Tensor Algebra and Tensor Analysis for Engineers

Mathematical Engineering

Series Editors:

Prof. Dr. Claus Hillermeier, Munich, Germany (volume editor)
Prof. Dr.-Ing. Jörg Schröder, Essen, Germany
Prof. Dr.-Ing. Bernhard Weigand, Stuttgart, Germany

For further volumes:
http://www.springer.com/series/8445

Mikhail Itskov

Tensor Algebra and Tensor Analysis for Engineers

With Applications to Continuum Mechanics

3rd Edition

Springer

Prof. Dr.-Ing. Mikhail Itskov
Department of Continuum Mechanics
RWTH Aachen University
Eilfschornsteinstr. 18
D 52062 Aachen
Germany

ISSN 2192-4732 ISSN 2192-4740 (electronic)
ISBN 978-3-642-44818-8 ISBN 978-3-642-30879-6 (eBook)
DOI 10.1007/978-3-642-30879-6
Springer Heidelberg New York Dordrecht London

Printed on acid-free paper

Springer is part of Springer Science+Business Media (www.springer.com)

Моим родителям

Preface to the Third Edition

This edition is enriched by new examples, problems, and solutions, in particular concerned with simple shear. I have also added an example with the derivation of constitutive relations and tangent moduli for hyperelastic materials with the isochoric-volumetric split of the strain energy function. Besides, Chap. 2 has some new figures illustrating spherical coordinates. These figures have again been prepared by Uwe Navrath. I also gratefully acknowledge Khiêm Ngoc Vu for careful proofreading of the manuscript. At this opportunity, I would also like to thank the Springer-Verlag and in particular Jan-Philip Schmidt for the fast and friendly support in getting this edition published.

Aachen, Germany Mikhail Itskov

Preface to the Second Edition

This second edition has a number of additional examples and exercises. In response to comments and questions of students using this book, solutions of many exercises have been improved for a better understanding. Some changes and enhancements are concerned with the treatment of skew-symmetric and rotation tensors in the first chapter. Besides, the text and formulae have been thoroughly reexamined and improved where necessary.

Aachen, Germany | Mikhail Itskov

Preface to the First Edition

Like many other textbooks, the present one is based on a lecture course given by the author for master's students of the RWTH Aachen University. In spite of a somewhat difficult matter those students were able to endure and, as far as I know, are still fine. I wish the same for the reader of this book.

Although this book can be referred to as a textbook, one finds only little plain text inside. I have tried to explain the matter in a brief way, nevertheless going into detail where necessary. I have also avoided tedious introductions and lengthy remarks about the significance of one topic or another. A reader interested in tensor algebra and tensor analysis but preferring, however, words instead of equations can close this book immediately after having read the preface.

The reader is assumed to be familiar with the basics of matrix algebra and continuum mechanics and is encouraged to solve at least some of numerous exercises accompanying every chapter. Having read many other texts on mathematics and mechanics, I was always upset vainly looking for solutions to the exercises which seemed to be most interesting for me. For this reason, all the exercises here are supplied with solutions, amounting to a substantial part of the book. Without doubt, this part facilitates a deeper understanding of the subject.

As a research work, this book is open for discussion which will certainly contribute to improving the text for further editions. In this sense, I am very grateful for comments, suggestions, and constructive criticism from the reader. I already expect such criticism, for example, with respect to the list of references which might be far from being complete. Indeed, throughout the book I only quote the sources indispensable to follow the exposition and notation. For this reason, I apologize to colleagues whose valuable contributions to the matter are not cited.

Finally, a word of acknowledgment is appropriate. I would like to thank Uwe Navrath for having prepared most of the figures for the book. Further, I am grateful to Alexander Ehret who taught me first steps as well as some "dirty" tricks in LaTeX, which were absolutely necessary to bring the manuscript to a printable form. He and Tran Dinh Tuyen are also acknowledged for careful proofreading and critical comments to an earlier version of the book. My special thanks go to the

Springer-Verlag and in particular to Eva Hestermann-Beyerle and Monika Lempe for their friendly support in getting this book published.

Aachen, Germany Mikhail Itskov

Contents

Chapter 1
Vectors and Tensors in a Finite-Dimensional Space

1.1 Notion of the Vector Space

We start with the definition of the vector space over the field of real numbers $\mathbb{R}$.

Definition 1.1. A vector space is a set $\mathbb{V}$ of elements called vectors satisfying the following axioms.

A. To every pair, $\boldsymbol{x}$ and $\boldsymbol{y}$ of vectors in $\mathbb{V}$ there corresponds a vector $\boldsymbol{x} + \boldsymbol{y}$, called the sum of $\boldsymbol{x}$ and $\boldsymbol{y}$, such that

   (A.1) $\boldsymbol{x} + \boldsymbol{y} = \boldsymbol{y} + \boldsymbol{x}$ (addition is commutative),
   (A.2) $(\boldsymbol{x} + \boldsymbol{y}) + \boldsymbol{z} = \boldsymbol{x} + (\boldsymbol{y} + \boldsymbol{z})$ (addition is associative),
   (A.3) There exists in $\mathbb{V}$ a unique vector zero $\boldsymbol{0}$, such that $\boldsymbol{0} + \boldsymbol{x} = \boldsymbol{x}$, $\forall \boldsymbol{x} \in \mathbb{V}$,
   (A.4) To every vector $\boldsymbol{x}$ in $\mathbb{V}$ there corresponds a unique vector $-\boldsymbol{x}$ such that $\boldsymbol{x} + (-\boldsymbol{x}) = \boldsymbol{0}$.

B. To every pair α and $\boldsymbol{x}$, where α is a scalar real number and $\boldsymbol{x}$ is a vector in $\mathbb{V}$, there corresponds a vector $\alpha \boldsymbol{x}$, called the product of α and $\boldsymbol{x}$, such that

   (B.1) $\alpha(\beta \boldsymbol{x}) = (\alpha\beta)\boldsymbol{x}$ (multiplication by scalars is associative),
   (B.2) $1\boldsymbol{x} = \boldsymbol{x}$,
   (B.3) $\alpha(\boldsymbol{x} + \boldsymbol{y}) = \alpha\boldsymbol{x} + \alpha\boldsymbol{y}$ (multiplication by scalars is distributive with respect to vector addition),
   (B.4) $(\alpha + \beta)\boldsymbol{x} = \alpha\boldsymbol{x} + \beta\boldsymbol{x}$ (multiplication by scalars is distributive with respect to scalar addition), $\forall \alpha, \beta \in \mathbb{R}$, $\forall \boldsymbol{x}, \boldsymbol{y} \in \mathbb{V}$.

Examples of Vector Spaces.

(1) The set of all real numbers $\mathbb{R}$.
(2) The set of all directional arrows in two or three dimensions. Applying the usual definitions for summation, multiplication by a scalar, the negative and zero vector (Fig. 1.1) one can easily see that the above axioms hold for directional arrows.

M. Itskov, *Tensor Algebra and Tensor Analysis for Engineers*, Mathematical Engineering,
DOI 10.1007/978-3-642-30879-6_1,

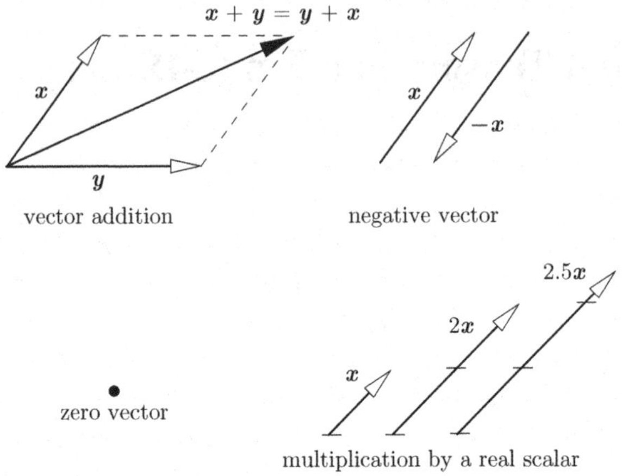

Fig. 1.1 Geometric illustration of vector axioms in two dimensions

(3) The set of all n-tuples of real numbers $\mathbb{R}$:

$$\boldsymbol{a} = \begin{Bmatrix} a_1 \\ a_2 \\ . \\ . \\ a_n \end{Bmatrix}.$$

Indeed, the axioms (A) and (B) apply to the n-tuples if one defines addition, multiplication by a scalar and finally the zero tuple, respectively, by

$$\boldsymbol{a} + \boldsymbol{b} = \begin{Bmatrix} a_1 + b_1 \\ a_2 + b_2 \\ . \\ . \\ a_n + b_n \end{Bmatrix}, \quad \alpha\boldsymbol{a} = \begin{Bmatrix} \alpha a_1 \\ \alpha a_2 \\ . \\ . \\ \alpha a_n \end{Bmatrix}, \quad \boldsymbol{0} = \begin{Bmatrix} 0 \\ 0 \\ . \\ . \\ 0 \end{Bmatrix}.$$

(4) The set of all real-valued functions defined on a real line.

1.2 Basis and Dimension of the Vector Space

Definition 1.2. A set of vectors $\boldsymbol{x}_1, \boldsymbol{x}_2, \ldots, \boldsymbol{x}_n$ is called linearly dependent if there exists a set of corresponding scalars $\alpha_1, \alpha_2, \ldots, \alpha_n \in \mathbb{R}$, not all zero, such that

$$\sum_{i=1}^{n} \alpha_i \boldsymbol{x}_i = \boldsymbol{0}. \tag{1.1}$$

Otherwise, the vectors $\boldsymbol{x}_1, \boldsymbol{x}_2, \ldots, \boldsymbol{x}_n$ are called linearly independent. In this case, none of the vectors $\boldsymbol{x}_i$ is the zero vector (Exercise 1.2).

Definition 1.3. The vector

$$\boldsymbol{x} = \sum_{i=1}^{n} \alpha_i \boldsymbol{x}_i \tag{1.2}$$

is called linear combination of the vectors $\boldsymbol{x}_1, \boldsymbol{x}_2, \ldots, \boldsymbol{x}_n$, where $\alpha_i \in \mathbb{R}(i = 1, 2, \ldots, n)$.

Theorem 1.1. *The set of n non-zero vectors $\boldsymbol{x}_1, \boldsymbol{x}_2, \ldots, \boldsymbol{x}_n$ is linearly dependent if and only if some vector $\boldsymbol{x}_k$ $(2 \le k \le n)$ is a linear combination of the preceding ones $\boldsymbol{x}_i$ $(i = 1, \ldots, k-1)$.*

Proof. If the vectors $\boldsymbol{x}_1, \boldsymbol{x}_2, \ldots, \boldsymbol{x}_n$ are linearly dependent, then

$$\sum_{i=1}^{n} \alpha_i \boldsymbol{x}_i = \boldsymbol{0},$$

where not all α_i are zero. Let α_k $(2 \le k \le n)$ be the last non-zero number, so that $\alpha_i = 0\,(i = k+1, \ldots, n)$. Then,

$$\sum_{i=1}^{k} \alpha_i \boldsymbol{x}_i = \boldsymbol{0} \Rightarrow \boldsymbol{x}_k = \sum_{i=1}^{k-1} \frac{-\alpha_i}{\alpha_k} \boldsymbol{x}_i.$$

Thereby, the case $k = 1$ is avoided because $\alpha_1 \boldsymbol{x}_1 = \boldsymbol{0}$ implies that $\boldsymbol{x}_1 = \boldsymbol{0}$ (Exercise 1.1). Thus, the sufficiency is proved. The necessity is evident.

Definition 1.4. A basis in a vector space $\mathbb{V}$ is a set $\mathcal{G} \subset \mathbb{V}$ of linearly independent vectors such that every vector in $\mathbb{V}$ is a linear combination of elements of $\mathcal{G}$. A vector space $\mathbb{V}$ is finite-dimensional if it has a finite basis.

Within this book, we restrict our attention to finite-dimensional vector spaces. Although one can find for a finite-dimensional vector space an infinite number of bases, they all have the same number of vectors.

Theorem 1.2. *All the bases of a finite-dimensional vector space $\mathbb{V}$ contain the same number of vectors.*

Proof. Let $\mathcal{G} = \{\boldsymbol{g}_1, \boldsymbol{g}_2, \ldots, \boldsymbol{g}_n\}$ and $\mathcal{F} = \{\boldsymbol{f}_1, \boldsymbol{f}_2, \ldots, \boldsymbol{f}_m\}$ be two arbitrary bases of $\mathbb{V}$ with different numbers of elements, say $m > n$. Then, every vector in $\mathbb{V}$ is a linear combination of the following vectors:

$$\boldsymbol{f}_1, \boldsymbol{g}_1, \boldsymbol{g}_2, \ldots, \boldsymbol{g}_n. \tag{1.3}$$

These vectors are non-zero and linearly dependent. Thus, according to Theorem 1.1 we can find such a vector $\boldsymbol{g}_k$, which is a linear combination of the preceding ones. Excluding this vector we obtain the set $\mathcal{G}'$ by

$$\boldsymbol{f}_1, \boldsymbol{g}_1, \boldsymbol{g}_2, \ldots, \boldsymbol{g}_{k-1}, \boldsymbol{g}_{k+1}, \ldots, \boldsymbol{g}_n$$

again with the property that every vector in $\mathbb{V}$ is a linear combination of the elements of $\mathcal{G}'$. Now, we consider the following vectors

$$\boldsymbol{f}_1, \boldsymbol{f}_2, \boldsymbol{g}_1, \boldsymbol{g}_2, \ldots, \boldsymbol{g}_{k-1}, \boldsymbol{g}_{k+1}, \ldots, \boldsymbol{g}_n$$

and repeat the excluding procedure just as before. We see that none of the vectors $\boldsymbol{f}_i$ can be eliminated in this way because they are linearly independent. As soon as all $\boldsymbol{g}_i$ $(i = 1, 2, \ldots, n)$ are exhausted we conclude that the vectors

$$\boldsymbol{f}_1, \boldsymbol{f}_2, \ldots, \boldsymbol{f}_{n+1}$$

are linearly dependent. This contradicts, however, the previous assumption that they belong to the basis $\mathcal{F}$.

Definition 1.5. The dimension of a finite-dimensional vector space $\mathbb{V}$ is the number of elements in a basis of $\mathbb{V}$.

Theorem 1.3. *Every set $\mathcal{F} = \{\boldsymbol{f}_1, \boldsymbol{f}_2, \ldots, \boldsymbol{f}_n\}$ of linearly independent vectors in an n-dimensional vectors space $\mathbb{V}$ forms a basis of $\mathbb{V}$. Every set of more than n vectors is linearly dependent.*

Proof. The proof of this theorem is similar to the preceding one. Let $\mathcal{G} = \{\boldsymbol{g}_1, \boldsymbol{g}_2, \ldots, \boldsymbol{g}_n\}$ be a basis of $\mathbb{V}$. Then, the vectors (1.3) are linearly dependent and non-zero. Excluding a vector $\boldsymbol{g}_k$ we obtain a set of vectors, say $\mathcal{G}'$, with the property that every vector in $\mathbb{V}$ is a linear combination of the elements of $\mathcal{G}'$. Repeating this procedure we finally end up with the set $\mathcal{F}$ with the same property. Since the vectors $\boldsymbol{f}_i$ $(i = 1, 2, \ldots, n)$ are linearly independent they form a basis of $\mathbb{V}$. Any further vectors in $\mathbb{V}$, say $\boldsymbol{f}_{n+1}, \boldsymbol{f}_{n+2}, \ldots$ are thus linear combinations of $\mathcal{F}$. Hence, any set of more than n vectors is linearly dependent.

Theorem 1.4. *Every set $\mathcal{F} = \{\boldsymbol{f}_1, \boldsymbol{f}_2, \ldots, \boldsymbol{f}_m\}$ of linearly independent vectors in an n-dimensional vector space $\mathbb{V}$ can be extended to a basis.*

Proof. If $m = n$, then $\mathcal{F}$ is already a basis according to Theorem 1.3. If $m < n$, then we try to find $n - m$ vectors $\boldsymbol{f}_{m+1}, \boldsymbol{f}_{m+2}, \ldots, \boldsymbol{f}_n$, such that all the vectors $\boldsymbol{f}_i$, that is, $\boldsymbol{f}_1, \boldsymbol{f}_2, \ldots, \boldsymbol{f}_m, \boldsymbol{f}_{m+1}, \ldots, \boldsymbol{f}_n$ are linearly independent and consequently form a basis. Let us assume, on the contrary, that only $k < n - m$ such vectors can be found. In this case, for all $\boldsymbol{x} \in \mathbb{V}$ there exist scalars $\alpha, \alpha_1, \alpha_2, \ldots, \alpha_{m+k}$, not all zero, such that

$$\alpha \boldsymbol{x} + \alpha_1 \boldsymbol{f}_1 + \alpha_2 \boldsymbol{f}_2 + \ldots + \alpha_{m+k} \boldsymbol{f}_{m+k} = \boldsymbol{0},$$

where $\alpha \neq 0$ since otherwise the vectors $\boldsymbol{f}_i$ $(i = 1, 2, \ldots, m + k)$ would be linearly dependent. Thus, all the vectors $\boldsymbol{x}$ of $\mathbb{V}$ are linear combinations of $\boldsymbol{f}_i$ $(i = 1, 2, \ldots, m + k)$. Then, the dimension of $\mathbb{V}$ is $m + k < n$, which contradicts the assumption of this theorem.

1.3 Components of a Vector, Summation Convention

Let $\mathcal{G} = \{\boldsymbol{g}_1, \boldsymbol{g}_2, \ldots, \boldsymbol{g}_n\}$ be a basis of an n-dimensional vector space $\mathbb{V}$. Then,

$$\boldsymbol{x} = \sum_{i=1}^{n} x^i \boldsymbol{g}_i, \quad \forall \boldsymbol{x} \in \mathbb{V}. \tag{1.4}$$

Theorem 1.5. *The representation (1.4) with respect to a given basis $\mathcal{G}$ is unique.*

Proof. Let

$$\boldsymbol{x} = \sum_{i=1}^{n} x^i \boldsymbol{g}_i \quad \text{and} \quad \boldsymbol{x} = \sum_{i=1}^{n} y^i \boldsymbol{g}_i$$

be two different representations of a vector $\boldsymbol{x}$, where not all scalar coefficients x^i and y^i $(i = 1, 2, \ldots, n)$ are pairwise identical. Then,

$$\boldsymbol{0} = \boldsymbol{x} + (-\boldsymbol{x}) = \boldsymbol{x} + (-1)\,\boldsymbol{x} = \sum_{i=1}^{n} x^i \boldsymbol{g}_i + \sum_{i=1}^{n} \left(-y^i\right) \boldsymbol{g}_i = \sum_{i=1}^{n} \left(x^i - y^i\right) \boldsymbol{g}_i,$$

where we use the identity $-\boldsymbol{x} = (-1)\,\boldsymbol{x}$ (Exercise 1.1). Thus, either the numbers x^i and y^i are pairwise equal $x^i = y^i$ $(i = 1, 2, \ldots, n)$ or the vectors $\boldsymbol{g}_i$ are linearly dependent. The latter one is likewise impossible because these vectors form a basis of $\mathbb{V}$.

The scalar numbers x^i $(i = 1, 2, \ldots, n)$ in the representation (1.4) are called components of the vector $\boldsymbol{x}$ with respect to the basis $\mathcal{G} = \{\boldsymbol{g}_1, \boldsymbol{g}_2, \ldots, \boldsymbol{g}_n\}$.

The summation of the form (1.4) is often used in tensor algebra. For this reason it is usually represented without the summation symbol in a short form by

$$\boldsymbol{x} = \sum_{i=1}^{n} x^i \boldsymbol{g}_i = x^i \boldsymbol{g}_i \tag{1.5}$$

referred to as Einstein's summation convention. Accordingly, the summation is implied if an index appears twice in a multiplicative term, once as a superscript and once as a subscript. Such a repeated index (called dummy index) takes the values from 1 to n (the dimension of the vector space in consideration). The sense of the index changes (from superscript to subscript or vice versa) if it appears under the fraction bar.

1.4 Scalar Product, Euclidean Space, Orthonormal Basis

The scalar product plays an important role in vector and tensor algebra. The properties of the vector space essentially depend on whether and how the scalar product is defined in this space.

Definition 1.6. The scalar (inner) product is a real-valued function $\boldsymbol{x} \cdot \boldsymbol{y}$ of two vectors $\boldsymbol{x}$ and $\boldsymbol{y}$ in a vector space $\mathbb{V}$, satisfying the following conditions.

C. (C.1) $\boldsymbol{x} \cdot \boldsymbol{y} = \boldsymbol{y} \cdot \boldsymbol{x}$ (commutative rule),
(C.2) $\boldsymbol{x} \cdot (\boldsymbol{y} + \boldsymbol{z}) = \boldsymbol{x} \cdot \boldsymbol{y} + \boldsymbol{x} \cdot \boldsymbol{z}$ (distributive rule),
(C.3) $\alpha (\boldsymbol{x} \cdot \boldsymbol{y}) = (\alpha \boldsymbol{x}) \cdot \boldsymbol{y} = \boldsymbol{x} \cdot (\alpha \boldsymbol{y})$ (associative rule for the multiplication by a scalar), $\forall \alpha \in \mathbb{R}$, $\forall \boldsymbol{x}, \boldsymbol{y}, \boldsymbol{z} \in \mathbb{V}$,
(C.4) $\boldsymbol{x} \cdot \boldsymbol{x} \geq 0 \;\; \forall \boldsymbol{x} \in \mathbb{V}$, $\;\; \boldsymbol{x} \cdot \boldsymbol{x} = 0$ if and only if $\boldsymbol{x} = \boldsymbol{0}$.

An n-dimensional vector space furnished by the scalar product with properties (C.1)–(C.4) is called Euclidean space $\mathbb{E}^n$. On the basis of this scalar product one defines the Euclidean length (also called norm) of a vector $\boldsymbol{x}$ by

$$\|\boldsymbol{x}\| = \sqrt{\boldsymbol{x} \cdot \boldsymbol{x}}. \tag{1.6}$$

A vector whose length is equal to 1 is referred to as unit vector.

Definition 1.7. Two vectors $\boldsymbol{x}$ and $\boldsymbol{y}$ are called orthogonal (perpendicular), denoted by $\boldsymbol{x} \perp \boldsymbol{y}$, if

$$\boldsymbol{x} \cdot \boldsymbol{y} = 0. \tag{1.7}$$

Of special interest is the so-called orthonormal basis of the Euclidean space.

Definition 1.8. A basis $\mathcal{E} = \{\boldsymbol{e}_1, \boldsymbol{e}_2, \ldots, \boldsymbol{e}_n\}$ of an n-dimensional Euclidean space $\mathbb{E}^n$ is called orthonormal if

$$\boldsymbol{e}_i \cdot \boldsymbol{e}_j = \delta_{ij}, \quad i, j = 1, 2, \ldots, n, \tag{1.8}$$

where

$$\delta_{ij} = \delta^{ij} = \delta^i_j = \begin{cases} 1 & \text{for } i = j, \\ 0 & \text{for } i \neq j \end{cases} \tag{1.9}$$

denotes the Kronecker delta.

Thus, the elements of an orthonormal basis represent pairwise orthogonal unit vectors. Of particular interest is the question of the existence of an orthonormal basis. Now, we are going to demonstrate that every set of $m \leq n$ linearly independent vectors in $\mathbb{E}^n$ can be orthogonalized and normalized by means of a linear transformation (Gram-Schmidt procedure). In other words, starting from linearly independent vectors $\boldsymbol{x}_1, \boldsymbol{x}_2, \ldots, \boldsymbol{x}_m$ one can always construct their linear combinations $\boldsymbol{e}_1, \boldsymbol{e}_2, \ldots, \boldsymbol{e}_m$ such that $\boldsymbol{e}_i \cdot \boldsymbol{e}_j = \delta_{ij}$ $(i, j = 1, 2, \ldots, m)$. Indeed, since the vectors $\boldsymbol{x}_i$ $(i = 1, 2, \ldots, m)$ are linearly independent they are all non-zero (see Exercise 1.2). Thus, we can define the first unit vector by

$$\boldsymbol{e}_1 = \frac{\boldsymbol{x}_1}{\|\boldsymbol{x}_1\|}. \tag{1.10}$$

Next, we consider the vector

$$\boldsymbol{e}'_2 = \boldsymbol{x}_2 - (\boldsymbol{x}_2 \cdot \boldsymbol{e}_1)\, \boldsymbol{e}_1 \tag{1.11}$$

orthogonal to $\boldsymbol{e}_1$. This holds for the unit vector $\boldsymbol{e}_2 = \boldsymbol{e}'_2 / \left\|\boldsymbol{e}'_2\right\|$ as well. It is also seen that $\left\|\boldsymbol{e}'_2\right\| = \sqrt{\boldsymbol{e}'_2 \cdot \boldsymbol{e}'_2} \neq 0$ because otherwise $\boldsymbol{e}'_2 = \boldsymbol{0}$ and thus $\boldsymbol{x}_2 = (\boldsymbol{x}_2 \cdot \boldsymbol{e}_1)\, \boldsymbol{e}_1 = (\boldsymbol{x}_2 \cdot \boldsymbol{e}_1)\, \|\boldsymbol{x}_1\|^{-1} \boldsymbol{x}_1$. However, the latter result contradicts the fact that the vectors $\boldsymbol{x}_1$ and $\boldsymbol{x}_2$ are linearly independent.

Further, we proceed to construct the vectors

$$\boldsymbol{e}'_3 = \boldsymbol{x}_3 - (\boldsymbol{x}_3 \cdot \boldsymbol{e}_2)\, \boldsymbol{e}_2 - (\boldsymbol{x}_3 \cdot \boldsymbol{e}_1)\, \boldsymbol{e}_1, \quad \boldsymbol{e}_3 = \frac{\boldsymbol{e}'_3}{\left\|\boldsymbol{e}'_3\right\|} \tag{1.12}$$

orthogonal to $\boldsymbol{e}_1$ and $\boldsymbol{e}_2$. Repeating this procedure we finally obtain the set of orthonormal vectors $\boldsymbol{e}_1, \boldsymbol{e}_2, \ldots, \boldsymbol{e}_m$. Since these vectors are non-zero and mutually orthogonal, they are linearly independent (see Exercise 1.6). In the case $m = n$, this set represents, according to Theorem 1.3, the orthonormal basis (1.8) in $\mathbb{E}^n$.

With respect to an orthonormal basis the scalar product of two vectors $\boldsymbol{x} = x^i \boldsymbol{e}_i$ and $\boldsymbol{y} = y^i \boldsymbol{e}_i$ in $\mathbb{E}^n$ takes the form

$$\boldsymbol{x} \cdot \boldsymbol{y} = x^1 y^1 + x^2 y^2 + \ldots + x^n y^n. \tag{1.13}$$

For the length of the vector $\boldsymbol{x}$ (1.6) we thus obtain the Pythagoras formula

$$\|\boldsymbol{x}\| = \sqrt{x^1 x^1 + x^2 x^2 + \ldots + x^n x^n}, \quad \boldsymbol{x} \in \mathbb{E}^n. \tag{1.14}$$

1.5 Dual Bases

Definition 1.9. Let $\mathcal{G} = \{\boldsymbol{g}_1, \boldsymbol{g}_2, \ldots, \boldsymbol{g}_n\}$ be a basis in the n-dimensional Euclidean space $\mathbb{E}^n$. Then, a basis $\mathcal{G}' = \left\{\boldsymbol{g}^1, \boldsymbol{g}^2, \ldots, \boldsymbol{g}^n\right\}$ of $\mathbb{E}^n$ is called dual to $\mathcal{G}$, if

$$\boldsymbol{g}_i \cdot \boldsymbol{g}^j = \delta_i^j, \quad i, j = 1, 2, \ldots, n. \tag{1.15}$$

In the following we show that a set of vectors $\mathcal{G}' = \{g^1, g^2, \ldots, g^n\}$ satisfying the conditions (1.15) always exists, is unique and forms a basis in $\mathbb{E}^n$.

Let $\mathcal{E} = \{e_1, e_2, \ldots, e_n\}$ be an orthonormal basis in $\mathbb{E}^n$. Since $\mathcal{G}$ also represents a basis, we can write

$$e_i = \alpha_i^j g_j, \quad g_i = \beta_i^j e_j, \quad i = 1, 2, \ldots, n, \tag{1.16}$$

where α_i^j and β_i^j $(i = 1, 2, \ldots, n)$ denote the components of e_i and g_i, respectively. Inserting the first relation (1.16) into the second one yields

$$g_i = \beta_i^j \alpha_j^k g_k, \quad \Rightarrow \quad \mathbf{0} = \left(\beta_i^j \alpha_j^k - \delta_i^k\right) g_k, \quad i = 1, 2, \ldots, n. \tag{1.17}$$

Since the vectors g_i are linearly independent we obtain

$$\beta_i^j \alpha_j^k = \delta_i^k, \quad i, k = 1, 2, \ldots, n. \tag{1.18}$$

Let further

$$g^i = \alpha_j^i e^j, \quad i = 1, 2, \ldots, n, \tag{1.19}$$

where and henceforth we set $e^j = e_j$ $(j = 1, 2, \ldots, n)$ in order to take the advantage of Einstein's summation convention. By virtue of (1.8), (1.16) and (1.18) one finally finds

$$g_i \cdot g^j = \left(\beta_i^k e_k\right) \cdot \left(\alpha_l^j e^l\right) = \beta_i^k \alpha_l^j \delta_k^l = \beta_i^k \alpha_k^j = \delta_i^j, \quad i, j = 1, 2, \ldots, n. \tag{1.20}$$

Next, we show that the vectors g^i $(i = 1, 2, \ldots, n)$ (1.19) are linearly independent and for this reason form a basis of $\mathbb{E}^n$. Assume on the contrary that

$$a_i g^i = \mathbf{0},$$

where not all scalars a_i $(i = 1, 2, \ldots, n)$ are zero. Multiplying both sides of this relation scalarly by the vectors g_j $(j = 1, 2, \ldots, n)$ leads to a contradiction. Indeed, using (1.167) (see Exercise 1.5) we obtain

$$0 = a_i g^i \cdot g_j = a_i \delta_j^i = a_j, \quad j = 1, 2, \ldots, n.$$

The next important question is whether the dual basis is unique. Let $\mathcal{G}' = \{g^1, g^2, \ldots, g^n\}$ and $\mathcal{H}' = \{h^1, h^2, \ldots, h^n\}$ be two arbitrary non-coinciding bases in $\mathbb{E}^n$, both dual to $\mathcal{G} = \{g_1, g_2, \ldots, g_n\}$. Then,

$$h^i = h_j^i g^j, \quad i = 1, 2, \ldots, n.$$

Forming the scalar product with the vectors $\boldsymbol{g}_j$ $(j = 1, 2, \ldots, n)$ we can conclude that the bases $\mathcal{G}'$ and $\mathcal{H}'$ coincide:

$$\delta^i_j = \boldsymbol{h}^i \cdot \boldsymbol{g}_j = \left(h^i_k \boldsymbol{g}^k\right) \cdot \boldsymbol{g}_j = h^i_k \delta^k_j = h^i_j \quad \Rightarrow \quad \boldsymbol{h}^i = \boldsymbol{g}^i, \quad i = 1, 2, \ldots, n.$$

Thus, we have proved the following theorem.

Theorem 1.6. *To every basis in an Euclidean space* $\mathbb{E}^n$ *there exists a unique dual basis.*

Relation (1.19) enables to determine the dual basis. However, it can also be obtained without any orthonormal basis. Indeed, let $\boldsymbol{g}^i$ be a basis dual to $\boldsymbol{g}_i$ $(i = 1, 2, \ldots, n)$. Then

$$\boldsymbol{g}^i = g^{ij} \boldsymbol{g}_j, \quad \boldsymbol{g}_i = g_{ij} \boldsymbol{g}^j, \quad i = 1, 2, \ldots, n. \tag{1.21}$$

Inserting the second relation (1.21) into the first one yields

$$\boldsymbol{g}^i = g^{ij} g_{jk} \boldsymbol{g}^k, \quad i = 1, 2, \ldots, n. \tag{1.22}$$

Multiplying scalarly with the vectors $\boldsymbol{g}_l$ we have by virtue of (1.15)

$$\delta^i_l = g^{ij} g_{jk} \delta^k_l = g^{ij} g_{jl}, \quad i, l = 1, 2, \ldots, n. \tag{1.23}$$

Thus, we see that the matrices $\left[g_{kj}\right]$ and $\left[g^{kj}\right]$ are inverse to each other such that

$$\left[g^{kj}\right] = \left[g_{kj}\right]^{-1}. \tag{1.24}$$

Now, multiplying scalarly the first and second relation (1.21) by the vectors $\boldsymbol{g}^j$ and $\boldsymbol{g}_j$ $(j = 1, 2, \ldots, n)$, respectively, we obtain with the aid of (1.15) the following important identities:

$$g^{ij} = g^{ji} = \boldsymbol{g}^i \cdot \boldsymbol{g}^j, \quad g_{ij} = g_{ji} = \boldsymbol{g}_i \cdot \boldsymbol{g}_j, \quad i, j = 1, 2, \ldots, n. \tag{1.25}$$

By definition (1.8) the orthonormal basis in $\mathbb{E}^n$ is self-dual, so that

$$\boldsymbol{e}_i = \boldsymbol{e}^i, \quad \boldsymbol{e}_i \cdot \boldsymbol{e}^j = \delta^j_i, \quad i, j = 1, 2, \ldots, n. \tag{1.26}$$

With the aid of the dual bases one can represent an arbitrary vector in $\mathbb{E}^n$ by

$$\boldsymbol{x} = x^i \boldsymbol{g}_i = x_i \boldsymbol{g}^i, \quad \forall \boldsymbol{x} \in \mathbb{E}^n, \tag{1.27}$$

where

$$x^i = \boldsymbol{x} \cdot \boldsymbol{g}^i, \quad x_i = \boldsymbol{x} \cdot \boldsymbol{g}_i, \quad i = 1, 2, \ldots, n. \tag{1.28}$$

Indeed, using (1.15) we can write

$$\boldsymbol{x} \cdot \boldsymbol{g}^i = \left(x^j \boldsymbol{g}_j\right) \cdot \boldsymbol{g}^i = x^j \delta^i_j = x^i,$$

$$\boldsymbol{x} \cdot \boldsymbol{g}_i = \left(x_j \boldsymbol{g}^j\right) \cdot \boldsymbol{g}_i = x_j \delta^j_i = x_i, \quad i = 1, 2, \ldots, n.$$

The components of a vector with respect to the dual bases are suitable for calculating the scalar product. For example, for two arbitrary vectors $\boldsymbol{x} = x^i \boldsymbol{g}_i = x_i \boldsymbol{g}^i$ and $\boldsymbol{y} = y^i \boldsymbol{g}_i = y_i \boldsymbol{g}^i$ we obtain

$$\boldsymbol{x} \cdot \boldsymbol{y} = x^i y^j g_{ij} = x_i y_j g^{ij} = x^i y_i = x_i y^i. \tag{1.29}$$

The length of the vector $\boldsymbol{x}$ can thus be written by

$$\|x\| = \sqrt{x_i x_j g^{ij}} = \sqrt{x^i x^j g_{ij}} = \sqrt{x_i x^i}. \tag{1.30}$$

Example 1.1. Dual basis in $\mathbb{E}^3$. Let $\mathcal{G} = \{\boldsymbol{g}_1, \boldsymbol{g}_2, \boldsymbol{g}_3\}$ be a basis of the three-dimensional Euclidean space and

$$g = [\boldsymbol{g}_1 \boldsymbol{g}_2 \boldsymbol{g}_3], \tag{1.31}$$

where $[\bullet \bullet \bullet]$ denotes the mixed product of vectors. It is defined by

$$[\boldsymbol{abc}] = (\boldsymbol{a} \times \boldsymbol{b}) \cdot \boldsymbol{c} = (\boldsymbol{b} \times \boldsymbol{c}) \cdot \boldsymbol{a} = (\boldsymbol{c} \times \boldsymbol{a}) \cdot \boldsymbol{b}, \tag{1.32}$$

where "×" denotes the vector (also called cross or outer) product of vectors. Consider the following set of vectors:

$$\boldsymbol{g}^1 = g^{-1} \boldsymbol{g}_2 \times \boldsymbol{g}_3, \quad \boldsymbol{g}^2 = g^{-1} \boldsymbol{g}_3 \times \boldsymbol{g}_1, \quad \boldsymbol{g}^3 = g^{-1} \boldsymbol{g}_1 \times \boldsymbol{g}_2. \tag{1.33}$$

It is seen that the vectors (1.33) satisfy conditions (1.15), are linearly independent (Exercise 1.11) and consequently form the basis dual to $\boldsymbol{g}_i$ $(i = 1, 2, 3)$. Further, it can be shown that

$$g^2 = \left|g_{ij}\right|, \tag{1.34}$$

where $|\bullet|$ denotes the determinant of the matrix $[\bullet]$. Indeed, with the aid of $(1.16)_2$ we obtain

$$g = [\boldsymbol{g}_1 \boldsymbol{g}_2 \boldsymbol{g}_3] = \left[\beta^i_1 \boldsymbol{e}_i \beta^j_2 \boldsymbol{e}_j \beta^k_3 \boldsymbol{e}_k\right]$$

$$= \beta^i_1 \beta^j_2 \beta^k_3 \left[\boldsymbol{e}_i \boldsymbol{e}_j \boldsymbol{e}_k\right] = \beta^i_1 \beta^j_2 \beta^k_3 e_{ijk} = \left|\beta^i_j\right|, \tag{1.35}$$

where e_{ijk} denotes the permutation symbol (also called Levi-Civita symbol). It is defined by

$$e_{ijk} = e^{ijk} = \left[\boldsymbol{e}_i \boldsymbol{e}_j \boldsymbol{e}_k \right] = \begin{cases} 1 & \text{if } ijk \text{ is an even permutation of 123,} \\ -1 & \text{if } ijk \text{ is an odd permutation of 123,} \\ 0 & \text{otherwise,} \end{cases} \tag{1.36}$$

where the orthonormal vectors $\boldsymbol{e}_1$, $\boldsymbol{e}_2$ and $\boldsymbol{e}_3$ are numerated in such a way that they form a right-handed system. In this case, $[\boldsymbol{e}_1 \boldsymbol{e}_2 \boldsymbol{e}_3] = 1$.

On the other hand, we can write again using $(1.16)_2$

$$g_{ij} = \boldsymbol{g}_i \cdot \boldsymbol{g}_j = \sum_{k=1}^{3} \beta_i^k \beta_j^k.$$

The latter sum can be represented as a product of two matrices so that

$$\left[g_{ij} \right] = \left[\beta_i^j \right] \left[\beta_i^j \right]^{\mathrm{T}}. \tag{1.37}$$

Since the determinant of the matrix product is equal to the product of the matrix determinants we finally have

$$\left| g_{ij} \right| = \left| \beta_i^j \right|^2 = g^2. \tag{1.38}$$

With the aid of the permutation symbol (1.36) one can write

$$\left[\boldsymbol{g}_i \boldsymbol{g}_j \boldsymbol{g}_k \right] = e_{ijk}\, g, \quad i, j, k = 1, 2, 3, \tag{1.39}$$

which by $(1.28)_2$ yields an alternative representation of the identities (1.33) as

$$\boldsymbol{g}_i \times \boldsymbol{g}_j = e_{ijk}\, g\, \boldsymbol{g}^k, \quad i, j = 1, 2, 3. \tag{1.40}$$

Similarly to (1.35) one can also show that (see Exercise 1.12)

$$\left[\boldsymbol{g}^1 \boldsymbol{g}^2 \boldsymbol{g}^3 \right] = g^{-1} \tag{1.41}$$

and

$$\left| g^{ij} \right| = g^{-2}. \tag{1.42}$$

Thus,

$$\left[\boldsymbol{g}^i \boldsymbol{g}^j \boldsymbol{g}^k\right] = \frac{e^{ijk}}{g}, \quad i, j, k = 1, 2, 3, \tag{1.43}$$

which yields by analogy with (1.40)

$$\boldsymbol{g}^i \times \boldsymbol{g}^j = \frac{e^{ijk}}{g} \boldsymbol{g}_k, \quad i, j = 1, 2, 3. \tag{1.44}$$

Relations (1.40) and (1.44) permit a useful representation of the vector product. Indeed, let $\boldsymbol{a} = a^i \boldsymbol{g}_i = a_i \boldsymbol{g}^i$ and $\boldsymbol{b} = b^j \boldsymbol{g}_j = b_j \boldsymbol{g}^j$ be two arbitrary vectors in $\mathbb{E}^3$. Then, in view of (1.32)

$$\boldsymbol{a} \times \boldsymbol{b} = \left(a^i \boldsymbol{g}_i\right) \times \left(b^j \boldsymbol{g}_j\right) = a^i b^j e_{ijk} g \boldsymbol{g}^k = g \begin{vmatrix} a^1 & a^2 & a^3 \\ b^1 & b^2 & b^3 \\ \boldsymbol{g}^1 & \boldsymbol{g}^2 & \boldsymbol{g}^3 \end{vmatrix},$$

$$\boldsymbol{a} \times \boldsymbol{b} = \left(a_i \boldsymbol{g}^i\right) \times \left(b_j \boldsymbol{g}^j\right) = a_i b_j e^{ijk} g^{-1} \boldsymbol{g}_k = \frac{1}{g} \begin{vmatrix} a_1 & a_2 & a_3 \\ b_1 & b_2 & b_3 \\ \boldsymbol{g}_1 & \boldsymbol{g}_2 & \boldsymbol{g}_3 \end{vmatrix}. \tag{1.45}$$

For the orthonormal basis in $\mathbb{E}^3$ relations (1.40) and (1.44) reduce to

$$\boldsymbol{e}_i \times \boldsymbol{e}_j = e_{ijk} \boldsymbol{e}^k = e^{ijk} \boldsymbol{e}_k, \quad i, j = 1, 2, 3, \tag{1.46}$$

so that the vector product (1.45) can be written by

$$\boldsymbol{a} \times \boldsymbol{b} = \begin{vmatrix} a_1 & a_2 & a_3 \\ b_1 & b_2 & b_3 \\ \boldsymbol{e}_1 & \boldsymbol{e}_2 & \boldsymbol{e}_3 \end{vmatrix}, \tag{1.47}$$

where $\boldsymbol{a} = a_i \boldsymbol{e}^i$ and $\boldsymbol{b} = b_j \boldsymbol{e}^j$.

1.6 Second-Order Tensor as a Linear Mapping

Let us consider a set $\mathbf{Lin}^n$ of all linear mappings of one vector into another one within $\mathbb{E}^n$. Such a mapping can be written as

$$\boldsymbol{y} = \mathbf{A}\boldsymbol{x}, \quad \boldsymbol{y} \in \mathbb{E}^n, \quad \forall \boldsymbol{x} \in \mathbb{E}^n, \quad \forall \mathbf{A} \in \mathbf{Lin}^n. \tag{1.48}$$

Elements of the set $\mathbf{Lin}^n$ are called second-order tensors or simply tensors. Linearity of the mapping (1.48) is expressed by the following relations:

$$\mathbf{A}\left(\boldsymbol{x} + \boldsymbol{y}\right) = \mathbf{A}\boldsymbol{x} + \mathbf{A}\boldsymbol{y}, \quad \forall \boldsymbol{x}, \boldsymbol{y} \in \mathbb{E}^n, \quad \forall \mathbf{A} \in \mathbf{Lin}^n, \tag{1.49}$$

$$\mathbf{A}(\alpha \boldsymbol{x}) = \alpha(\mathbf{A}\boldsymbol{x}), \quad \forall \boldsymbol{x} \in \mathbb{E}^n, \quad \forall \alpha \in \mathbb{R}, \quad \forall \mathbf{A} \in \mathbf{Lin}^n. \tag{1.50}$$

Further, we define the product of a tensor by a scalar number $\alpha \in \mathbb{R}$ as

$$(\alpha \mathbf{A})\boldsymbol{x} = \alpha(\mathbf{A}\boldsymbol{x}) = \mathbf{A}(\alpha \boldsymbol{x}), \quad \forall \boldsymbol{x} \in \mathbb{E}^n \tag{1.51}$$

and the sum of two tensors **A** and **B** as

$$(\mathbf{A} + \mathbf{B})\boldsymbol{x} = \mathbf{A}\boldsymbol{x} + \mathbf{B}\boldsymbol{x}, \quad \forall \boldsymbol{x} \in \mathbb{E}^n. \tag{1.52}$$

Thus, properties (A.1), (A.2) and (B.1)–(B.4) apply to the set $\mathbf{Lin}^n$. Setting in (1.51) $\alpha = -1$ we obtain the negative tensor by

$$-\mathbf{A} = (-1)\mathbf{A}. \tag{1.53}$$

Further, we define a zero tensor $\mathbf{0}$ in the following manner

$$\mathbf{0}\boldsymbol{x} = \boldsymbol{0}, \quad \forall \boldsymbol{x} \in \mathbb{E}^n, \tag{1.54}$$

so that the elements of the set $\mathbf{Lin}^n$ also fulfill conditions (A.3) and (A.4) and accordingly form a vector space.

The properties of second-order tensors can thus be summarized by

$$\mathbf{A} + \mathbf{B} = \mathbf{B} + \mathbf{A}, \quad \text{(addition is commutative)}, \tag{1.55}$$

$$\mathbf{A} + (\mathbf{B} + \mathbf{C}) = (\mathbf{A} + \mathbf{B}) + \mathbf{C}, \quad \text{(addition is associative)}, \tag{1.56}$$

$$\mathbf{0} + \mathbf{A} = \mathbf{A}, \tag{1.57}$$

$$\mathbf{A} + (-\mathbf{A}) = \mathbf{0}, \tag{1.58}$$

$$\alpha(\beta\mathbf{A}) = (\alpha\beta)\mathbf{A}, \quad \text{(multiplication by scalars is associative)}, \tag{1.59}$$

$$1\mathbf{A} = \mathbf{A}, \tag{1.60}$$

$$\alpha(\mathbf{A} + \mathbf{B}) = \alpha\mathbf{A} + \alpha\mathbf{B}, \quad \text{(multiplication by scalars is distributive with respect to tensor addition)}, \tag{1.61}$$

$$(\alpha + \beta)\mathbf{A} = \alpha\mathbf{A} + \beta\mathbf{A}, \quad \text{(multiplication by scalars is distributive with respect to scalar addition)}, \quad \forall \mathbf{A}, \mathbf{B}, \mathbf{C} \in \mathbf{Lin}^n, \ \forall \alpha, \beta \in \mathbb{R}. \tag{1.62}$$

Example 1.2. Vector product in $\mathbb{E}^3$. The vector product of two vectors in $\mathbb{E}^3$ represents again a vector in $\mathbb{E}^3$

$$\boldsymbol{z} = \boldsymbol{w} \times \boldsymbol{x}, \quad \boldsymbol{z} \in \mathbb{E}^3, \quad \forall \boldsymbol{w}, \boldsymbol{x} \in \mathbb{E}^3. \tag{1.63}$$

According to (1.45) the mapping $\boldsymbol{x} \to \boldsymbol{z}$ is linear (Exercise 1.16) so that

$$\boldsymbol{w} \times (\alpha \boldsymbol{x}) = \alpha (\boldsymbol{w} \times \boldsymbol{x}),$$

$$\boldsymbol{w} \times (\boldsymbol{x} + \boldsymbol{y}) = \boldsymbol{w} \times \boldsymbol{x} + \boldsymbol{w} \times \boldsymbol{y}, \quad \forall \boldsymbol{w}, \boldsymbol{x}, \boldsymbol{y} \in \mathbb{E}^3, \quad \forall \alpha \in \mathbb{R}. \tag{1.64}$$

Thus, it can be described by means of a tensor of the second order by

$$\boldsymbol{w} \times \boldsymbol{x} = \mathbf{W}\boldsymbol{x}, \quad \mathbf{W} \in \mathbf{Lin}^3, \quad \forall \boldsymbol{x} \in \mathbb{E}^3. \tag{1.65}$$

The tensor which forms the vector product by a vector $\boldsymbol{w}$ according to (1.65) will be denoted in the following by $\hat{\boldsymbol{w}}$. Thus, we write

$$\boldsymbol{w} \times \boldsymbol{x} = \hat{\boldsymbol{w}}\boldsymbol{x}. \tag{1.66}$$

Clearly

$$\hat{\boldsymbol{0}} = \mathbf{0}. \tag{1.67}$$

Example 1.3. Representation of a rotation by a second-order tensor. A rotation of a vector $\boldsymbol{a}$ in $\mathbb{E}^3$ about an axis yields another vector $\boldsymbol{r}$ in $\mathbb{E}^3$. It can be shown that the mapping $\boldsymbol{a} \to \boldsymbol{r}(\boldsymbol{a})$ is linear such that

$$\boldsymbol{r}(\alpha \boldsymbol{a}) = \alpha \boldsymbol{r}(\boldsymbol{a}), \ \boldsymbol{r}(\boldsymbol{a} + \boldsymbol{b}) = \boldsymbol{r}(\boldsymbol{a}) + \boldsymbol{r}(\boldsymbol{b}), \ \forall \alpha \in \mathbb{R}, \ \forall \boldsymbol{a}, \boldsymbol{b} \in \mathbb{E}^3. \tag{1.68}$$

Thus, it can again be described by a second-order tensor as

$$\boldsymbol{r}(\boldsymbol{a}) = \mathbf{R}\boldsymbol{a}, \quad \forall \boldsymbol{a} \in \mathbb{E}^3, \quad \mathbf{R} \in \mathbf{Lin}^3. \tag{1.69}$$

This tensor $\mathbf{R}$ is referred to as rotation tensor.

Let us construct the rotation tensor which rotates an arbitrary vector $\boldsymbol{a} \in \mathbb{E}^3$ about an axis specified by a unit vector $\boldsymbol{e} \in \mathbb{E}^3$ (see Fig. 1.2). Decomposing the vector $\boldsymbol{a}$ by $\boldsymbol{a} = \boldsymbol{a}^* + \boldsymbol{x}$ in two vectors along and perpendicular to the rotation axis we can write

$$\boldsymbol{r}(\boldsymbol{a}) = \boldsymbol{a}^* + \boldsymbol{x} \cos\omega + \boldsymbol{y} \sin\omega = \boldsymbol{a}^* + \left(\boldsymbol{a} - \boldsymbol{a}^*\right) \cos\omega + \boldsymbol{y} \sin\omega, \tag{1.70}$$

where ω denotes the rotation angle. By virtue of the geometric identities

$$\boldsymbol{a}^* = (\boldsymbol{a} \cdot \boldsymbol{e})\, \boldsymbol{e} = (\boldsymbol{e} \otimes \boldsymbol{e})\, \boldsymbol{a}, \quad \boldsymbol{y} = \boldsymbol{e} \times \boldsymbol{x} = \boldsymbol{e} \times \left(\boldsymbol{a} - \boldsymbol{a}^*\right) = \boldsymbol{e} \times \boldsymbol{a} = \hat{\boldsymbol{e}}\boldsymbol{a}, \tag{1.71}$$

where "$\otimes$" denotes the so-called tensor product (1.80) (see Sect. 1.7), we obtain

$$\boldsymbol{r}(\boldsymbol{a}) = \cos\omega \boldsymbol{a} + \sin\omega \hat{\boldsymbol{e}}\boldsymbol{a} + (1 - \cos\omega)(\boldsymbol{e} \otimes \boldsymbol{e})\, \boldsymbol{a}. \tag{1.72}$$

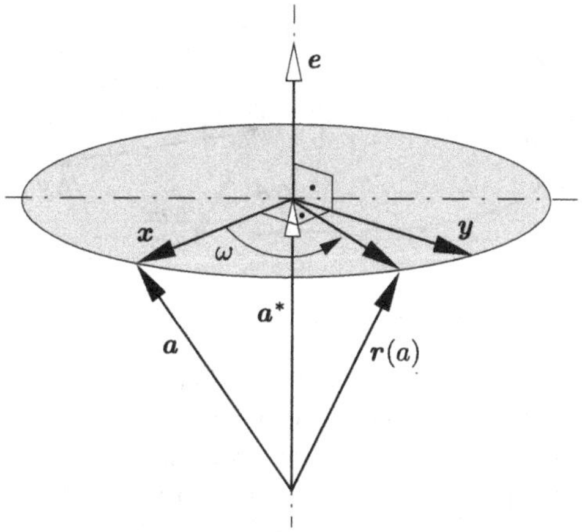

Fig. 1.2 Finite rotation of a vector in $\mathbb{E}^3$

Thus the rotation tensor can be given by

$$\mathbf{R} = \cos\omega \mathbf{I} + \sin\omega \hat{\boldsymbol{e}} + (1 - \cos\omega)\, \boldsymbol{e} \otimes \boldsymbol{e}, \tag{1.73}$$

where $\mathbf{I}$ denotes the so-called identity tensor (1.89) (see Sect. 1.7).

Another useful representation for the rotation tensor can be obtained utilizing the fact that $\boldsymbol{x} = \boldsymbol{y} \times \boldsymbol{e} = -\boldsymbol{e} \times \boldsymbol{y}$. Indeed, rewriting (1.70) by

$$\boldsymbol{r}\,(\boldsymbol{a}) = \boldsymbol{a} + \boldsymbol{x}\,(\cos\omega - 1) + \boldsymbol{y} \sin\omega \tag{1.74}$$

and keeping $(1.71)_2$ in mind we receive

$$\boldsymbol{r}\,(\boldsymbol{a}) = \boldsymbol{a} + \sin\omega \hat{\boldsymbol{e}} \boldsymbol{a} + (1 - \cos\omega)\,(\hat{\boldsymbol{e}})^2\, \boldsymbol{a}. \tag{1.75}$$

This leads to the expression for the rotation tensor

$$\mathbf{R} = \mathbf{I} + \sin\omega \hat{\boldsymbol{e}} + (1 - \cos\omega)\,(\hat{\boldsymbol{e}})^2 \tag{1.76}$$

known as the Euler-Rodrigues formula (see, e.g., [9]).

Example 1.4. The Cauchy stress tensor as a linear mapping of the unit surface normal into the Cauchy stress vector. Let us consider a body B in the current configuration at a time t. In order to define the stress in some point P let us further imagine a smooth surface going through P and separating B into two parts (Fig. 1.3). Then, one can define a force $\Delta \boldsymbol{p}$ and a couple $\Delta \boldsymbol{m}$ resulting from the forces exerted by the (hidden) material on one side of the surface ΔA and acting on the material on the other side of this surface. Let the area ΔA tend to zero keeping P as inner point. A basic postulate of continuum mechanics is that the limit

$$\boldsymbol{t} = \lim_{\Delta A \to 0} \frac{\Delta \boldsymbol{p}}{\Delta A}$$

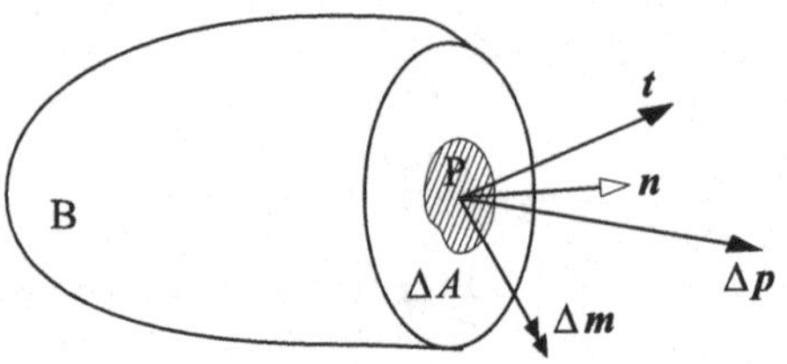

Fig. 1.3 Cauchy stress vector

exists and is final. The so-defined vector $\boldsymbol{t}$ is called Cauchy stress vector. Cauchy's fundamental postulate states that the vector $\boldsymbol{t}$ depends on the surface only through the outward unit normal $\boldsymbol{n}$. In other words, the Cauchy stress vector is the same for all surfaces through P which have $\boldsymbol{n}$ as the normal in P. Further, according to Cauchy's theorem the mapping $\boldsymbol{n} \to \boldsymbol{t}$ is linear provided $\boldsymbol{t}$ is a continuous function of the position vector $\boldsymbol{x}$ at P. Hence, this mapping can be described by a second-order tensor $\boldsymbol{\sigma}$ called the Cauchy stress tensor so that

$$\boldsymbol{t} = \boldsymbol{\sigma}\boldsymbol{n}. \tag{1.77}$$

On the basis of the "right" mapping (1.48) we can also define the "left" one by the following condition

$$(\boldsymbol{y}\mathbf{A}) \cdot \boldsymbol{x} = \boldsymbol{y} \cdot (\mathbf{A}\boldsymbol{x}), \quad \forall \boldsymbol{x} \in \mathbb{E}^n, \quad \mathbf{A} \in \mathbf{Lin}^n. \tag{1.78}$$

First, it should be shown that for all $\boldsymbol{y} \in \mathbb{E}^n$ there exists a unique vector $\boldsymbol{y}\mathbf{A} \in \mathbb{E}^n$ satisfying the condition (1.78) for all $\boldsymbol{x} \in \mathbb{E}^n$. Let $\mathcal{G} = \{\boldsymbol{g}_1, \boldsymbol{g}_2, \ldots, \boldsymbol{g}_n\}$ and $\mathcal{G}' = \{\boldsymbol{g}^1, \boldsymbol{g}^2, \ldots, \boldsymbol{g}^n\}$ be dual bases in $\mathbb{E}^n$. Then, we can represent two arbitrary vectors $\boldsymbol{x}, \boldsymbol{y} \in \mathbb{E}^n$, by $\boldsymbol{x} = x_i \boldsymbol{g}^i$ and $\boldsymbol{y} = y_i \boldsymbol{g}^i$. Now, consider the vector

$$\boldsymbol{y}\mathbf{A} = y_i \left[\boldsymbol{g}^i \cdot \left(\mathbf{A}\boldsymbol{g}^j\right)\right] \boldsymbol{g}_j.$$

It holds: $(\boldsymbol{y}\mathbf{A}) \cdot \boldsymbol{x} = y_i x_j \left[\boldsymbol{g}^i \cdot \left(\mathbf{A}\boldsymbol{g}^j\right)\right]$. On the other hand, we obtain the same result also by

$$\boldsymbol{y} \cdot (\mathbf{A}\boldsymbol{x}) = \boldsymbol{y} \cdot \left(x_j \mathbf{A}\boldsymbol{g}^j\right) = y_i x_j \left[\boldsymbol{g}^i \cdot \left(\mathbf{A}\boldsymbol{g}^j\right)\right].$$

Further, we show that the vector $\boldsymbol{y}\mathbf{A}$, satisfying condition (1.78) for all $\boldsymbol{x} \in \mathbb{E}^n$, is unique. Conversely, let $\boldsymbol{a}, \boldsymbol{b} \in \mathbb{E}^n$ be two such vectors. Then, we have

$$\boldsymbol{a} \cdot \boldsymbol{x} = \boldsymbol{b} \cdot \boldsymbol{x} \Rightarrow (\boldsymbol{a} - \boldsymbol{b}) \cdot \boldsymbol{x} = 0, \ \forall \boldsymbol{x} \in \mathbb{E}^n \Rightarrow (\boldsymbol{a} - \boldsymbol{b}) \cdot (\boldsymbol{a} - \boldsymbol{b}) = 0,$$

which by axiom (C.4) implies that $\boldsymbol{a} = \boldsymbol{b}$.

Since the order of mappings in (1.78) is irrelevant we can write them without brackets and dots as follows

$$\boldsymbol{y} \cdot (\mathbf{A}\boldsymbol{x}) = (\boldsymbol{y}\mathbf{A}) \cdot \boldsymbol{x} = \boldsymbol{y}\mathbf{A}\boldsymbol{x}. \tag{1.79}$$

1.7 Tensor Product, Representation of a Tensor with Respect to a Basis

The tensor product plays an important role since it enables to construct a second-order tensor from two vectors. In order to define the tensor product we consider two vectors $\boldsymbol{a}, \boldsymbol{b} \in \mathbb{E}^n$. An arbitrary vector $\boldsymbol{x} \in \mathbb{E}^n$ can be mapped into another vector $\boldsymbol{a}(\boldsymbol{b} \cdot \boldsymbol{x}) \in \mathbb{E}^n$. This mapping is denoted by symbol "$\otimes$" as $\boldsymbol{a} \otimes \boldsymbol{b}$. Thus,

$$(\boldsymbol{a} \otimes \boldsymbol{b})\boldsymbol{x} = \boldsymbol{a}(\boldsymbol{b} \cdot \boldsymbol{x}), \quad \boldsymbol{a}, \boldsymbol{b} \in \mathbb{E}^n, \ \forall \boldsymbol{x} \in \mathbb{E}^n. \tag{1.80}$$

It can be shown that the mapping (1.80) fulfills the conditions (1.49)–(1.51) and for this reason is linear. Indeed, by virtue of (B.1), (B.4), (C.2) and (C.3) we can write

$$\begin{aligned}(\boldsymbol{a} \otimes \boldsymbol{b})(\boldsymbol{x} + \boldsymbol{y}) &= \boldsymbol{a}[\boldsymbol{b} \cdot (\boldsymbol{x} + \boldsymbol{y})] = \boldsymbol{a}(\boldsymbol{b} \cdot \boldsymbol{x} + \boldsymbol{b} \cdot \boldsymbol{y}) \\ &= (\boldsymbol{a} \otimes \boldsymbol{b})\boldsymbol{x} + (\boldsymbol{a} \otimes \boldsymbol{b})\boldsymbol{y},\end{aligned} \tag{1.81}$$

$$\begin{aligned}(\boldsymbol{a} \otimes \boldsymbol{b})(\alpha\boldsymbol{x}) &= \boldsymbol{a}[\boldsymbol{b} \cdot (\alpha\boldsymbol{x})] = \alpha(\boldsymbol{b} \cdot \boldsymbol{x})\boldsymbol{a} \\ &= \alpha(\boldsymbol{a} \otimes \boldsymbol{b})\boldsymbol{x}, \quad \boldsymbol{a}, \boldsymbol{b} \in \mathbb{E}^n, \ \forall \boldsymbol{x}, \boldsymbol{y} \in \mathbb{E}^n, \ \forall \alpha \in \mathbb{R}.\end{aligned} \tag{1.82}$$

Thus, the tensor product of two vectors represents a second-order tensor. Further, it holds

$$\boldsymbol{c} \otimes (\boldsymbol{a} + \boldsymbol{b}) = \boldsymbol{c} \otimes \boldsymbol{a} + \boldsymbol{c} \otimes \boldsymbol{b}, \quad (\boldsymbol{a} + \boldsymbol{b}) \otimes \boldsymbol{c} = \boldsymbol{a} \otimes \boldsymbol{c} + \boldsymbol{b} \otimes \boldsymbol{c}, \tag{1.83}$$

$$(\alpha\boldsymbol{a}) \otimes (\beta\boldsymbol{b}) = \alpha\beta(\boldsymbol{a} \otimes \boldsymbol{b}), \quad \boldsymbol{a}, \boldsymbol{b}, \boldsymbol{c} \in \mathbb{E}^n, \ \forall \alpha, \beta \in \mathbb{R}. \tag{1.84}$$

Indeed, mapping an arbitrary vector $\boldsymbol{x} \in \mathbb{E}^n$ by both sides of these relations and using (1.52) and (1.80) we obtain

$$\begin{aligned}\boldsymbol{c} \otimes (\boldsymbol{a} + \boldsymbol{b})\boldsymbol{x} &= \boldsymbol{c}(\boldsymbol{a} \cdot \boldsymbol{x} + \boldsymbol{b} \cdot \boldsymbol{x}) = \boldsymbol{c}(\boldsymbol{a} \cdot \boldsymbol{x}) + \boldsymbol{c}(\boldsymbol{b} \cdot \boldsymbol{x}) \\ &= (\boldsymbol{c} \otimes \boldsymbol{a})\boldsymbol{x} + (\boldsymbol{c} \otimes \boldsymbol{b})\boldsymbol{x} = (\boldsymbol{c} \otimes \boldsymbol{a} + \boldsymbol{c} \otimes \boldsymbol{b})\boldsymbol{x},\end{aligned}$$

$$\begin{aligned}[(\boldsymbol{a} + \boldsymbol{b}) \otimes \boldsymbol{c}]\boldsymbol{x} &= (\boldsymbol{a} + \boldsymbol{b})(\boldsymbol{c} \cdot \boldsymbol{x}) = \boldsymbol{a}(\boldsymbol{c} \cdot \boldsymbol{x}) + \boldsymbol{b}(\boldsymbol{c} \cdot \boldsymbol{x}) \\ &= (\boldsymbol{a} \otimes \boldsymbol{c})\boldsymbol{x} + (\boldsymbol{b} \otimes \boldsymbol{c})\boldsymbol{x} = (\boldsymbol{a} \otimes \boldsymbol{c} + \boldsymbol{b} \otimes \boldsymbol{c})\boldsymbol{x},\end{aligned}$$

$$\begin{aligned}(\alpha\boldsymbol{a}) \otimes (\beta\boldsymbol{b})\boldsymbol{x} &= (\alpha\boldsymbol{a})(\beta\boldsymbol{b} \cdot \boldsymbol{x}) \\ &= \alpha\beta\boldsymbol{a}(\boldsymbol{b} \cdot \boldsymbol{x}) = \alpha\beta(\boldsymbol{a} \otimes \boldsymbol{b})\boldsymbol{x}, \quad \forall \boldsymbol{x} \in \mathbb{E}^n.\end{aligned}$$

For the "left" mapping by the tensor $\boldsymbol{a} \otimes \boldsymbol{b}$ we obtain from (1.78) (see Exercise 1.21)

$$\boldsymbol{y}\,(\boldsymbol{a} \otimes \boldsymbol{b}) = (\boldsymbol{y} \cdot \boldsymbol{a})\,\boldsymbol{b}, \quad \forall \boldsymbol{y} \in \mathbb{E}^n. \tag{1.85}$$

We have already seen that the set of all second-order tensors $\mathbf{Lin}^n$ represents a vector space. In the following, we show that a basis of $\mathbf{Lin}^n$ can be constructed with the aid of the tensor product (1.80).

Theorem 1.7. *Let $\mathcal{F} = \{\boldsymbol{f}_1, \boldsymbol{f}_2, \ldots, \boldsymbol{f}_n\}$ and $\mathcal{G} = \{\boldsymbol{g}_1, \boldsymbol{g}_2, \ldots, \boldsymbol{g}_n\}$ be two arbitrary bases of $\mathbb{E}^n$. Then, the tensors $\boldsymbol{f}_i \otimes \boldsymbol{g}_j$ $(i, j = 1, 2, \ldots, n)$ represent a basis of $\mathbf{Lin}^n$. The dimension of the vector space $\mathbf{Lin}^n$ is thus n^2.*

Proof. First, we prove that every tensor in $\mathbf{Lin}^n$ represents a linear combination of the tensors $\boldsymbol{f}_i \otimes \boldsymbol{g}_j$ $(i, j = 1, 2, \ldots, n)$. Indeed, let $\mathbf{A} \in \mathbf{Lin}^n$ be an arbitrary second-order tensor. Consider the following linear combination

$$\mathbf{A}' = \left(\boldsymbol{f}^i \mathbf{A} \boldsymbol{g}^j\right) \boldsymbol{f}_i \otimes \boldsymbol{g}_j,$$

where the vectors $\boldsymbol{f}^i$ and $\boldsymbol{g}^i$ $(i = 1, 2, \ldots, n)$ form the bases dual to $\mathcal{F}$ and $\mathcal{G}$, respectively. The tensors $\mathbf{A}$ and $\mathbf{A}'$ coincide if and only if

$$\mathbf{A}'\boldsymbol{x} = \mathbf{A}\boldsymbol{x}, \quad \forall \boldsymbol{x} \in \mathbb{E}^n. \tag{1.86}$$

Let $\boldsymbol{x} = x_j \boldsymbol{g}^j$. Then

$$\mathbf{A}'\boldsymbol{x} = \left(\boldsymbol{f}^i \mathbf{A} \boldsymbol{g}^j\right) \boldsymbol{f}_i \otimes \boldsymbol{g}_j \left(x_k \boldsymbol{g}^k\right) = \left(\boldsymbol{f}^i \mathbf{A} \boldsymbol{g}^j\right) \boldsymbol{f}_i x_k \delta_j^k = x_j \left(\boldsymbol{f}^i \mathbf{A} \boldsymbol{g}^j\right) \boldsymbol{f}_i.$$

On the other hand, $\mathbf{A}\boldsymbol{x} = x_j \mathbf{A} \boldsymbol{g}^j$. By virtue of (1.27) and (1.28) we can represent the vectors $\mathbf{A}\boldsymbol{g}^j$ $(j = 1, 2, \ldots, n)$ with respect to the basis $\mathcal{F}$ by $\mathbf{A}\boldsymbol{g}^j = \left[\boldsymbol{f}^i \cdot \left(\mathbf{A}\boldsymbol{g}^j\right)\right] \boldsymbol{f}_i = \left(\boldsymbol{f}^i \mathbf{A} \boldsymbol{g}^j\right) \boldsymbol{f}_i$ $(j = 1, 2, \ldots, n)$. Hence,

$$\mathbf{A}\boldsymbol{x} = x_j \left(\boldsymbol{f}^i \mathbf{A} \boldsymbol{g}^j\right) \boldsymbol{f}_i.$$

Thus, it is seen that condition (1.86) is satisfied for all $\boldsymbol{x} \in \mathbb{E}^n$. Finally, we show that the tensors $\boldsymbol{f}_i \otimes \boldsymbol{g}_j$ $(i, j = 1, 2, \ldots, n)$ are linearly independent. Otherwise, there would exist scalars α^{ij} $(i, j = 1, 2, \ldots, n)$, not all zero, such that

$$\alpha^{ij} \boldsymbol{f}_i \otimes \boldsymbol{g}_j = \mathbf{0}.$$

The right mapping of $\boldsymbol{g}^k$ $(k = 1, 2, \ldots, n)$ by this tensor equality yields then: $\alpha^{ik} \boldsymbol{f}_i = \boldsymbol{0}$ $(k = 1, 2, \ldots, n)$. This contradicts, however, the fact that the vectors $\boldsymbol{f}_k$ $(k = 1, 2, \ldots, n)$ form a basis and are therefore linearly independent.

For the representation of second-order tensors we will in the following use primarily the bases $\boldsymbol{g}_i \otimes \boldsymbol{g}_j$, $\boldsymbol{g}^i \otimes \boldsymbol{g}^j$, $\boldsymbol{g}^i \otimes \boldsymbol{g}_j$ or $\boldsymbol{g}_i \otimes \boldsymbol{g}^j$ $(i, j = 1, 2, \ldots, n)$. With respect to these bases a tensor $\mathbf{A} \in \mathbf{Lin}^n$ is written as

$$\mathbf{A} = \mathrm{A}^{ij} \boldsymbol{g}_i \otimes \boldsymbol{g}_j = \mathrm{A}_{ij} \boldsymbol{g}^i \otimes \boldsymbol{g}^j = \mathrm{A}^{i}_{\cdot j} \boldsymbol{g}_i \otimes \boldsymbol{g}^j = \mathrm{A}_{i\cdot}^{\ j} \boldsymbol{g}^i \otimes \boldsymbol{g}_j \tag{1.87}$$

with the components (see Exercise 1.22)

$$\mathrm{A}^{ij} = \boldsymbol{g}^i \mathbf{A} \boldsymbol{g}^j, \quad \mathrm{A}_{ij} = \boldsymbol{g}_i \mathbf{A} \boldsymbol{g}_j,$$
$$\mathrm{A}^{i}_{\cdot j} = \boldsymbol{g}^i \mathbf{A} \boldsymbol{g}_j, \quad \mathrm{A}_{i\cdot}^{\ j} = \boldsymbol{g}_i \mathbf{A} \boldsymbol{g}^j, \quad i, j = 1, 2, \ldots, n. \tag{1.88}$$

Note, that the subscript dot indicates the position of the above index. For example, for the components $\mathrm{A}^{i}_{\cdot j}$, i is the first index while for the components $\mathrm{A}_{j\cdot}^{\ i}$, i is the second index.

Of special importance is the so-called identity tensor **I**. It is defined by

$$\mathbf{I}\boldsymbol{x} = \boldsymbol{x}, \quad \forall \boldsymbol{x} \in \mathbb{E}^n. \tag{1.89}$$

With the aid of (1.25), (1.87) and (1.88) the components of the identity tensor can be expressed by

$$\mathrm{I}^{ij} = \boldsymbol{g}^i \mathbf{I} \boldsymbol{g}^j = \boldsymbol{g}^i \cdot \boldsymbol{g}^j = g^{ij}, \quad \mathrm{I}_{ij} = \boldsymbol{g}_i \mathbf{I} \boldsymbol{g}_j = \boldsymbol{g}_i \cdot \boldsymbol{g}_j = g_{ij},$$

$$\mathrm{I}^{i}_{\cdot j} = \mathrm{I}_{i\cdot}^{\ j} = \mathrm{I}^{i}_{j} = \boldsymbol{g}^i \mathbf{I} \boldsymbol{g}_j = \boldsymbol{g}_i \mathbf{I} \boldsymbol{g}^j = \boldsymbol{g}^i \cdot \boldsymbol{g}_j = \boldsymbol{g}_i \cdot \boldsymbol{g}^j = \delta^i_j, \tag{1.90}$$

where $i, j = 1, 2, \ldots, n$. Thus,

$$\mathbf{I} = g_{ij} \boldsymbol{g}^i \otimes \boldsymbol{g}^j = g^{ij} \boldsymbol{g}_i \otimes \boldsymbol{g}_j = \boldsymbol{g}^i \otimes \boldsymbol{g}_i = \boldsymbol{g}_i \otimes \boldsymbol{g}^i. \tag{1.91}$$

It is seen that the components $(1.90)_{1,2}$ of the identity tensor are given by relation (1.25). In view of (1.30) they characterize metric properties of the Euclidean space and are referred to as metric coefficients. For this reason, the identity tensor is frequently called metric tensor. With respect to an orthonormal basis relation (1.91) reduces to

$$\mathbf{I} = \sum_{i=1}^{n} \boldsymbol{e}_i \otimes \boldsymbol{e}_i. \tag{1.92}$$

1.8 Change of the Basis, Transformation Rules

Now, we are going to clarify how the vector and tensor components transform with the change of the basis. Let $\boldsymbol{x}$ be a vector and **A** a second-order tensor. According to (1.27) and (1.87)

$$\boldsymbol{x} = x^i \boldsymbol{g}_i = x_i \boldsymbol{g}^i, \tag{1.93}$$

$$\mathbf{A} = \mathrm{A}^{ij} \boldsymbol{g}_i \otimes \boldsymbol{g}_j = \mathrm{A}_{ij} \boldsymbol{g}^i \otimes \boldsymbol{g}^j = \mathrm{A}^{i}_{\cdot j} \boldsymbol{g}_i \otimes \boldsymbol{g}^j = \mathrm{A}_{i\cdot}^{\ j} \boldsymbol{g}^i \otimes \boldsymbol{g}_j. \tag{1.94}$$

With the aid of (1.21) and (1.28) we can write

$$x^i = \boldsymbol{x} \cdot \boldsymbol{g}^i = \boldsymbol{x} \cdot \left(g^{ij} \boldsymbol{g}_j\right) = x_j g^{ji}, \quad x_i = \boldsymbol{x} \cdot \boldsymbol{g}_i = \boldsymbol{x} \cdot \left(g_{ij} \boldsymbol{g}^j\right) = x^j g_{ji}, \tag{1.95}$$

where $i = 1, 2, \ldots, n$. Similarly we obtain by virtue of (1.88)

$$\begin{aligned} \mathrm{A}^{ij} = \boldsymbol{g}^i \mathbf{A} \boldsymbol{g}^j &= \boldsymbol{g}^i \mathbf{A} \left(g^{jk} \boldsymbol{g}_k\right) \\ &= \left(g^{il} \boldsymbol{g}_l\right) \mathbf{A} \left(g^{jk} \boldsymbol{g}_k\right) = \mathrm{A}^i_{\cdot k} g^{kj} = g^{il} \mathrm{A}_{lk} g^{kj}, \end{aligned} \tag{1.96}$$

$$\begin{aligned} \mathrm{A}_{ij} = \boldsymbol{g}_i \mathbf{A} \boldsymbol{g}_j &= \boldsymbol{g}_i \mathbf{A} \left(g_{jk} \boldsymbol{g}^k\right) \\ &= \left(g_{il} \boldsymbol{g}^l\right) \mathbf{A} \left(g_{jk} \boldsymbol{g}^k\right) = \mathrm{A}_{i\cdot}^{\ k} g_{kj} = g_{il} \mathrm{A}^{lk} g_{kj}, \end{aligned} \tag{1.97}$$

where $i, j = 1, 2, \ldots, n$. The transformation rules (1.95)–(1.97) hold not only for dual bases. Indeed, let $\boldsymbol{g}_i$ and $\bar{\boldsymbol{g}}_i$ $(i = 1, 2, \ldots, n)$ be two arbitrary bases in $\mathbb{E}^n$, so that

$$\boldsymbol{x} = x^i \boldsymbol{g}_i = \bar{x}^i \bar{\boldsymbol{g}}_i, \tag{1.98}$$

$$\mathbf{A} = \mathrm{A}^{ij} \boldsymbol{g}_i \otimes \boldsymbol{g}_j = \bar{\mathrm{A}}^{ij} \bar{\boldsymbol{g}}_i \otimes \bar{\boldsymbol{g}}_j. \tag{1.99}$$

By means of the relations

$$\boldsymbol{g}_i = a_i^j \bar{\boldsymbol{g}}_j, \quad i = 1, 2, \ldots, n \tag{1.100}$$

one thus obtains

$$\boldsymbol{x} = x^i \boldsymbol{g}_i = x^i a_i^j \bar{\boldsymbol{g}}_j \quad \Rightarrow \quad \bar{x}^j = x^i a_i^j, \quad j = 1, 2, \ldots, n, \tag{1.101}$$

$$\begin{aligned} \mathbf{A} = \mathrm{A}^{ij} \boldsymbol{g}_i \otimes \boldsymbol{g}_j &= \mathrm{A}^{ij} \left(a_i^k \bar{\boldsymbol{g}}_k\right) \otimes \left(a_j^l \bar{\boldsymbol{g}}_l\right) = \mathrm{A}^{ij} a_i^k a_j^l \bar{\boldsymbol{g}}_k \otimes \bar{\boldsymbol{g}}_l \\ &\Rightarrow \bar{\mathrm{A}}^{kl} = \mathrm{A}^{ij} a_i^k a_j^l, \quad k, l = 1, 2, \ldots, n. \end{aligned} \tag{1.102}$$

1.9 Special Operations with Second-Order Tensors

In Sect. 1.6 we have seen that the set $\mathbf{Lin}^n$ represents a finite-dimensional vector space. Its elements are second-order tensors that can be treated as vectors in $\mathbb{E}^{n^2}$ with all the operations specific for vectors such as summation, multiplication by a scalar or a scalar product (the latter one will be defined for second-order tensors in Sect. 1.10). However, in contrast to conventional vectors in the Euclidean space, for second-order tensors one can additionally define some special operations as for example composition, transposition or inversion.

Composition (simple contraction). Let $\mathbf{A}, \mathbf{B} \in \mathbf{Lin}^n$ be two second-order tensors. The tensor $\mathbf{C} = \mathbf{AB}$ is called composition of $\mathbf{A}$ and $\mathbf{B}$ if

$$\mathbf{C}\boldsymbol{x} = \mathbf{A}(\mathbf{B}\boldsymbol{x}), \quad \forall \boldsymbol{x} \in \mathbb{E}^n. \tag{1.103}$$

For the left mapping (1.78) one can write

$$\boldsymbol{y}(\mathbf{AB}) = (\boldsymbol{y}\mathbf{A})\mathbf{B}, \quad \forall \boldsymbol{y} \in \mathbb{E}^n. \tag{1.104}$$

In order to prove the last relation we use again (1.78) and (1.103):

$$\begin{aligned} \boldsymbol{y}(\mathbf{AB})\boldsymbol{x} &= \boldsymbol{y} \cdot [(\mathbf{AB})\boldsymbol{x}] = \boldsymbol{y} \cdot [\mathbf{A}(\mathbf{B}\boldsymbol{x})] \\ &= (\boldsymbol{y}\mathbf{A}) \cdot (\mathbf{B}\boldsymbol{x}) = [(\boldsymbol{y}\mathbf{A})\mathbf{B}] \cdot \boldsymbol{x}, \quad \forall \boldsymbol{x} \in \mathbb{E}^n. \end{aligned}$$

The composition of tensors (1.103) is generally not commutative so that $\mathbf{AB} \neq \mathbf{BA}$. Two tensors $\mathbf{A}$ and $\mathbf{B}$ are called commutative if on the contrary $\mathbf{AB} = \mathbf{BA}$. Besides, the composition of tensors is characterized by the following properties (see Exercise 1.26):

$$\mathbf{A0} = \mathbf{0A} = \mathbf{0}, \quad \mathbf{AI} = \mathbf{IA} = \mathbf{A}, \tag{1.105}$$

$$\mathbf{A}(\mathbf{B} + \mathbf{C}) = \mathbf{AB} + \mathbf{AC}, \quad (\mathbf{B} + \mathbf{C})\mathbf{A} = \mathbf{BA} + \mathbf{CA}, \tag{1.106}$$

$$\mathbf{A}(\mathbf{BC}) = (\mathbf{AB})\mathbf{C}. \tag{1.107}$$

For example, the distributive rule $(1.106)_1$ can be proved as follows

$$\begin{aligned} [\mathbf{A}(\mathbf{B} + \mathbf{C})]\boldsymbol{x} &= \mathbf{A}[(\mathbf{B} + \mathbf{C})\boldsymbol{x}] = \mathbf{A}(\mathbf{B}\boldsymbol{x} + \mathbf{C}\boldsymbol{x}) = \mathbf{A}(\mathbf{B}\boldsymbol{x}) + \mathbf{A}(\mathbf{C}\boldsymbol{x}) \\ &= (\mathbf{AB})\boldsymbol{x} + (\mathbf{AC})\boldsymbol{x} = (\mathbf{AB} + \mathbf{AC})\boldsymbol{x}, \quad \forall \boldsymbol{x} \in \mathbb{E}^n. \end{aligned}$$

For the tensor product (1.80) the composition (1.103) yields

$$(\boldsymbol{a} \otimes \boldsymbol{b})(\boldsymbol{c} \otimes \boldsymbol{d}) = (\boldsymbol{b} \cdot \boldsymbol{c})\boldsymbol{a} \otimes \boldsymbol{d}, \quad \boldsymbol{a}, \boldsymbol{b}, \boldsymbol{c}, \boldsymbol{d} \in \mathbb{E}^n. \tag{1.108}$$

Indeed, by virtue of (1.80), (1.82) and (1.103)

$$\begin{aligned} (\boldsymbol{a} \otimes \boldsymbol{b})(\boldsymbol{c} \otimes \boldsymbol{d})\boldsymbol{x} &= (\boldsymbol{a} \otimes \boldsymbol{b})[(\boldsymbol{c} \otimes \boldsymbol{d})\boldsymbol{x}] = (\boldsymbol{d} \cdot \boldsymbol{x})(\boldsymbol{a} \otimes \boldsymbol{b})\boldsymbol{c} \\ &= (\boldsymbol{d} \cdot \boldsymbol{x})(\boldsymbol{b} \cdot \boldsymbol{c})\boldsymbol{a} = (\boldsymbol{b} \cdot \boldsymbol{c})(\boldsymbol{a} \otimes \boldsymbol{d})\boldsymbol{x} \\ &= [(\boldsymbol{b} \cdot \boldsymbol{c})\boldsymbol{a} \otimes \boldsymbol{d}]\boldsymbol{x}, \quad \forall \boldsymbol{x} \in \mathbb{E}^n. \end{aligned}$$

Thus, we can write

$$\mathbf{AB} = \mathrm{A}^{ik}\mathrm{B}_{k.}^{\ j}\boldsymbol{g}_i \otimes \boldsymbol{g}_j = \mathrm{A}_{ik}\mathrm{B}^{kj}\boldsymbol{g}^i \otimes \boldsymbol{g}_j$$
$$= \mathrm{A}_{.k}^{i}\mathrm{B}_{.j}^{k}\boldsymbol{g}_i \otimes \boldsymbol{g}^j = \mathrm{A}_{i.}^{\ k}\mathrm{B}_{kj}\boldsymbol{g}^i \otimes \boldsymbol{g}^j, \qquad (1.109)$$

where **A** and **B** are given in the form (1.87).

Powers, polynomials and functions of second-order tensors. On the basis of the composition (1.103) one defines by

$$\mathbf{A}^m = \underbrace{\mathbf{AA}\ldots\mathbf{A}}_{m \text{ times}}, \quad m = 1, 2, 3\ldots, \quad \mathbf{A}^0 = \mathbf{I} \qquad (1.110)$$

powers (monomials) of second-order tensors characterized by the following evident properties

$$\mathbf{A}^k\mathbf{A}^l = \mathbf{A}^{k+l}, \quad \left(\mathbf{A}^k\right)^l = \mathbf{A}^{kl}, \qquad (1.111)$$

$$(\alpha\mathbf{A})^k = \alpha^k\mathbf{A}^k, \quad k, l = 0, 1, 2\ldots \qquad (1.112)$$

With the aid of the tensor powers a polynomial of **A** can be defined by

$$g(\mathbf{A}) = a_0\mathbf{I} + a_1\mathbf{A} + a_2\mathbf{A}^2 + \ldots + a_m\mathbf{A}^m = \sum_{k=0}^{m} a_k\mathbf{A}^k. \qquad (1.113)$$

$g(\mathbf{A})$: $\mathbf{Lin}^n \mapsto \mathbf{Lin}^n$ represents a tensor function mapping one second-order tensor into another one within $\mathbf{Lin}^n$. By this means one can define various tensor functions. Of special interest is the exponential one

$$\exp(\mathbf{A}) = \sum_{k=0}^{\infty} \frac{\mathbf{A}^k}{k!} \qquad (1.114)$$

given by the infinite power series.

Transposition. The transposed tensor $\mathbf{A}^{\mathrm{T}}$ is defined by:

$$\mathbf{A}^{\mathrm{T}}\boldsymbol{x} = \boldsymbol{x}\mathbf{A}, \quad \forall \boldsymbol{x} \in \mathbb{E}^n, \qquad (1.115)$$

so that one can also write

$$\mathbf{A}\boldsymbol{y} = \boldsymbol{y}\mathbf{A}^{\mathrm{T}}, \quad \boldsymbol{x}\mathbf{A}\boldsymbol{y} = \boldsymbol{y}\mathbf{A}^{\mathrm{T}}\boldsymbol{x}, \quad \forall \boldsymbol{x}, \boldsymbol{y} \in \mathbb{E}^n. \qquad (1.116)$$

Indeed,

$$\boldsymbol{x} \cdot (\mathbf{A}\boldsymbol{y}) = (\boldsymbol{x}\mathbf{A}) \cdot \boldsymbol{y} = \boldsymbol{y} \cdot \left(\mathbf{A}^{\mathrm{T}}\boldsymbol{x}\right) = \boldsymbol{y}\mathbf{A}^{\mathrm{T}}\boldsymbol{x} = \boldsymbol{x} \cdot \left(\boldsymbol{y}\mathbf{A}^{\mathrm{T}}\right), \ \forall \boldsymbol{x}, \boldsymbol{y} \in \mathbb{E}^n.$$

Consequently,

$$\left(\mathbf{A}^{\mathrm{T}}\right)^{\mathrm{T}} = \mathbf{A}. \tag{1.117}$$

Transposition represents a linear operation over a second-order tensor since

$$(\mathbf{A} + \mathbf{B})^{\mathrm{T}} = \mathbf{A}^{\mathrm{T}} + \mathbf{B}^{\mathrm{T}} \tag{1.118}$$

and

$$(\alpha \mathbf{A})^{\mathrm{T}} = \alpha \mathbf{A}^{\mathrm{T}}, \quad \forall \alpha \in \mathbb{R}. \tag{1.119}$$

The composition of second-order tensors is transposed by

$$(\mathbf{AB})^{\mathrm{T}} = \mathbf{B}^{\mathrm{T}} \mathbf{A}^{\mathrm{T}}. \tag{1.120}$$

Indeed, in view of (1.104) and (1.115)

$$(\mathbf{AB})^{\mathrm{T}} \boldsymbol{x} = \boldsymbol{x} (\mathbf{AB}) = (\boldsymbol{x} \mathbf{A}) \mathbf{B} = \mathbf{B}^{\mathrm{T}} (\boldsymbol{x} \mathbf{A}) = \mathbf{B}^{\mathrm{T}} \mathbf{A}^{\mathrm{T}} \boldsymbol{x}, \quad \forall \boldsymbol{x} \in \mathbb{E}^n.$$

For the tensor product of two vectors $\boldsymbol{a}, \boldsymbol{b} \in \mathbb{E}^n$ we further obtain by use of (1.80) and (1.85)

$$(\boldsymbol{a} \otimes \boldsymbol{b})^{\mathrm{T}} = \boldsymbol{b} \otimes \boldsymbol{a}. \tag{1.121}$$

This ensures the existence and uniqueness of the transposed tensor. Indeed, every tensor $\mathbf{A}$ in $\mathbf{Lin}^n$ can be represented with respect to the tensor product of the basis vectors in $\mathbb{E}^n$ in the form (1.87). Hence, considering (1.121) we have

$$\mathbf{A}^{\mathrm{T}} = \mathrm{A}^{ij} \boldsymbol{g}_j \otimes \boldsymbol{g}_i = \mathrm{A}_{ij} \boldsymbol{g}^j \otimes \boldsymbol{g}^i = \mathrm{A}^{i}_{\cdot j} \boldsymbol{g}^j \otimes \boldsymbol{g}_i = \mathrm{A}_{i \cdot}^{\ j} \boldsymbol{g}_j \otimes \boldsymbol{g}^i, \tag{1.122}$$

or

$$\mathbf{A}^{\mathrm{T}} = \mathrm{A}^{ji} \boldsymbol{g}_i \otimes \boldsymbol{g}_j = \mathrm{A}_{ji} \boldsymbol{g}^i \otimes \boldsymbol{g}^j = \mathrm{A}^{j}_{\cdot i} \boldsymbol{g}^i \otimes \boldsymbol{g}_j = \mathrm{A}_{j \cdot}^{\ i} \boldsymbol{g}_i \otimes \boldsymbol{g}^j. \tag{1.123}$$

Comparing the latter result with the original representation (1.87) one observes that the components of the transposed tensor can be expressed by

$$\left(\mathbf{A}^{\mathrm{T}}\right)_{ij} = \mathrm{A}_{ji}, \quad \left(\mathbf{A}^{\mathrm{T}}\right)^{ij} = \mathrm{A}^{ji}, \tag{1.124}$$

$$\left(\mathbf{A}^{\mathrm{T}}\right)_{i \cdot}^{\ j} = \mathrm{A}^{j}_{\cdot i} = g^{jk} \mathrm{A}_{k \cdot}^{\ l} g_{li}, \quad \left(\mathbf{A}^{\mathrm{T}}\right)^{i}_{\cdot j} = \mathrm{A}_{j \cdot}^{\ i} = g_{jk} \mathrm{A}^{k}_{\cdot l} g^{li}. \tag{1.125}$$

For example, the last relation results from (1.88) and (1.116) within the following steps

$$\left(\mathbf{A}^{\mathrm{T}}\right)^{i}_{\cdot j} = \boldsymbol{g}^i \mathbf{A}^{\mathrm{T}} \boldsymbol{g}_j = \boldsymbol{g}_j \mathbf{A} \boldsymbol{g}^i = \boldsymbol{g}_j \left(\mathrm{A}^{k}_{\cdot l} \boldsymbol{g}_k \otimes \boldsymbol{g}^l\right) \boldsymbol{g}^i = g_{jk} \mathrm{A}^{k}_{\cdot l} g^{li}.$$

According to (1.124) the homogeneous (covariant or contravariant) components of the transposed tensor can simply be obtained by reflecting the matrix of the original components from the main diagonal. It does not, however, hold for the mixed components (1.125).

The transposition operation (1.115) gives rise to the definition of symmetric $\mathbf{M}^{\mathrm{T}} = \mathbf{M}$ and skew-symmetric second-order tensors $\mathbf{W}^{\mathrm{T}} = -\mathbf{W}$.

Obviously, the identity tensor is symmetric

$$\mathbf{I}^{\mathrm{T}} = \mathbf{I}. \tag{1.126}$$

Indeed,

$$\boldsymbol{x}\mathbf{I}\boldsymbol{y} = \boldsymbol{x} \cdot \boldsymbol{y} = \boldsymbol{y} \cdot \boldsymbol{x} = \boldsymbol{y}\mathbf{I}\boldsymbol{x} = \boldsymbol{x}\mathbf{I}^{\mathrm{T}}\boldsymbol{y}, \quad \forall \boldsymbol{x}, \boldsymbol{y} \in \mathbb{E}^n.$$

One can easily show that the tensor $\hat{\boldsymbol{w}}$ (1.66) is skew-symmetric so that

$$\hat{\boldsymbol{w}}^{\mathrm{T}} = -\hat{\boldsymbol{w}}. \tag{1.127}$$

Indeed, by virtue of (1.32) and (1.116) on can write

$$\boldsymbol{x}\hat{\boldsymbol{w}}^{\mathrm{T}}\boldsymbol{y} = \boldsymbol{y}\hat{\boldsymbol{w}}\boldsymbol{x} = \boldsymbol{y} \cdot (\boldsymbol{w} \times \boldsymbol{x}) = [\boldsymbol{y}\boldsymbol{w}\boldsymbol{x}] = -[\boldsymbol{x}\boldsymbol{w}\boldsymbol{y}]$$

$$= -\boldsymbol{x} \cdot (\boldsymbol{w} \times \boldsymbol{y}) = \boldsymbol{x}(-\hat{\boldsymbol{w}})\boldsymbol{y}, \quad \forall \boldsymbol{x}, \boldsymbol{y} \in \mathbb{E}^3.$$

Inversion. Let

$$\boldsymbol{y} = \mathbf{A}\boldsymbol{x}. \tag{1.128}$$

A tensor $\mathbf{A} \in \mathbf{Lin}^n$ is referred to as invertible if there exists a tensor $\mathbf{A}^{-1} \in \mathbf{Lin}^n$ satisfying the condition

$$\boldsymbol{x} = \mathbf{A}^{-1}\boldsymbol{y}, \quad \forall \boldsymbol{x} \in \mathbb{E}^n. \tag{1.129}$$

The tensor $\mathbf{A}^{-1}$ is called inverse of $\mathbf{A}$. The set of all invertible tensors $\mathbf{Inv}^n = \{\mathbf{A} \in \mathbf{Lin}^n : \exists \mathbf{A}^{-1}\}$ forms a subset of all second-order tensors $\mathbf{Lin}^n$.

Inserting (1.128) into (1.129) yields

$$\boldsymbol{x} = \mathbf{A}^{-1}\boldsymbol{y} = \mathbf{A}^{-1}(\mathbf{A}\boldsymbol{x}) = \left(\mathbf{A}^{-1}\mathbf{A}\right)\boldsymbol{x}, \quad \forall \boldsymbol{x} \in \mathbb{E}^n$$

and consequently

$$\mathbf{A}^{-1}\mathbf{A} = \mathbf{I}. \tag{1.130}$$

Theorem 1.8. *A tensor* $\mathbf{A}$ *is invertible if and only if* $\mathbf{A}\boldsymbol{x} = \boldsymbol{0}$ *implies that* $\boldsymbol{x} = \boldsymbol{0}$.

Proof. First we prove the sufficiency. To this end, we map the vector equation $\mathbf{A}\boldsymbol{x} = \boldsymbol{0}$ by $\mathbf{A}^{-1}$. According to (1.130) it yields: $\boldsymbol{0} = \mathbf{A}^{-1}\mathbf{A}\boldsymbol{x} = \mathbf{I}\boldsymbol{x} = \boldsymbol{x}$. To prove the necessity we consider a basis $\mathcal{G} = \{\boldsymbol{g}_1, \boldsymbol{g}_2, \ldots, \boldsymbol{g}_n\}$ in $\mathbb{E}^n$. It can be shown that the vectors $\boldsymbol{h}_i = \mathbf{A}\boldsymbol{g}_i$ $(i = 1, 2, \ldots, n)$ form likewise a basis of $\mathbb{E}^n$. Conversely, let these vectors be linearly dependent so that $a^i\boldsymbol{h}_i = \boldsymbol{0}$, where not all scalars a^i $(i = 1, 2, \ldots, n)$ are zero. Then, $\boldsymbol{0} = a^i\boldsymbol{h}_i = a^i\mathbf{A}\boldsymbol{g}_i = \mathbf{A}\boldsymbol{a}$, where $\boldsymbol{a} = a^i\boldsymbol{g}_i \neq \boldsymbol{0}$, which contradicts the assumption of the theorem. Now, consider the tensor $\mathbf{A}' = \boldsymbol{g}_i \otimes \boldsymbol{h}^i$, where the vectors $\boldsymbol{h}^i$ are dual to $\boldsymbol{h}_i$ $(i = 1, 2, \ldots, n)$. One can show that this tensor is inverse to $\mathbf{A}$, such that $\mathbf{A}' = \mathbf{A}^{-1}$. Indeed, let $\boldsymbol{x} = x^i \boldsymbol{g}_i$ be an arbitrary vector in $\mathbb{E}^n$. Then, $\boldsymbol{y} = \mathbf{A}\boldsymbol{x} = x^i \mathbf{A}\boldsymbol{g}_i = x^i \boldsymbol{h}_i$ and therefore $\mathbf{A}'\boldsymbol{y} = \boldsymbol{g}_i \otimes \boldsymbol{h}^i \left(x^j \boldsymbol{h}_j\right) = \boldsymbol{g}_i x^j \delta^i_j = x^i \boldsymbol{g}_i = \boldsymbol{x}$.

Conversely, it can be shown that an invertible tensor $\mathbf{A}$ is inverse to $\mathbf{A}^{-1}$ and consequently

$$\mathbf{A}\mathbf{A}^{-1} = \mathbf{I}. \tag{1.131}$$

For the proof we again consider the bases $\boldsymbol{g}_i$ and $\mathbf{A}\boldsymbol{g}_i$ $(i = 1, 2, \ldots, n)$. Let $\boldsymbol{y} = y^i \mathbf{A}\boldsymbol{g}_i$ be an arbitrary vector in $\mathbb{E}^n$. Let further $\boldsymbol{x} = \mathbf{A}^{-1}\boldsymbol{y} = y^i \boldsymbol{g}_i$ in view of (1.130). Then, $\mathbf{A}\boldsymbol{x} = y^i \mathbf{A}\boldsymbol{g}_i = \boldsymbol{y}$ which implies that the tensor $\mathbf{A}$ is inverse to $\mathbf{A}^{-1}$.

Relation (1.131) implies the uniqueness of the inverse. Indeed, if $\mathbf{A}^{-1}$ and $\widetilde{\mathbf{A}^{-1}}$ are two distinct tensors both inverse to $\mathbf{A}$ then there exists at least one vector $\boldsymbol{y} \in \mathbb{E}^n$ such that $\mathbf{A}^{-1}\boldsymbol{y} \neq \widetilde{\mathbf{A}^{-1}}\boldsymbol{y}$. Mapping both sides of this vector inequality by $\mathbf{A}$ and taking (1.131) into account we immediately come to the contradiction.

By means of (1.120), (1.126) and (1.131) we can write (see Exercise 1.39)

$$\left(\mathbf{A}^{-1}\right)^{\mathrm{T}} = \left(\mathbf{A}^{\mathrm{T}}\right)^{-1} = \mathbf{A}^{-\mathrm{T}}. \tag{1.132}$$

The composition of two arbitrary invertible tensors $\mathbf{A}$ and $\mathbf{B}$ is inverted by

$$(\mathbf{A}\mathbf{B})^{-1} = \mathbf{B}^{-1}\mathbf{A}^{-1}. \tag{1.133}$$

Indeed, let

$$\boldsymbol{y} = \mathbf{A}\mathbf{B}\boldsymbol{x}.$$

Mapping both sides of this vector identity by $\mathbf{A}^{-1}$ and then by $\mathbf{B}^{-1}$, we obtain with the aid of (1.130)

$$\boldsymbol{x} = \mathbf{B}^{-1}\mathbf{A}^{-1}\boldsymbol{y}, \quad \forall \boldsymbol{x} \in \mathbb{E}^n.$$

On the basis of transposition and inversion one defines the so-called orthogonal tensors. They do not change after consecutive transposition and inversion and form the following subset of $\mathbf{Lin}^n$:

$$\mathbf{Orth}^n = \left\{\mathbf{Q} \in \mathbf{Lin}^n : \mathbf{Q} = \mathbf{Q}^{-\mathrm{T}}\right\}. \tag{1.134}$$

For orthogonal tensors we can write in view of (1.130) and (1.131)

$$\mathbf{Q}\mathbf{Q}^{\mathrm{T}} = \mathbf{Q}^{\mathrm{T}}\mathbf{Q} = \mathbf{I}, \quad \forall \mathbf{Q} \in \mathbf{Orth}^n. \tag{1.135}$$

For example, one can show that the rotation tensor (1.73) is orthogonal. To this end, we complete the vector $\boldsymbol{e}$ defining the rotation axis (Fig. 1.2) to an orthonormal basis $\{\boldsymbol{e}, \boldsymbol{q}, \boldsymbol{p}\}$ such that $\boldsymbol{e} = \boldsymbol{q} \times \boldsymbol{p}$. Then, using the vector identity (see Exercise 1.15)

$$\boldsymbol{p}\left(\boldsymbol{q} \cdot \boldsymbol{x}\right) - \boldsymbol{q}\left(\boldsymbol{p} \cdot \boldsymbol{x}\right) = \left(\boldsymbol{q} \times \boldsymbol{p}\right) \times \boldsymbol{x}, \quad \forall \boldsymbol{x} \in \mathbb{E}^3 \tag{1.136}$$

we can write

$$\hat{e} = \boldsymbol{p} \otimes \boldsymbol{q} - \boldsymbol{q} \otimes \boldsymbol{p}. \tag{1.137}$$

The rotation tensor (1.73) takes thus the form

$$\mathbf{R} = \cos\omega \mathbf{I} + \sin\omega \left(\boldsymbol{p} \otimes \boldsymbol{q} - \boldsymbol{q} \otimes \boldsymbol{p}\right) + (1 - \cos\omega)\left(\boldsymbol{e} \otimes \boldsymbol{e}\right). \tag{1.138}$$

Hence,

$$\begin{aligned}\mathbf{R}\mathbf{R}^{\mathrm{T}} &= \left[\cos\omega \mathbf{I} + \sin\omega \left(\boldsymbol{p} \otimes \boldsymbol{q} - \boldsymbol{q} \otimes \boldsymbol{p}\right) + (1 - \cos\omega)\left(\boldsymbol{e} \otimes \boldsymbol{e}\right)\right] \\ &\quad \left[\cos\omega \mathbf{I} - \sin\omega \left(\boldsymbol{p} \otimes \boldsymbol{q} - \boldsymbol{q} \otimes \boldsymbol{p}\right) + (1 - \cos\omega)\left(\boldsymbol{e} \otimes \boldsymbol{e}\right)\right] \\ &= \cos^2\omega \mathbf{I} + \sin^2\omega \left(\boldsymbol{e} \otimes \boldsymbol{e}\right) + \sin^2\omega \left(\boldsymbol{p} \otimes \boldsymbol{p} + \boldsymbol{q} \otimes \boldsymbol{q}\right) = \mathbf{I}.\end{aligned}$$

Alternatively one can express the transposed rotation tensor (1.73) by

$$\begin{aligned}\mathbf{R}^{\mathrm{T}} &= \cos\omega \mathbf{I} + \sin\omega \hat{e}^{\mathrm{T}} + (1 - \cos\omega)\, \boldsymbol{e} \otimes \boldsymbol{e} \\ &= \cos(-\omega)\, \mathbf{I} + \sin(-\omega)\, \hat{e} + \left[1 - \cos(-\omega)\right] \boldsymbol{e} \otimes \boldsymbol{e}\end{aligned} \tag{1.139}$$

taking (1.121), (1.126) and (1.127) into account. Thus, $\mathbf{R}^{\mathrm{T}}$ (1.139) describes the rotation about the same axis $\boldsymbol{e}$ by the angle $-\omega$, which likewise implies that $\mathbf{R}^{\mathrm{T}}\mathbf{R}\boldsymbol{x} = \boldsymbol{x}, \ \forall \boldsymbol{x} \in \mathbb{E}^3$.

It is interesting that the exponential function (1.114) of a skew-symmetric tensors represents an orthogonal tensor. Indeed, keeping in mind that a skew-symmetric tensor $\mathbf{W}$ commutes with its transposed counterpart $\mathbf{W}^{\mathrm{T}} = -\mathbf{W}$ and using the identities $\exp(\mathbf{A} + \mathbf{B}) = \exp(\mathbf{A}) \exp(\mathbf{B})$ for commutative tensors (Exercise 1.29) and $\left(\mathbf{A}^k\right)^{\mathrm{T}} = \left(\mathbf{A}^{\mathrm{T}}\right)^k$ for integer k (Exercise 1.37) we can write

$$\begin{aligned}\mathbf{I} &= \exp(\mathbf{0}) = \exp(\mathbf{W} - \mathbf{W}) = \exp\left(\mathbf{W} + \mathbf{W}^{\mathrm{T}}\right) \\ &= \exp(\mathbf{W}) \exp\left(\mathbf{W}^{\mathrm{T}}\right) = \exp(\mathbf{W}) \left[\exp(\mathbf{W})\right]^{\mathrm{T}},\end{aligned} \tag{1.140}$$

where $\mathbf{W}$ denotes an arbitrary skew-symmetric tensor.

1.10 Scalar Product of Second-Order Tensors

Consider two second-order tensors $\boldsymbol{a} \otimes \boldsymbol{b}$ and $\boldsymbol{c} \otimes \boldsymbol{d}$ given in terms of the tensor product (1.80). Their scalar product can be defined in the following manner:

$$(\boldsymbol{a} \otimes \boldsymbol{b}) : (\boldsymbol{c} \otimes \boldsymbol{d}) = (\boldsymbol{a} \cdot \boldsymbol{c})\, (\boldsymbol{b} \cdot \boldsymbol{d}), \quad \boldsymbol{a}, \boldsymbol{b}, \boldsymbol{c}, \boldsymbol{d} \in \mathbb{E}^n. \tag{1.141}$$

It leads to the following identity (Exercise 1.41):

$$\boldsymbol{c} \otimes \boldsymbol{d} : \mathbf{A} = \boldsymbol{c}\mathbf{A}\boldsymbol{d} = \boldsymbol{d}\mathbf{A}^{\mathrm{T}}\boldsymbol{c}. \tag{1.142}$$

For two arbitrary tensors **A** and **B** given in the form (1.87) we thus obtain

$$\mathbf{A} : \mathbf{B} = \mathrm{A}_{ij}\mathrm{B}^{ij} = \mathrm{A}^{ij}\mathrm{B}_{ij} = \mathrm{A}^{i}_{\cdot j}\mathrm{B}_{i\cdot}^{\ j} = \mathrm{A}_{i\cdot}^{\ j}\mathrm{B}^{i}_{\cdot j}. \tag{1.143}$$

Similar to vectors the scalar product of tensors is a real function characterized by the following properties (see Exercise 1.42)

D. (D.1) $\mathbf{A} : \mathbf{B} = \mathbf{B} : \mathbf{A}$ (commutative rule),
(D.2) $\mathbf{A} : (\mathbf{B} + \mathbf{C}) = \mathbf{A} : \mathbf{B} + \mathbf{A} : \mathbf{C}$ (distributive rule),
(D.3) $\alpha\,(\mathbf{A} : \mathbf{B}) = (\alpha\mathbf{A}) : \mathbf{B} = \mathbf{A} : (\alpha\mathbf{B})$ (associative rule for multiplication by a scalar), $\forall \mathbf{A}, \mathbf{B} \in \mathbf{Lin}^n$, $\forall \alpha \in \mathbb{R}$,
(D.4) $\mathbf{A} : \mathbf{A} \geq 0 \ \forall \mathbf{A} \in \mathbf{Lin}^n$, $\mathbf{A} : \mathbf{A} = 0$ if and only if $\mathbf{A} = \mathbf{0}$.

We prove for example the property (D.4). To this end, we represent an arbitrary tensor **A** with respect to an orthonormal basis of $\mathbf{Lin}^n$ as: $\mathbf{A} = \mathrm{A}^{ij}\boldsymbol{e}_i \otimes \boldsymbol{e}_j = \mathrm{A}_{ij}\boldsymbol{e}^i \otimes \boldsymbol{e}^j$, where $\mathrm{A}^{ij} = \mathrm{A}_{ij}$, $(i, j = 1, 2, \ldots, n)$, since $\boldsymbol{e}^i = \boldsymbol{e}_i$ $(i = 1, 2, \ldots, n)$ form an orthonormal basis of $\mathbb{E}^n$ (1.8). Keeping (1.143) in mind we then obtain:

$$\mathbf{A} : \mathbf{A} = \mathrm{A}^{ij}\mathrm{A}_{ij} = \sum_{i,j=1}^{n} \mathrm{A}^{ij}\mathrm{A}^{ij} = \sum_{i,j=1}^{n} \left(\mathrm{A}^{ij}\right)^2 \geq 0.$$

Using this important property one can define the norm of a second-order tensor by:

$$\|\mathbf{A}\| = (\mathbf{A} : \mathbf{A})^{1/2}, \quad \mathbf{A} \in \mathbf{Lin}^n. \tag{1.144}$$

For the scalar product of tensors one of which is given by a composition we can write

$$\mathbf{A} : (\mathbf{BC}) = \left(\mathbf{B}^{\mathrm{T}}\mathbf{A}\right) : \mathbf{C} = \left(\mathbf{A}\mathbf{C}^{\mathrm{T}}\right) : \mathbf{B}. \tag{1.145}$$

We prove this identity first for the tensor products:

$$\begin{aligned}(\boldsymbol{a} \otimes \boldsymbol{b}) : [(\boldsymbol{c} \otimes \boldsymbol{d})(\boldsymbol{e} \otimes \boldsymbol{f})] &= (\boldsymbol{d} \cdot \boldsymbol{e})[(\boldsymbol{a} \otimes \boldsymbol{b}) : (\boldsymbol{c} \otimes \boldsymbol{f})] \\ &= (\boldsymbol{d} \cdot \boldsymbol{e})(\boldsymbol{a} \cdot \boldsymbol{c})(\boldsymbol{b} \cdot \boldsymbol{f}),\end{aligned}$$

$$\begin{aligned}\left[(\boldsymbol{c} \otimes \boldsymbol{d})^{\mathrm{T}}(\boldsymbol{a} \otimes \boldsymbol{b})\right] : (\boldsymbol{e} \otimes \boldsymbol{f}) &= [(\boldsymbol{d} \otimes \boldsymbol{c})(\boldsymbol{a} \otimes \boldsymbol{b})] : (\boldsymbol{e} \otimes \boldsymbol{f}) \\ &= (\boldsymbol{a} \cdot \boldsymbol{c})[(\boldsymbol{d} \otimes \boldsymbol{b}) : (\boldsymbol{e} \otimes \boldsymbol{f})] \\ &= (\boldsymbol{d} \cdot \boldsymbol{e})(\boldsymbol{a} \cdot \boldsymbol{c})(\boldsymbol{b} \cdot \boldsymbol{f}),\end{aligned}$$

$$\left[(\boldsymbol{a} \otimes \boldsymbol{b})(\boldsymbol{e} \otimes \boldsymbol{f})^{\mathrm{T}}\right] : (\boldsymbol{c} \otimes \boldsymbol{d}) = [(\boldsymbol{a} \otimes \boldsymbol{b})(\boldsymbol{f} \otimes \boldsymbol{e})] : (\boldsymbol{c} \otimes \boldsymbol{d})$$
$$= (\boldsymbol{b} \cdot \boldsymbol{f})[(\boldsymbol{a} \otimes \boldsymbol{e}) : (\boldsymbol{c} \otimes \boldsymbol{d})]$$
$$= (\boldsymbol{d} \cdot \boldsymbol{e})(\boldsymbol{a} \cdot \boldsymbol{c})(\boldsymbol{b} \cdot \boldsymbol{f}).$$

For three arbitrary tensors **A**, **B** and **C** given in the form (1.87) we can write in view of (1.109), (1.125) and (1.143)

$$\mathrm{A}^{i}_{\cdot j}\left(\mathrm{B}^{\ k}_{i\cdot}\mathrm{C}^{\ j}_{k\cdot}\right) = \left(\mathrm{B}^{\ k}_{i\cdot}\mathrm{A}^{i}_{\cdot j}\right)\mathrm{C}^{\ j}_{k\cdot} = \left[\left(\mathbf{B}^{\mathrm{T}}\right)^{k}_{\cdot i}\mathrm{A}^{i}_{\cdot j}\right]\mathrm{C}^{\ j}_{k\cdot},$$

$$\mathrm{A}^{i}_{\cdot j}\left(\mathrm{B}^{\ k}_{i\cdot}\mathrm{C}^{\ j}_{k\cdot}\right) = \left(\mathrm{A}^{i}_{\cdot j}\mathrm{C}^{\ j}_{k\cdot}\right)\mathrm{B}^{\ k}_{i\cdot} = \left[\mathrm{A}^{i}_{\cdot j}\left(\mathbf{C}^{\mathrm{T}}\right)^{j}_{\cdot k}\right]\mathrm{B}^{\ k}_{i\cdot}. \tag{1.146}$$

Similarly we can prove that

$$\mathbf{A} : \mathbf{B} = \mathbf{A}^{\mathrm{T}} : \mathbf{B}^{\mathrm{T}}. \tag{1.147}$$

On the basis of the scalar product one defines the trace of second-order tensors by:

$$\mathrm{tr}\mathbf{A} = \mathbf{A} : \mathbf{I}. \tag{1.148}$$

For the tensor product (1.80) the trace (1.148) yields in view of (1.142)

$$\mathrm{tr}\,(\boldsymbol{a} \otimes \boldsymbol{b}) = \boldsymbol{a} \cdot \boldsymbol{b}. \tag{1.149}$$

With the aid of the relation (1.145) we further write

$$\mathrm{tr}\,(\mathbf{AB}) = \mathbf{A} : \mathbf{B}^{\mathrm{T}} = \mathbf{A}^{\mathrm{T}} : \mathbf{B}. \tag{1.150}$$

In view of (D.1) this also implies that

$$\mathrm{tr}\,(\mathbf{AB}) = \mathrm{tr}\,(\mathbf{BA}). \tag{1.151}$$

1.11 Decompositions of Second-Order Tensors

Additive decomposition into a symmetric and a skew-symmetric part. Every second-order tensor can be decomposed additively into a symmetric and a skew-symmetric part by

$$\mathbf{A} = \mathrm{sym}\mathbf{A} + \mathrm{skew}\mathbf{A}, \tag{1.152}$$

where

$$\mathrm{sym}\mathbf{A} = \frac{1}{2}\left(\mathbf{A} + \mathbf{A}^{\mathrm{T}}\right), \quad \mathrm{skew}\mathbf{A} = \frac{1}{2}\left(\mathbf{A} - \mathbf{A}^{\mathrm{T}}\right). \tag{1.153}$$

Symmetric and skew-symmetric tensors form subsets of $\mathbf{Lin}^n$ defined respectively by

$$\mathbf{Sym}^n = \left\{\mathbf{M} \in \mathbf{Lin}^n : \mathbf{M} = \mathbf{M}^{\mathrm{T}}\right\}, \tag{1.154}$$

$$\mathbf{Skew}^n = \left\{\mathbf{W} \in \mathbf{Lin}^n : \mathbf{W} = -\mathbf{W}^{\mathrm{T}}\right\}. \tag{1.155}$$

One can easily show that these subsets represent vector spaces and can be referred to as subspaces of $\mathbf{Lin}^n$. Indeed, the axioms (A.1)–(A.4) and (B.1)–(B.4) including operations with the zero tensor are valid both for symmetric and skew-symmetric tensors. The zero tensor is the only linear mapping that is both symmetric and skew-symmetric such that $\mathbf{Sym}^n \cap \mathbf{Skew}^n = \mathbf{0}$.

For every symmetric tensor $\mathbf{M} = \mathrm{M}^{ij} \boldsymbol{g}_i \otimes \boldsymbol{g}_j$ it follows from (1.124) that $\mathrm{M}^{ij} = \mathrm{M}^{ji}$ $(i \neq j,\ i, j = 1, 2, \ldots, n)$. Thus, we can write

$$\mathbf{M} = \sum_{i=1}^{n} \mathrm{M}^{ii} \boldsymbol{g}_i \otimes \boldsymbol{g}_i + \sum_{\substack{i,j=1 \\ i>j}}^{n} \mathrm{M}^{ij} \left(\boldsymbol{g}_i \otimes \boldsymbol{g}_j + \boldsymbol{g}_j \otimes \boldsymbol{g}_i\right), \quad \mathbf{M} \in \mathbf{Sym}^n. \tag{1.156}$$

Similarly we can write for a skew-symmetric tensor

$$\mathbf{W} = \sum_{\substack{i,j=1 \\ i>j}}^{n} \mathrm{W}^{ij} \left(\boldsymbol{g}_i \otimes \boldsymbol{g}_j - \boldsymbol{g}_j \otimes \boldsymbol{g}_i\right), \quad \mathbf{W} \in \mathbf{Skew}^n \tag{1.157}$$

taking into account that $\mathrm{W}^{ii} = 0$ and $\mathrm{W}^{ij} = -\mathrm{W}^{ji}$ $(i \neq j,\ i, j = 1, 2, \ldots, n)$. Therefore, the basis of $\mathbf{Sym}^n$ is formed by n tensors $\boldsymbol{g}_i \otimes \boldsymbol{g}_i$ and $\frac{1}{2}n\,(n-1)$ tensors $\boldsymbol{g}_i \otimes \boldsymbol{g}_j + \boldsymbol{g}_j \otimes \boldsymbol{g}_i$, while the basis of $\mathbf{Skew}^n$ consists of $\frac{1}{2}n\,(n-1)$ tensors $\boldsymbol{g}_i \otimes \boldsymbol{g}_j - \boldsymbol{g}_j \otimes \boldsymbol{g}_i$, where $i > j = 1, 2, \ldots, n$. Thus, the dimensions of $\mathbf{Sym}^n$ and $\mathbf{Skew}^n$ are $\frac{1}{2}n\,(n+1)$ and $\frac{1}{2}n\,(n-1)$, respectively. It follows from (1.152) that any basis of $\mathbf{Skew}^n$ complements any basis of $\mathbf{Sym}^n$ to a basis of $\mathbf{Lin}^n$.

Taking (1.40) and (1.169) into account a skew symmetric tensor (1.157) can be represented in three-dimensional space by

$$\begin{aligned}\mathbf{W} &= \sum_{\substack{i,j=1 \\ i>j}}^{3} \mathrm{W}^{ij} \left(\boldsymbol{g}_i \otimes \boldsymbol{g}_j - \boldsymbol{g}_j \otimes \boldsymbol{g}_i\right) \\ &= \sum_{\substack{i,j=1 \\ i>j}}^{3} \mathrm{W}^{ij} \,\widehat{\boldsymbol{g}_j \times \boldsymbol{g}_i} = \hat{\boldsymbol{w}}, \quad \mathbf{W} \in \mathbf{Skew}^3,\end{aligned} \tag{1.158}$$

where

$$\boldsymbol{w} = \sum_{\substack{i,j=1 \\ i>j}}^{3} \mathrm{W}^{ij} \boldsymbol{g}_j \times \boldsymbol{g}_i = \frac{1}{2} \mathrm{W}^{ij} \boldsymbol{g}_j \times \boldsymbol{g}_i = \frac{1}{2} \boldsymbol{g}_j \times \left(\mathbf{W} \boldsymbol{g}^j \right)$$

$$= \frac{1}{2} \mathrm{W}^{ij} e_{jik}\, g\, \boldsymbol{g}^k = g \left(\mathrm{W}^{32} \boldsymbol{g}^1 + \mathrm{W}^{13} \boldsymbol{g}^2 + \mathrm{W}^{21} \boldsymbol{g}^3 \right). \tag{1.159}$$

Thus, every skew-symmetric tensor in three-dimensional space describes a cross product by a vector $\boldsymbol{w}$ (1.159) called axial vector. One immediately observes that

$$\mathbf{W}\boldsymbol{w} = \boldsymbol{0}, \quad \mathbf{W} \in \mathrm{Skew}^3. \tag{1.160}$$

Obviously, symmetric and skew-symmetric tensors are mutually orthogonal such that (see Exercise 1.46)

$$\mathbf{M} : \mathbf{W} = 0, \quad \forall \mathbf{M} \in \mathbf{Sym}^n, \ \forall \mathbf{W} \in \mathbf{Skew}^n. \tag{1.161}$$

Spaces characterized by this property are called orthogonal.

Additive decomposition into a spherical and a deviatoric part. For every second-order tensor $\mathbf{A}$ we can write

$$\mathbf{A} = \mathrm{sph}\mathbf{A} + \mathrm{dev}\mathbf{A}, \tag{1.162}$$

where

$$\mathrm{sph}\mathbf{A} = \frac{1}{n} \mathrm{tr}\left(\mathbf{A}\right) \mathbf{I}, \quad \mathrm{dev}\mathbf{A} = \mathbf{A} - \frac{1}{n} \mathrm{tr}\left(\mathbf{A}\right) \mathbf{I} \tag{1.163}$$

denote its spherical and deviatoric part, respectively. Thus, every spherical tensor $\mathbf{S}$ can be represented by $\mathbf{S} = \alpha \mathbf{I}$, where α is a scalar number. In turn, every deviatoric tensor $\mathbf{D}$ is characterized by the condition $\mathrm{tr}\mathbf{D} = 0$. Just like symmetric and skew-symmetric tensors, spherical and deviatoric tensors form orthogonal subspaces of $\mathbf{Lin}^n$.

1.12 Tensors of Higher Orders

Similarly to second-order tensors we can define tensors of higher orders. For example, a third-order tensor can be defined as a linear mapping from $\mathbb{E}^n$ to $\mathbf{Lin}^n$. Thus, we can write

$$\mathbf{Y} = \mathsf{A}\boldsymbol{x}, \quad \mathbf{Y} \in \mathbf{Lin}^n, \quad \forall \boldsymbol{x} \in \mathbb{E}^n, \quad \forall \mathsf{A} \in \mathsf{Lin}^n, \tag{1.164}$$

where $\mathbf{Lin}^n$ denotes the set of all linear mappings of vectors in $\mathbb{E}^n$ into second-order tensors in $\mathbf{Lin}^n$. The tensors of the third order can likewise be represented with respect to a basis in $\mathbf{Lin}^n$ e.g. by

$$\mathbf{A} = \mathsf{A}^{ijk} \boldsymbol{g}_i \otimes \boldsymbol{g}_j \otimes \boldsymbol{g}_k = \mathsf{A}_{ijk} \boldsymbol{g}^i \otimes \boldsymbol{g}^j \otimes \boldsymbol{g}^k$$

$$= \mathsf{A}^i_{\cdot jk} \boldsymbol{g}_i \otimes \boldsymbol{g}^j \otimes \boldsymbol{g}^k = \mathsf{A}_{i\cdot k}^{\cdot j} \boldsymbol{g}^i \otimes \boldsymbol{g}_j \otimes \boldsymbol{g}^k. \quad (1.165)$$

For the components of the tensor $\mathbf{A}$ (1.165) we can thus write by analogy with (1.146)

$$\mathsf{A}^{ijk} = \mathsf{A}^{ij}_{\cdot\cdot s} g^{sk} = \mathsf{A}^i_{\cdot st} g^{sj} g^{tk} = \mathsf{A}_{rst} g^{ri} g^{sj} g^{tk},$$

$$\mathsf{A}_{ijk} = \mathsf{A}^r_{\cdot jk} g_{ri} = \mathsf{A}^{rs}_{\cdot\cdot k} g_{ri} g_{sj} = \mathsf{A}^{rst} g_{ri} g_{sj} g_{tk}. \quad (1.166)$$

Exercises

1.1. Prove that if $\boldsymbol{x} \in \mathbb{V}$ is a vector and $\alpha \in \mathbb{R}$ is a scalar, then the following identities hold.
(a) $-\boldsymbol{0} = \boldsymbol{0}$, (b) $\alpha\boldsymbol{0} = \boldsymbol{0}$, (c) $0\boldsymbol{x} = \boldsymbol{0}$, (d) $-\boldsymbol{x} = (-1)\,\boldsymbol{x}$, (e) if $\alpha\boldsymbol{x} = \boldsymbol{0}$, then either $\alpha = 0$ or $\boldsymbol{x} = \boldsymbol{0}$ or both.

1.2. Prove that $\boldsymbol{x}_i \neq \boldsymbol{0}\,(i = 1, 2, \ldots, n)$ for linearly independent vectors $\boldsymbol{x}_1, \boldsymbol{x}_2, \ldots, \boldsymbol{x}_n$. In other words, linearly independent vectors are all non-zero.

1.3. Prove that any non-empty subset of linearly independent vectors $\boldsymbol{x}_1, \boldsymbol{x}_2, \ldots, \boldsymbol{x}_n$ is also linearly independent.

1.4. Write out in full the following expressions for $n = 3$: (a) $\delta^i_j a^j$, (b) $\delta_{ij} x^i x^j$, (c) δ^i_i, (d) $\dfrac{\partial f_i}{\partial x^j} dx^j$.

1.5. Prove that

$$\boldsymbol{0} \cdot \boldsymbol{x} = 0, \ \forall \boldsymbol{x} \in \mathbb{E}^n. \quad (1.167)$$

1.6. Prove that a set of mutually orthogonal non-zero vectors is always linearly independent.

1.7. Prove the so-called parallelogram law: $\|\boldsymbol{x} + \boldsymbol{y}\|^2 = \|\boldsymbol{x}\|^2 + 2\boldsymbol{x} \cdot \boldsymbol{y} + \|\boldsymbol{y}\|^2$.

1.8. Let $\mathcal{G} = \{\boldsymbol{g}_1, \boldsymbol{g}_2, \ldots, \boldsymbol{g}_n\}$ be a basis in $\mathbb{E}^n$ and $\boldsymbol{a} \in \mathbb{E}^n$ be a vector. Prove that $\boldsymbol{a} \cdot \boldsymbol{g}_i = 0 \ (i = 1, 2, \ldots, n)$ if and only if $\boldsymbol{a} = \boldsymbol{0}$.

1.9. Prove that $\boldsymbol{a} = \boldsymbol{b}$ if and only if $\boldsymbol{a} \cdot \boldsymbol{x} = \boldsymbol{b} \cdot \boldsymbol{x}, \ \forall \boldsymbol{x} \in \mathbb{E}^n$.

1.10. (a) Construct an orthonormal set of vectors orthogonalizing and normalizing (with the aid of the procedure described in Sect. 1.4) the following linearly independent vectors:

$$g_1 = \begin{Bmatrix} 1 \\ 1 \\ 0 \end{Bmatrix}, \quad g_2 = \begin{Bmatrix} 2 \\ 1 \\ -2 \end{Bmatrix}, \quad g_3 = \begin{Bmatrix} 4 \\ 2 \\ 1 \end{Bmatrix},$$

where the components are given with respect to an orthonormal basis.

(b) Construct a basis in $\mathbb{E}^3$ dual to the given above utilizing relations $(1.16)_2$, (1.18) and (1.19).

(c) As an alternative, construct a basis in $\mathbb{E}^3$ dual to the given above by means of $(1.21)_1$, (1.24) and $(1.25)_2$.

(d) Calculate again the vectors $\boldsymbol{g}^i$ dual to $\boldsymbol{g}_i$ $(i = 1, 2, 3)$ by using relations (1.33) and (1.35). Compare the result with the solution of problem (b).

1.11. Verify that the vectors (1.33) are linearly independent.

1.12. Prove identities (1.41) and (1.42) by means of (1.18), (1.19) and (1.24), respectively.

1.13. Prove relations (1.40) and (1.44) by using (1.39) and (1.43), respectively.

1.14. Verify the following identities involving the permutation symbol (1.36) for $n = 3$: (a) $\delta^{ij} e_{ijk} = 0$, (b) $e^{ikm} e_{jkm} = 2\delta^i_j$, (c) $e^{ijk} e_{ijk} = 6$, (d) $e^{ijm} e_{klm} = \delta^i_k \delta^j_l - \delta^i_l \delta^j_k$.

1.15. Prove the following identities

$$(\boldsymbol{a} \times \boldsymbol{b}) \times \boldsymbol{c} = (\boldsymbol{a} \cdot \boldsymbol{c})\, \boldsymbol{b} - (\boldsymbol{b} \cdot \boldsymbol{c})\, \boldsymbol{a}, \tag{1.168}$$

$$\widehat{\boldsymbol{a} \times \boldsymbol{b}} = \boldsymbol{b} \otimes \boldsymbol{a} - \boldsymbol{a} \otimes \boldsymbol{b}, \quad \forall \boldsymbol{a}, \boldsymbol{b}, \boldsymbol{c} \in \mathbb{E}^3. \tag{1.169}$$

1.16. Prove relations (1.64) using (1.45).

1.17. Prove that $\mathbf{A}\boldsymbol{0} = \boldsymbol{0}\mathbf{A} = \boldsymbol{0}$, $\forall \mathbf{A} \in \mathbf{Lin}^n$.

1.18. Prove that $0\mathbf{A} = \mathbf{0}$, $\forall \mathbf{A} \in \mathbf{Lin}^n$.

1.19. Prove formula (1.58), where the negative tensor $-\mathbf{A}$ is defined by (1.53).

1.20. Prove that not every second order tensor in $\mathbf{Lin}^n$ can be represented as a tensor product of two vectors $\boldsymbol{a}, \boldsymbol{b} \in \mathbb{E}^n$ as $\boldsymbol{a} \otimes \boldsymbol{b}$.

1.21. Prove relation (1.85).

1.22. Prove (1.88) using (1.87) and (1.15).

1.23. Evaluate the tensor $\mathbf{W} = \hat{\boldsymbol{w}} = \boldsymbol{w}\times$, where $\boldsymbol{w} = w^i \boldsymbol{g}_i$.

1.24. Evaluate components of the tensor describing a rotation about the axis $\boldsymbol{e}_3$ by the angle α.

1.25. Let $\mathbf{A} = \mathrm{A}^{ij} \boldsymbol{g}_i \otimes \boldsymbol{g}_j$, where

$$\left[\mathrm{A}^{ij}\right] = \begin{bmatrix} 0 & -1 & 0 \\ 0 & 0 & 0 \\ 1 & 0 & 0 \end{bmatrix}$$

and the vectors $\boldsymbol{g}_i$ $(i = 1, 2, 3)$ are given in Exercise 1.10. Evaluate the components A_{ij}, $\mathrm{A}^{i}_{\cdot j}$ and $\mathrm{A}_{i\cdot}^{\ j}$.

1.26. Prove identities (1.105) and (1.107).

1.27. Let $\mathbf{A} = \mathrm{A}^{i}_{\cdot j} \boldsymbol{g}_i \otimes \boldsymbol{g}^j$, $\mathbf{B} = \mathrm{B}^{i}_{\cdot j} \boldsymbol{g}_i \otimes \boldsymbol{g}^j$, $\mathbf{C} = \mathrm{C}^{i}_{\cdot j} \boldsymbol{g}_i \otimes \boldsymbol{g}^j$ and $\mathbf{D} = \mathrm{D}^{i}_{\cdot j} \boldsymbol{g}_i \otimes \boldsymbol{g}^j$, where

$$\left[\mathrm{A}^{i}_{\cdot j}\right] = \begin{bmatrix} 0 & 2 & 0 \\ 0 & 0 & 0 \\ 0 & 0 & 0 \end{bmatrix}, \quad \left[\mathrm{B}^{i}_{\cdot j}\right] = \begin{bmatrix} 0 & 0 & 0 \\ 0 & 0 & 0 \\ 0 & 0 & 1 \end{bmatrix}, \quad \left[\mathrm{C}^{i}_{\cdot j}\right] = \begin{bmatrix} 1 & 2 & 3 \\ 0 & 0 & 0 \\ 0 & 1 & 0 \end{bmatrix},$$

$$\left[\mathrm{D}^{i}_{\cdot j}\right] = \begin{bmatrix} 1 & 0 & 0 \\ 0 & 1/2 & 0 \\ 0 & 0 & 10 \end{bmatrix}.$$

Find commutative pairs of tensors.

1.28. Let $\mathbf{A}$ and $\mathbf{B}$ be two commutative tensors. Write out in full $(\mathbf{A} + \mathbf{B})^k$, where $k = 2, 3, \ldots$

1.29. Prove that

$$\exp(\mathbf{A} + \mathbf{B}) = \exp(\mathbf{A}) \exp(\mathbf{B}), \tag{1.170}$$

where $\mathbf{A}$ and $\mathbf{B}$ commute.

1.30. Evaluate $\exp(\mathbf{0})$ and $\exp(\mathbf{I})$.

1.31. Prove that $\exp(-\mathbf{A}) \exp(\mathbf{A}) = \exp(\mathbf{A}) \exp(-\mathbf{A}) = \mathbf{I}$.

1.32. Prove that $\exp(k\mathbf{A}) = [\exp(\mathbf{A})]^k$ for all integer k.

1.33. Prove that $\exp(\mathbf{A} + \mathbf{B}) = \exp(\mathbf{A}) + \exp(\mathbf{B}) - \mathbf{I}$ if $\mathbf{AB} = \mathbf{BA} = \mathbf{0}$.

1.34. Prove that $\exp\left(\mathbf{QAQ}^{\mathrm{T}}\right) = \mathbf{Q} \exp(\mathbf{A}) \mathbf{Q}^{\mathrm{T}}, \ \forall \mathbf{Q} \in \mathbf{Orth}^n$.

1.35. Compute the exponential of the tensors $\mathbf{D} = \mathrm{D}^{i}_{\cdot j} \boldsymbol{g}_i \otimes \boldsymbol{g}^j$, $\mathbf{E} = \mathrm{E}^{i}_{\cdot j} \boldsymbol{g}_i \otimes \boldsymbol{g}^j$ and $\mathbf{F} = \mathrm{F}^{i}_{\cdot j} \boldsymbol{g}_i \otimes \boldsymbol{g}^j$, where

$$\left[\mathrm{D}^{i}_{\cdot j}\right] = \begin{bmatrix} 2 & 0 & 0 \\ 0 & 3 & 0 \\ 0 & 0 & 1 \end{bmatrix}, \quad \left[\mathrm{E}^{i}_{\cdot j}\right] = \begin{bmatrix} 0 & 1 & 0 \\ 0 & 0 & 0 \\ 0 & 0 & 0 \end{bmatrix}, \quad \left[\mathrm{F}^{i}_{\cdot j}\right] = \begin{bmatrix} 0 & 2 & 0 \\ 0 & 0 & 0 \\ 0 & 0 & 1 \end{bmatrix}.$$

1.36. Prove that $(\mathbf{ABCD})^{\mathrm{T}} = \mathbf{D}^{\mathrm{T}}\mathbf{C}^{\mathrm{T}}\mathbf{B}^{\mathrm{T}}\mathbf{A}^{\mathrm{T}}$.

1.37. Verify that $\left(\mathbf{A}^k\right)^{\mathrm{T}} = \left(\mathbf{A}^{\mathrm{T}}\right)^k$, where $k = 1, 2, 3, \ldots$

1.38. Evaluate the components B^{ij}, B_{ij}, $\mathrm{B}^{i}_{\cdot j}$ and $\mathrm{B}_{i\cdot}^{\ j}$ of the tensor $\mathbf{B} = \mathbf{A}^{\mathrm{T}}$, where $\mathbf{A}$ is defined in Exercise 1.25.

1.39. Prove relation (1.132).

1.40. Verify that $\left(\mathbf{A}^{-1}\right)^k = \left(\mathbf{A}^k\right)^{-1} = \mathbf{A}^{-k}$, where $k = 1, 2, 3, \ldots$

1.41. Prove identity (1.142) using (1.87) and (1.141).

1.42. Prove by means of (1.141)–(1.143) the properties of the scalar product (D.1)–(D.3).

1.43. Verify that $[(\boldsymbol{a} \otimes \boldsymbol{b})(\boldsymbol{c} \otimes \boldsymbol{d})] : \mathbf{I} = (\boldsymbol{a} \cdot \boldsymbol{d})(\boldsymbol{b} \cdot \boldsymbol{c})$.

1.44. Express $\mathrm{tr}\mathbf{A}$ in terms of the components $\mathrm{A}^{i}_{\cdot j}$, A_{ij}, A^{ij}.

1.45. Let $\mathbf{W} = \mathrm{W}^{ij} \boldsymbol{g}_i \otimes \boldsymbol{g}_j$, where

$$\left[\mathrm{W}^{ij}\right] = \begin{bmatrix} 0 & -1 & -3 \\ 1 & 0 & 1 \\ 3 & -1 & 0 \end{bmatrix}$$

and the vectors $\boldsymbol{g}_i$ $(i = 1, 2, 3)$ are given in Exercise 1.10. Calculate the axial vector of $\mathbf{W}$.

1.46. Prove that $\mathbf{M} : \mathbf{W} = 0$, where $\mathbf{M}$ is a symmetric tensor and $\mathbf{W}$ a skew-symmetric tensor.

1.47. Evaluate $\mathrm{tr}\mathbf{W}^k$, where $\mathbf{W}$ is a skew-symmetric tensor and $k = 1, 3, 5, \ldots$

1.48. Verify that $\mathrm{sym}(\mathrm{skew}\mathbf{A}) = \mathrm{skew}(\mathrm{sym}\mathbf{A}) = \mathbf{0}$, $\forall \mathbf{A} \in \mathbf{Lin}^n$.

1.49. Prove that $\mathrm{sph}(\mathrm{dev}\mathbf{A}) = \mathrm{dev}(\mathrm{sph}\mathbf{A}) = \mathbf{0}$, $\forall \mathbf{A} \in \mathbf{Lin}^n$.

Chapter 2
Vector and Tensor Analysis in Euclidean Space

2.1 Vector- and Tensor-Valued Functions, Differential Calculus

In the following we consider a vector-valued function $\boldsymbol{x}(t)$ and a tensor-valued function $\mathbf{A}(t)$ of a real variable t. Henceforth, we assume that these functions are continuous such that

$$\lim_{t \to t_0} [\boldsymbol{x}(t) - \boldsymbol{x}(t_0)] = \boldsymbol{0}, \quad \lim_{t \to t_0} [\mathbf{A}(t) - \mathbf{A}(t_0)] = \mathbf{0} \tag{2.1}$$

for all t_0 within the definition domain. The functions $\boldsymbol{x}(t)$ and $\mathbf{A}(t)$ are called differentiable if the following limits

$$\frac{\mathrm{d}\boldsymbol{x}}{\mathrm{d}t} = \lim_{s \to 0} \frac{\boldsymbol{x}(t+s) - \boldsymbol{x}(t)}{s}, \quad \frac{\mathrm{d}\mathbf{A}}{\mathrm{d}t} = \lim_{s \to 0} \frac{\mathbf{A}(t+s) - \mathbf{A}(t)}{s} \tag{2.2}$$

exist and are finite. They are referred to as the derivatives of the vector- and tensor-valued functions $\boldsymbol{x}(t)$ and $\mathbf{A}(t)$, respectively.

For differentiable vector- and tensor-valued functions the usual rules of differentiation hold.

1. Product of a scalar function with a vector- or tensor-valued function:

$$\frac{\mathrm{d}}{\mathrm{d}t} [u(t)\, \boldsymbol{x}(t)] = \frac{\mathrm{d}u}{\mathrm{d}t} \boldsymbol{x}(t) + u(t) \frac{\mathrm{d}\boldsymbol{x}}{\mathrm{d}t}, \tag{2.3}$$

$$\frac{\mathrm{d}}{\mathrm{d}t} [u(t)\, \mathbf{A}(t)] = \frac{\mathrm{d}u}{\mathrm{d}t} \mathbf{A}(t) + u(t) \frac{\mathrm{d}\mathbf{A}}{\mathrm{d}t}. \tag{2.4}$$

2. Mapping of a vector-valued function by a tensor-valued function:

$$\frac{\mathrm{d}}{\mathrm{d}t} [\mathbf{A}(t)\, \boldsymbol{x}(t)] = \frac{\mathrm{d}\mathbf{A}}{\mathrm{d}t} \boldsymbol{x}(t) + \mathbf{A}(t) \frac{\mathrm{d}\boldsymbol{x}}{\mathrm{d}t}. \tag{2.5}$$

M. Itskov, *Tensor Algebra and Tensor Analysis for Engineers*, Mathematical Engineering,
DOI 10.1007/978-3-642-30879-6_2,

3. Scalar product of two vector- or tensor-valued functions:

$$\frac{\mathrm{d}}{\mathrm{d}t}\left[\boldsymbol{x}(t)\cdot\boldsymbol{y}(t)\right]=\frac{\mathrm{d}\boldsymbol{x}}{\mathrm{d}t}\cdot\boldsymbol{y}(t)+\boldsymbol{x}(t)\cdot\frac{\mathrm{d}\boldsymbol{y}}{\mathrm{d}t},\tag{2.6}$$

$$\frac{\mathrm{d}}{\mathrm{d}t}\left[\mathbf{A}(t):\mathbf{B}(t)\right]=\frac{\mathrm{d}\mathbf{A}}{\mathrm{d}t}:\mathbf{B}(t)+\mathbf{A}(t):\frac{\mathrm{d}\mathbf{B}}{\mathrm{d}t}.\tag{2.7}$$

4. Tensor product of two vector-valued functions:

$$\frac{\mathrm{d}}{\mathrm{d}t}\left[\boldsymbol{x}(t)\otimes\boldsymbol{y}(t)\right]=\frac{\mathrm{d}\boldsymbol{x}}{\mathrm{d}t}\otimes\boldsymbol{y}(t)+\boldsymbol{x}(t)\otimes\frac{\mathrm{d}\boldsymbol{y}}{\mathrm{d}t}.\tag{2.8}$$

5. Composition of two tensor-valued functions:

$$\frac{\mathrm{d}}{\mathrm{d}t}\left[\mathbf{A}(t)\,\mathbf{B}(t)\right]=\frac{\mathrm{d}\mathbf{A}}{\mathrm{d}t}\mathbf{B}(t)+\mathbf{A}(t)\,\frac{\mathrm{d}\mathbf{B}}{\mathrm{d}t}.\tag{2.9}$$

6. Chain rule:

$$\frac{\mathrm{d}}{\mathrm{d}t}\boldsymbol{x}\left[u(t)\right]=\frac{\mathrm{d}\boldsymbol{x}}{\mathrm{d}u}\frac{\mathrm{d}u}{\mathrm{d}t},\quad\frac{\mathrm{d}}{\mathrm{d}t}\mathbf{A}\left[u(t)\right]=\frac{\mathrm{d}\mathbf{A}}{\mathrm{d}u}\frac{\mathrm{d}u}{\mathrm{d}t}.\tag{2.10}$$

7. Chain rule for functions of several arguments:

$$\frac{\mathrm{d}}{\mathrm{d}t}\boldsymbol{x}\left[u(t),v(t)\right]=\frac{\partial\boldsymbol{x}}{\partial u}\frac{\mathrm{d}u}{\mathrm{d}t}+\frac{\partial\boldsymbol{x}}{\partial v}\frac{\mathrm{d}v}{\mathrm{d}t},\tag{2.11}$$

$$\frac{\mathrm{d}}{\mathrm{d}t}\mathbf{A}\left[u(t),v(t)\right]=\frac{\partial\mathbf{A}}{\partial u}\frac{\mathrm{d}u}{\mathrm{d}t}+\frac{\partial\mathbf{A}}{\partial v}\frac{\mathrm{d}v}{\mathrm{d}t},\tag{2.12}$$

where $\partial/\partial u$ denotes the partial derivative. It is defined for vector and tensor valued functions in the standard manner by

$$\frac{\partial\boldsymbol{x}(u,v)}{\partial u}=\lim_{s\to 0}\frac{\boldsymbol{x}(u+s,v)-\boldsymbol{x}(u,v)}{s},\tag{2.13}$$

$$\frac{\partial\mathbf{A}(u,v)}{\partial u}=\lim_{s\to 0}\frac{\mathbf{A}(u+s,v)-\mathbf{A}(u,v)}{s}.\tag{2.14}$$

The above differentiation rules can be verified with the aid of elementary differential calculus. For example, for the derivative of the composition of two second-order tensors (2.9) we proceed as follows. Let us define two tensor-valued functions by

$$\mathbf{O}_1(s)=\frac{\mathbf{A}(t+s)-\mathbf{A}(t)}{s}-\frac{\mathrm{d}\mathbf{A}}{\mathrm{d}t},\quad\mathbf{O}_2(s)=\frac{\mathbf{B}(t+s)-\mathbf{B}(t)}{s}-\frac{\mathrm{d}\mathbf{B}}{\mathrm{d}t}.\tag{2.15}$$

Bearing the definition of the derivative (2.2) in mind we have

$$\lim_{s\to 0}\mathbf{O}_1(s)=\mathbf{0},\quad\lim_{s\to 0}\mathbf{O}_2(s)=\mathbf{0}.$$

Then,

$$\frac{\mathrm{d}}{\mathrm{d}t}\left[\mathbf{A}(t)\,\mathbf{B}(t)\right] = \lim_{s\to 0}\frac{\mathbf{A}(t+s)\,\mathbf{B}(t+s) - \mathbf{A}(t)\,\mathbf{B}(t)}{s}$$
$$= \lim_{s\to 0}\frac{1}{s}\left\{\left[\mathbf{A}(t) + s\frac{\mathrm{d}\mathbf{A}}{\mathrm{d}t} + s\mathbf{O}_1(s)\right]\left[\mathbf{B}(t) + s\frac{\mathrm{d}\mathbf{B}}{\mathrm{d}t} + s\mathbf{O}_2(s)\right]\right.$$
$$\left. - \mathbf{A}(t)\,\mathbf{B}(t)\right\}$$
$$= \lim_{s\to 0}\left\{\left[\frac{\mathrm{d}\mathbf{A}}{\mathrm{d}t} + \mathbf{O}_1(s)\right]\mathbf{B}(t) + \mathbf{A}(t)\left[\frac{\mathrm{d}\mathbf{B}}{\mathrm{d}t} + \mathbf{O}_2(s)\right]\right\}$$
$$+ \lim_{s\to 0} s\left[\frac{\mathrm{d}\mathbf{A}}{\mathrm{d}t} + \mathbf{O}_1(s)\right]\left[\frac{\mathrm{d}\mathbf{B}}{\mathrm{d}t} + \mathbf{O}_2(s)\right] = \frac{\mathrm{d}\mathbf{A}}{\mathrm{d}t}\mathbf{B}(t) + \mathbf{A}(t)\frac{\mathrm{d}\mathbf{B}}{\mathrm{d}t}.$$

2.2 Coordinates in Euclidean Space, Tangent Vectors

Definition 2.1. A coordinate system is a one to one correspondence between vectors in the n-dimensional Euclidean space $\mathbb{E}^n$ and a set of n real numbers $(x^1, x^2, \ldots, x^n)$. These numbers are called coordinates of the corresponding vectors.

Thus, we can write

$$x^i = x^i(\boldsymbol{r}) \quad \Leftrightarrow \quad \boldsymbol{r} = \boldsymbol{r}\left(x^1, x^2, \ldots, x^n\right), \tag{2.16}$$

where $\boldsymbol{r} \in \mathbb{E}^n$ and $x^i \in \mathbb{R}$ $(i = 1, 2, \ldots, n)$. Henceforth, we assume that the functions $x^i = x^i(\boldsymbol{r})$ and $\boldsymbol{r} = \boldsymbol{r}\left(x^1, x^2, \ldots, x^n\right)$ are sufficiently differentiable.

Example 2.1. Cylindrical coordinates in $\mathbb{E}^3$. The cylindrical coordinates (Fig. 2.1) are defined by

$$\boldsymbol{r} = \boldsymbol{r}(\varphi, z, r) = r\cos\varphi\,\boldsymbol{e}_1 + r\sin\varphi\,\boldsymbol{e}_2 + z\boldsymbol{e}_3 \tag{2.17}$$

and

$$r = \sqrt{(\boldsymbol{r}\cdot\boldsymbol{e}_1)^2 + (\boldsymbol{r}\cdot\boldsymbol{e}_2)^2}, \quad z = \boldsymbol{r}\cdot\boldsymbol{e}_3,$$
$$\varphi = \begin{cases} \arccos\dfrac{\boldsymbol{r}\cdot\boldsymbol{e}_1}{r} & \text{if } \boldsymbol{r}\cdot\boldsymbol{e}_2 \geq 0, \\ 2\pi - \arccos\dfrac{\boldsymbol{r}\cdot\boldsymbol{e}_1}{r} & \text{if } \boldsymbol{r}\cdot\boldsymbol{e}_2 < 0, \end{cases} \tag{2.18}$$

where $\boldsymbol{e}_i$ $(i = 1, 2, 3)$ form an orthonormal basis in $\mathbb{E}^3$.

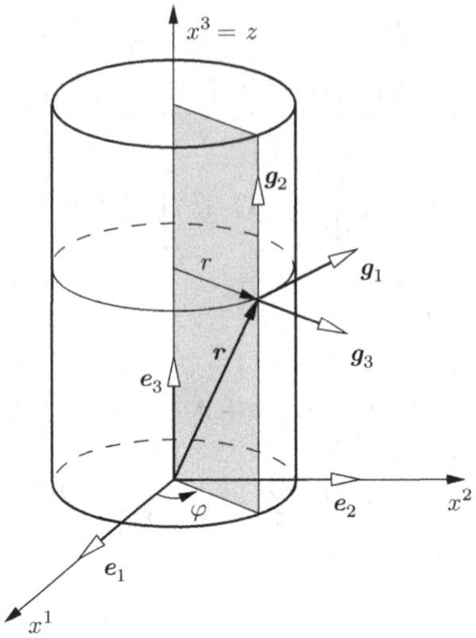

Fig. 2.1 Cylindrical coordinates in three-dimensional space

The vector components with respect to a fixed basis, say $\mathcal{H} = \{\boldsymbol{h}_1, \boldsymbol{h}_2, \ldots, \boldsymbol{h}_n\}$, obviously represent its coordinates. Indeed, according to Theorem 1.5 of the previous chapter the following correspondence is one to one

$$\boldsymbol{r} = x^i \boldsymbol{h}_i \quad \Leftrightarrow \quad x^i = \boldsymbol{r} \cdot \boldsymbol{h}^i, \quad i = 1, 2, \ldots, n, \tag{2.19}$$

where $\boldsymbol{r} \in \mathbb{E}^n$ and $\mathcal{H}' = \left\{\boldsymbol{h}^1, \boldsymbol{h}^2, \ldots, \boldsymbol{h}^n\right\}$ is the basis dual to $\mathcal{H}$. The components x^i $(2.19)_2$ are referred to as the linear coordinates of the vector $\boldsymbol{r}$.

The Cartesian coordinates result as a special case of the linear coordinates (2.19) where $\boldsymbol{h}_i = \boldsymbol{e}_i$ $(i = 1, 2, \ldots, n)$ so that

$$\boldsymbol{r} = x^i \boldsymbol{e}_i \quad \Leftrightarrow \quad x^i = \boldsymbol{r} \cdot \boldsymbol{e}_i, \quad i = 1, 2, \ldots, n. \tag{2.20}$$

Let $x^i = x^i(\boldsymbol{r})$ and $y^i = y^i(\boldsymbol{r})$ $(i = 1, 2, \ldots, n)$ be two arbitrary coordinate systems in $\mathbb{E}^n$. Since their correspondences are one to one, the functions

$$x^i = \hat{x}^i\left(y^1, y^2, \ldots, y^n\right) \Leftrightarrow y^i = \hat{y}^i\left(x^1, x^2, \ldots, x^n\right), \; i = 1, 2, \ldots, n \tag{2.21}$$

are invertible. These functions describe the transformation of the coordinate systems. Inserting one relation (2.21) into another one yields

$$y^i = \hat{y}^i\left(\hat{x}^1\left(y^1, y^2, \ldots, y^n\right), \right.$$
$$\left. \hat{x}^2\left(y^1, y^2, \ldots, y^n\right), \ldots, \hat{x}^n\left(y^1, y^2, \ldots, y^n\right)\right). \tag{2.22}$$

The further differentiation with respect to y^j delivers with the aid of the chain rule

$$\frac{\partial y^i}{\partial y^j} = \delta_{ij} = \frac{\partial y^i}{\partial x^k}\frac{\partial x^k}{\partial y^j}, \quad i,j = 1,2,\ldots,n. \tag{2.23}$$

The determinant of the matrix (2.23) takes the form

$$\left|\delta_{ij}\right| = 1 = \left|\frac{\partial y^i}{\partial x^k}\frac{\partial x^k}{\partial y^j}\right| = \left|\frac{\partial y^i}{\partial x^k}\right|\left|\frac{\partial x^k}{\partial y^j}\right|. \tag{2.24}$$

The determinant $\left|\partial y^i / \partial x^k\right|$ on the right hand side of (2.24) is referred to as Jacobian determinant of the coordinate transformation $y^i = \hat{y}^i\left(x^1, x^2, \ldots, x^n\right)$ $(i = 1, 2, \ldots, n)$. Thus, we have proved the following theorem.

Theorem 2.1. *If the transformation of the coordinates* $y^i = \hat{y}^i\left(x^1, x^2, \ldots, x^n\right)$ *admits an inverse form* $x^i = \hat{x}^i\left(y^1, y^2, \ldots, y^n\right)$ $(i = 1, 2, \ldots, n)$ *and if* J *and* K *are the Jacobians of these transformations then* $JK = 1$.

One of the important consequences of this theorem is that

$$J = \left|\frac{\partial y^i}{\partial x^k}\right| \neq 0. \tag{2.25}$$

Now, we consider an arbitrary curvilinear coordinate system

$$\theta^i = \theta^i(\boldsymbol{r}) \Leftrightarrow \boldsymbol{r} = \boldsymbol{r}\left(\theta^1, \theta^2, \ldots, \theta^n\right), \tag{2.26}$$

where $\boldsymbol{r} \in \mathbb{E}^n$ and $\theta^i \in \mathbb{R}$ $(i = 1, 2, \ldots, n)$. The equations

$$\theta^i = const, \; i = 1, 2, \ldots, k-1, k+1, \ldots, n \tag{2.27}$$

define a curve in $\mathbb{E}^n$ called θ^k-coordinate line. The vectors (see Fig. 2.2)

$$\boldsymbol{g}_k = \frac{\partial \boldsymbol{r}}{\partial \theta^k}, \quad k = 1, 2, \ldots, n \tag{2.28}$$

are called the tangent vectors to the corresponding θ^k-coordinate lines (2.27).

One can verify that the tangent vectors are linearly independent and form thus a basis of $\mathbb{E}^n$. Conversely, let the vectors (2.28) be linearly dependent. Then, there are scalars $\alpha^i \in \mathbb{R}$ $(i = 1, 2, \ldots, n)$, not all zero, such that $\alpha^i \boldsymbol{g}_i = \boldsymbol{0}$. Let further $x^i = x^i(\boldsymbol{r})$ $(i = 1, 2, \ldots, n)$ be linear coordinates in $\mathbb{E}^n$ with respect to a basis $\mathcal{H} = \{\boldsymbol{h}_1, \boldsymbol{h}_2, \ldots, \boldsymbol{h}_n\}$. Then,

$$\boldsymbol{0} = \alpha^i \boldsymbol{g}_i = \alpha^i \frac{\partial \boldsymbol{r}}{\partial \theta^i} = \alpha^i \frac{\partial \boldsymbol{r}}{\partial x^j}\frac{\partial x^j}{\partial \theta^i} = \alpha^i \frac{\partial x^j}{\partial \theta^i}\boldsymbol{h}_j.$$

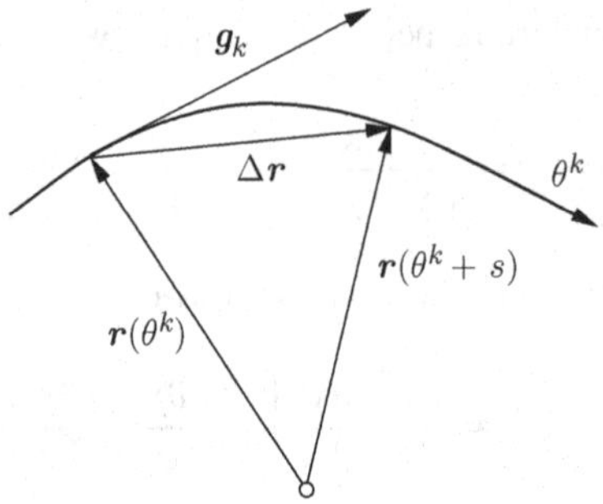

Fig. 2.2 Illustration of the tangent vectors

Since the basis vectors $\boldsymbol{h}_j$ $(j = 1, 2, \ldots, n)$ are linearly independent

$$\alpha^i \frac{\partial x^j}{\partial \theta^i} = 0, \quad j = 1, 2, \ldots, n.$$

This is a homogeneous linear equation system with a non-trivial solution α^i $(i = 1, 2, \ldots, n)$. Hence, $\left|\partial x^j / \partial \theta^i\right| = 0$, which obviously contradicts relation (2.25).

Example 2.2. Tangent vectors and metric coefficients of cylindrical coordinates in $\mathbb{E}^3$. By means of (2.17) and (2.28) we obtain

$$\begin{aligned}
\boldsymbol{g}_1 &= \frac{\partial \boldsymbol{r}}{\partial \varphi} = -r \sin\varphi \boldsymbol{e}_1 + r \cos\varphi \boldsymbol{e}_2, \\
\boldsymbol{g}_2 &= \frac{\partial \boldsymbol{r}}{\partial z} = \boldsymbol{e}_3, \\
\boldsymbol{g}_3 &= \frac{\partial \boldsymbol{r}}{\partial r} = \cos\varphi \boldsymbol{e}_1 + \sin\varphi \boldsymbol{e}_2.
\end{aligned} \tag{2.29}$$

The metric coefficients take by virtue of (1.24) and $(1.25)_2$ the form

$$\left[g_{ij}\right] = \left[\boldsymbol{g}_i \cdot \boldsymbol{g}_j\right] = \begin{bmatrix} r^2 & 0 & 0 \\ 0 & 1 & 0 \\ 0 & 0 & 1 \end{bmatrix}, \quad \left[g^{ij}\right] = \left[g_{ij}\right]^{-1} = \begin{bmatrix} r^{-2} & 0 & 0 \\ 0 & 1 & 0 \\ 0 & 0 & 1 \end{bmatrix}. \tag{2.30}$$

The dual basis results from $(1.21)_1$ by

$$\begin{aligned}
\boldsymbol{g}^1 &= \frac{1}{r^2} \boldsymbol{g}_1 = -\frac{1}{r} \sin\varphi \boldsymbol{e}_1 + \frac{1}{r} \cos\varphi \boldsymbol{e}_2, \\
\boldsymbol{g}^2 &= \boldsymbol{g}_2 = \boldsymbol{e}_3, \\
\boldsymbol{g}^3 &= \boldsymbol{g}_3 = \cos\varphi \boldsymbol{e}_1 + \sin\varphi \boldsymbol{e}_2.
\end{aligned} \tag{2.31}$$

2.3 Coordinate Transformation. Co-, Contra- and Mixed Variant Components

Let $\theta^i = \theta^i(\boldsymbol{r})$ and $\bar{\theta}^i = \bar{\theta}^i(\boldsymbol{r})$ $(i = 1, 2, \ldots, n)$ be two arbitrary coordinate systems in $\mathbb{E}^n$. It holds

$$\bar{\boldsymbol{g}}_i = \frac{\partial \boldsymbol{r}}{\partial \bar{\theta}^i} = \frac{\partial \boldsymbol{r}}{\partial \theta^j}\frac{\partial \theta^j}{\partial \bar{\theta}^i} = \boldsymbol{g}_j \frac{\partial \theta^j}{\partial \bar{\theta}^i}, \quad i = 1, 2, \ldots, n. \tag{2.32}$$

If $\boldsymbol{g}^i$ is the dual basis to $\boldsymbol{g}_i$ $(i = 1, 2, \ldots, n)$, then we can write

$$\bar{\boldsymbol{g}}^i = \boldsymbol{g}^j \frac{\partial \bar{\theta}^i}{\partial \theta^j}, \quad i = 1, 2, \ldots, n. \tag{2.33}$$

Indeed,

$$\bar{\boldsymbol{g}}^i \cdot \bar{\boldsymbol{g}}_j = \left(\boldsymbol{g}^k \frac{\partial \bar{\theta}^i}{\partial \theta^k}\right) \cdot \left(\boldsymbol{g}_l \frac{\partial \theta^l}{\partial \bar{\theta}^j}\right) = \boldsymbol{g}^k \cdot \boldsymbol{g}_l \left(\frac{\partial \bar{\theta}^i}{\partial \theta^k}\frac{\partial \theta^l}{\partial \bar{\theta}^j}\right)$$

$$= \delta^k_l \left(\frac{\partial \bar{\theta}^i}{\partial \theta^k}\frac{\partial \theta^l}{\partial \bar{\theta}^j}\right) = \frac{\partial \bar{\theta}^i}{\partial \theta^k}\frac{\partial \theta^k}{\partial \bar{\theta}^j} = \frac{\partial \bar{\theta}^i}{\partial \bar{\theta}^j} = \delta^i_j, \; i, j = 1, 2, \ldots, n. \tag{2.34}$$

One can observe the difference in the transformation of the dual vectors (2.32) and (2.33) which results from the change of the coordinate system. The transformation rules of the form (2.32) and (2.33) and the corresponding variables are referred to as covariant and contravariant, respectively. Covariant and contravariant variables are denoted by lower and upper indices, respectively.

The co- and contravariant rules can also be recognized in the transformation of the components of vectors and tensors if they are related to tangent vectors. Indeed, let

$$\boldsymbol{x} = x_i \boldsymbol{g}^i = x^i \boldsymbol{g}_i = \bar{x}_i \bar{\boldsymbol{g}}^i = \bar{x}^i \bar{\boldsymbol{g}}_i, \tag{2.35}$$

$$\mathbf{A} = \mathrm{A}_{ij} \boldsymbol{g}^i \otimes \boldsymbol{g}^j = \mathrm{A}^{ij} \boldsymbol{g}_i \otimes \boldsymbol{g}_j = \mathrm{A}^i_{\cdot j} \boldsymbol{g}_i \otimes \boldsymbol{g}^j$$

$$= \bar{\mathrm{A}}_{ij} \bar{\boldsymbol{g}}^i \otimes \bar{\boldsymbol{g}}^j = \bar{\mathrm{A}}^{ij} \bar{\boldsymbol{g}}_i \otimes \bar{\boldsymbol{g}}_j = \bar{\mathrm{A}}^i_{\cdot j} \bar{\boldsymbol{g}}_i \otimes \bar{\boldsymbol{g}}^j. \tag{2.36}$$

Then, by means of (1.28), (1.88), (2.32) and (2.33) we obtain

$$\bar{x}_i = \boldsymbol{x} \cdot \bar{\boldsymbol{g}}_i = \boldsymbol{x} \cdot \left(\boldsymbol{g}_j \frac{\partial \theta^j}{\partial \bar{\theta}^i}\right) = x_j \frac{\partial \theta^j}{\partial \bar{\theta}^i}, \tag{2.37}$$

$$\bar{x}^i = \boldsymbol{x} \cdot \bar{\boldsymbol{g}}^i = \boldsymbol{x} \cdot \left(\boldsymbol{g}^j \frac{\partial \bar{\theta}^i}{\partial \theta^j}\right) = x^j \frac{\partial \bar{\theta}^i}{\partial \theta^j}, \tag{2.38}$$

$$\bar{A}_{ij} = \bar{\boldsymbol{g}}_i \mathbf{A} \bar{\boldsymbol{g}}_j = \left(\boldsymbol{g}_k \frac{\partial \theta^k}{\partial \bar{\theta}^i} \right) \mathbf{A} \left(\boldsymbol{g}_l \frac{\partial \theta^l}{\partial \bar{\theta}^j} \right) = \frac{\partial \theta^k}{\partial \bar{\theta}^i} \frac{\partial \theta^l}{\partial \bar{\theta}^j} A_{kl}, \tag{2.39}$$

$$\bar{A}^{ij} = \bar{\boldsymbol{g}}^i \mathbf{A} \bar{\boldsymbol{g}}^j = \left(\boldsymbol{g}^k \frac{\partial \bar{\theta}^i}{\partial \theta^k} \right) \mathbf{A} \left(\boldsymbol{g}^l \frac{\partial \bar{\theta}^j}{\partial \theta^l} \right) = \frac{\partial \bar{\theta}^i}{\partial \theta^k} \frac{\partial \bar{\theta}^j}{\partial \theta^l} A^{kl}, \tag{2.40}$$

$$\bar{A}^{i}_{\cdot j} = \bar{\boldsymbol{g}}^i \mathbf{A} \bar{\boldsymbol{g}}_j = \left(\boldsymbol{g}^k \frac{\partial \bar{\theta}^i}{\partial \theta^k} \right) \mathbf{A} \left(\boldsymbol{g}_l \frac{\partial \theta^l}{\partial \bar{\theta}^j} \right) = \frac{\partial \bar{\theta}^i}{\partial \theta^k} \frac{\partial \theta^l}{\partial \bar{\theta}^j} A^{k}_{\cdot l}. \tag{2.41}$$

Accordingly, the vector and tensor components x_i, A_{ij} and x^i, A^{ij} are called covariant and contravariant, respectively. The tensor components $A^{i}_{\cdot j}$ are referred to as mixed variant. The transformation rules (2.37)–(2.41) can similarly be written for tensors of higher orders as well. For example, one obtains for third-order tensors

$$\bar{A}_{ijk} = \frac{\partial \theta^r}{\partial \bar{\theta}^i} \frac{\partial \theta^s}{\partial \bar{\theta}^j} \frac{\partial \theta^t}{\partial \bar{\theta}^k} A_{rst}, \quad \bar{A}^{ijk} = \frac{\partial \bar{\theta}^i}{\partial \theta^r} \frac{\partial \bar{\theta}^j}{\partial \theta^s} \frac{\partial \bar{\theta}^k}{\partial \theta^t} A^{rst}, \ldots \tag{2.42}$$

From the very beginning we have supplied coordinates with upper indices which imply the contravariant transformation rule. Indeed, let us consider the transformation of a coordinate system $\bar{\theta}^i = \bar{\theta}^i \left(\theta^1, \theta^2, \ldots, \theta^n \right)$ $(i = 1, 2, \ldots, n)$. It holds:

$$d\bar{\theta}^i = \frac{\partial \bar{\theta}^i}{\partial \theta^k} d\theta^k, \quad i = 1, 2, \ldots, n. \tag{2.43}$$

Thus, the differentials of the coordinates really transform according to the contravariant law (2.33).

Example 2.3. Transformation of linear coordinates into cylindrical ones (2.17). Let $x^i = x^i(\boldsymbol{r})$ be linear coordinates with respect to an orthonormal basis $\boldsymbol{e}_i$ $(i = 1, 2, 3)$ in $\mathbb{E}^3$:

$$x^i = \boldsymbol{r} \cdot \boldsymbol{e}_i \quad \Leftrightarrow \quad \boldsymbol{r} = x^i \boldsymbol{e}_i. \tag{2.44}$$

By means of (2.17) one can write

$$x^1 = r \cos\varphi, \quad x^2 = r \sin\varphi, \quad x^3 = z \tag{2.45}$$

and consequently

$$\begin{aligned}
&\frac{\partial x^1}{\partial \varphi} = -r \sin\varphi = -x^2, && \frac{\partial x^1}{\partial z} = 0, && \frac{\partial x^1}{\partial r} = \cos\varphi = \frac{x^1}{r}, \\
&\frac{\partial x^2}{\partial \varphi} = r \cos\varphi = x^1, && \frac{\partial x^2}{\partial z} = 0, && \frac{\partial x^2}{\partial r} = \sin\varphi = \frac{x^2}{r}, \\
&\frac{\partial x^3}{\partial \varphi} = 0, && \frac{\partial x^3}{\partial z} = 1, && \frac{\partial x^3}{\partial r} = 0.
\end{aligned} \tag{2.46}$$

The reciprocal derivatives can easily be obtained from (2.23) by inverting the matrix $\left[\frac{\partial x^i}{\partial \varphi}\ \frac{\partial x^i}{\partial z}\ \frac{\partial x^i}{\partial r}\right]$. This yields:

$$\begin{aligned}
&\frac{\partial \varphi}{\partial x^1} = -\frac{1}{r}\sin\varphi = -\frac{x^2}{r^2}, && \frac{\partial \varphi}{\partial x^2} = \frac{1}{r}\cos\varphi = \frac{x^1}{r^2}, && \frac{\partial \varphi}{\partial x^3} = 0,\\
&\frac{\partial z}{\partial x^1} = 0, && \frac{\partial z}{\partial x^2} = 0, && \frac{\partial z}{\partial x^3} = 1,\\
&\frac{\partial r}{\partial x^1} = \cos\varphi = \frac{x^1}{r}, && \frac{\partial r}{\partial x^2} = \sin\varphi = \frac{x^2}{r}, && \frac{\partial r}{\partial x^3} = 0.
\end{aligned} \tag{2.47}$$

2.4 Gradient, Covariant and Contravariant Derivatives

Let $\Phi = \Phi\left(\theta^1, \theta^2, \ldots, \theta^n\right)$, $\boldsymbol{x} = \boldsymbol{x}\left(\theta^1, \theta^2, \ldots, \theta^n\right)$ and $\mathbf{A} = \mathbf{A}\left(\theta^1, \theta^2, \ldots, \theta^n\right)$ be, respectively, a scalar-, a vector- and a tensor-valued differentiable function of the coordinates $\theta^i \in \mathbb{R}$ $(i = 1, 2, \ldots, n)$. Such functions of coordinates are generally referred to as fields, as for example, the scalar field, the vector field or the tensor field. Due to the one to one correspondence (2.26) these fields can alternatively be represented by

$$\Phi = \Phi(\boldsymbol{r}), \quad \boldsymbol{x} = \boldsymbol{x}(\boldsymbol{r}), \quad \mathbf{A} = \mathbf{A}(\boldsymbol{r}). \tag{2.48}$$

In the following we assume that the so-called directional derivatives of the functions (2.48)

$$\begin{aligned}
\left.\frac{\mathrm{d}}{\mathrm{d}s}\Phi(\boldsymbol{r} + s\boldsymbol{a})\right|_{s=0} &= \lim_{s\to 0}\frac{\Phi(\boldsymbol{r} + s\boldsymbol{a}) - \Phi(\boldsymbol{r})}{s},\\
\left.\frac{\mathrm{d}}{\mathrm{d}s}\boldsymbol{x}(\boldsymbol{r} + s\boldsymbol{a})\right|_{s=0} &= \lim_{s\to 0}\frac{\boldsymbol{x}(\boldsymbol{r} + s\boldsymbol{a}) - \boldsymbol{x}(\boldsymbol{r})}{s},\\
\left.\frac{\mathrm{d}}{\mathrm{d}s}\mathbf{A}(\boldsymbol{r} + s\boldsymbol{a})\right|_{s=0} &= \lim_{s\to 0}\frac{\mathbf{A}(\boldsymbol{r} + s\boldsymbol{a}) - \mathbf{A}(\boldsymbol{r})}{s}
\end{aligned} \tag{2.49}$$

exist for all $\boldsymbol{a} \in \mathbb{E}^n$. Further, one can show that the mappings $\boldsymbol{a} \to \frac{\mathrm{d}}{\mathrm{d}s}\Phi(\boldsymbol{r} + s\boldsymbol{a})\big|_{s=0}$, $\boldsymbol{a} \to \frac{\mathrm{d}}{\mathrm{d}s}\boldsymbol{x}(\boldsymbol{r} + s\boldsymbol{a})\big|_{s=0}$ and $\boldsymbol{a} \to \frac{\mathrm{d}}{\mathrm{d}s}\mathbf{A}(\boldsymbol{r} + s\boldsymbol{a})\big|_{s=0}$ are linear with respect to the vector $\boldsymbol{a}$. For example, we can write for the directional derivative of the scalar function $\Phi = \Phi(\boldsymbol{r})$

$$\left.\frac{\mathrm{d}}{\mathrm{d}s}\Phi\left[\boldsymbol{r} + s(\boldsymbol{a} + \boldsymbol{b})\right]\right|_{s=0} = \left.\frac{\mathrm{d}}{\mathrm{d}s}\Phi\left[\boldsymbol{r} + s_1\boldsymbol{a} + s_2\boldsymbol{b}\right]\right|_{s=0}, \tag{2.50}$$

where s_1 and s_2 are assumed to be functions of s such that $s_1 = s$ and $s_2 = s$. With the aid of the chain rule this delivers

$$\begin{aligned}
&\frac{\mathrm{d}}{\mathrm{d}s}\Phi\left[\boldsymbol{r}+s_1\boldsymbol{a}+s_2\boldsymbol{b}\right]\Big|_{s=0} \\
&= \left\{\frac{\partial}{\partial s_1}\Phi\left[\boldsymbol{r}+s_1\boldsymbol{a}+s_2\boldsymbol{b}\right]\frac{\mathrm{d}s_1}{\mathrm{d}s}+\frac{\partial}{\partial s_2}\Phi\left[\boldsymbol{r}+s_1\boldsymbol{a}+s_2\boldsymbol{b}\right]\frac{\mathrm{d}s_2}{\mathrm{d}s}\right\}\Big|_{s=0} \\
&= \frac{\partial}{\partial s_1}\Phi\left(\boldsymbol{r}+s_1\boldsymbol{a}+s_2\boldsymbol{b}\right)\Big|_{s_1=0,s_2=0}+\frac{\partial}{\partial s_2}\Phi\left(\boldsymbol{r}+s_1\boldsymbol{a}+s_2\boldsymbol{b}\right)\Big|_{s_1=0,s_2=0} \\
&= \frac{\mathrm{d}}{\mathrm{d}s}\Phi\left(\boldsymbol{r}+s\boldsymbol{a}\right)\Big|_{s=0}+\frac{\mathrm{d}}{\mathrm{d}s}\Phi\left(\boldsymbol{r}+s\boldsymbol{b}\right)\Big|_{s=0}
\end{aligned}$$

and finally

$$\frac{\mathrm{d}}{\mathrm{d}s}\Phi\left[\boldsymbol{r}+s\left(\boldsymbol{a}+\boldsymbol{b}\right)\right]\Big|_{s=0} = \frac{\mathrm{d}}{\mathrm{d}s}\Phi\left(\boldsymbol{r}+s\boldsymbol{a}\right)\Big|_{s=0}+\frac{\mathrm{d}}{\mathrm{d}s}\Phi\left(\boldsymbol{r}+s\boldsymbol{b}\right)\Big|_{s=0} \tag{2.51}$$

for all $\boldsymbol{a},\boldsymbol{b}\in\mathbb{E}^n$. In a similar fashion we can write

$$\begin{aligned}
\frac{\mathrm{d}}{\mathrm{d}s}\Phi\left(\boldsymbol{r}+s\alpha\boldsymbol{a}\right)\Big|_{s=0} &= \frac{\mathrm{d}}{\mathrm{d}\left(\alpha s\right)}\Phi\left(\boldsymbol{r}+s\alpha\boldsymbol{a}\right)\frac{\mathrm{d}\left(\alpha s\right)}{\mathrm{d}s}\Big|_{s=0} \\
&= \alpha\,\frac{\mathrm{d}}{\mathrm{d}s}\Phi\left(\boldsymbol{r}+s\boldsymbol{a}\right)\Big|_{s=0}, \quad \forall\boldsymbol{a}\in\mathbb{E}^n,\ \forall\alpha\in\mathbb{R}.
\end{aligned} \tag{2.52}$$

Representing $\boldsymbol{a}$ with respect to a basis as $\boldsymbol{a} = a^i\boldsymbol{g}_i$ we thus obtain

$$\begin{aligned}
\frac{\mathrm{d}}{\mathrm{d}s}\Phi\left(\boldsymbol{r}+s\boldsymbol{a}\right)\Big|_{s=0} &= \frac{\mathrm{d}}{\mathrm{d}s}\Phi\left(\boldsymbol{r}+sa^i\boldsymbol{g}_i\right)\Big|_{s=0} = a^i\,\frac{\mathrm{d}}{\mathrm{d}s}\Phi\left(\boldsymbol{r}+s\boldsymbol{g}_i\right)\Big|_{s=0} \\
&= \frac{\mathrm{d}}{\mathrm{d}s}\Phi\left(\boldsymbol{r}+s\boldsymbol{g}_i\right)\Big|_{s=0}\boldsymbol{g}^i\cdot\left(a^j\boldsymbol{g}_j\right),
\end{aligned} \tag{2.53}$$

where $\boldsymbol{g}^i$ form the basis dual to $\boldsymbol{g}_i$ $(i = 1, 2, \ldots, n)$. This result can finally be expressed by

$$\frac{\mathrm{d}}{\mathrm{d}s}\Phi\left(\boldsymbol{r}+s\boldsymbol{a}\right)\Big|_{s=0} = \mathrm{grad}\Phi\cdot\boldsymbol{a}, \quad \forall\boldsymbol{a}\in\mathbb{E}^n, \tag{2.54}$$

where the vector denoted by $\mathrm{grad}\Phi\in\mathbb{E}^n$ is referred to as gradient of the function $\Phi = \Phi\left(\boldsymbol{r}\right)$. According to (2.53) and (2.54) it can be represented by

$$\mathrm{grad}\Phi = \frac{\mathrm{d}}{\mathrm{d}s}\Phi\left(\boldsymbol{r}+s\boldsymbol{g}_i\right)\Big|_{s=0}\boldsymbol{g}^i. \tag{2.55}$$

Example 2.4. Gradient of the scalar function $\|\boldsymbol{r}\|$. Using the definition of the directional derivative (2.49) we can write

$$\begin{aligned}\frac{\mathrm{d}}{\mathrm{d}s}\|\boldsymbol{r}+s\boldsymbol{a}\|\Big|_{s=0} &= \frac{\mathrm{d}}{\mathrm{d}s}\sqrt{(\boldsymbol{r}+s\boldsymbol{a})\cdot(\boldsymbol{r}+s\boldsymbol{a})}\Big|_{s=0}\\ &= \frac{\mathrm{d}}{\mathrm{d}s}\sqrt{\boldsymbol{r}\cdot\boldsymbol{r}+2s(\boldsymbol{r}\cdot\boldsymbol{a})+s^2(\boldsymbol{a}\cdot\boldsymbol{a})}\Big|_{s=0}\\ &= \frac{1}{2}\frac{2(\boldsymbol{r}\cdot\boldsymbol{a})+2s(\boldsymbol{a}\cdot\boldsymbol{a})}{\sqrt{\boldsymbol{r}\cdot\boldsymbol{r}+2s(\boldsymbol{r}\cdot\boldsymbol{a})+s^2(\boldsymbol{a}\cdot\boldsymbol{a})}}\Big|_{s=0} = \frac{\boldsymbol{r}\cdot\boldsymbol{a}}{\|\boldsymbol{r}\|}.\end{aligned}$$

Comparing this result with (2.54) delivers

$$\operatorname{grad}\|\boldsymbol{r}\| = \frac{\boldsymbol{r}}{\|\boldsymbol{r}\|}. \tag{2.56}$$

Similarly to (2.54) one defines the gradient of the vector function $\boldsymbol{x}=\boldsymbol{x}(\boldsymbol{r})$ and the gradient of the tensor function $\mathbf{A}=\mathbf{A}(\boldsymbol{r})$:

$$\frac{\mathrm{d}}{\mathrm{d}s}\boldsymbol{x}(\boldsymbol{r}+s\boldsymbol{a})\Big|_{s=0} = (\operatorname{grad}\boldsymbol{x})\,\boldsymbol{a}, \quad \forall\boldsymbol{a}\in\mathbb{E}^n, \tag{2.57}$$

$$\frac{\mathrm{d}}{\mathrm{d}s}\mathbf{A}(\boldsymbol{r}+s\boldsymbol{a})\Big|_{s=0} = (\operatorname{grad}\mathbf{A})\,\boldsymbol{a}, \quad \forall\boldsymbol{a}\in\mathbb{E}^n. \tag{2.58}$$

Herein, grad$\boldsymbol{x}$ and grad$\mathbf{A}$ represent tensors of second and third order, respectively.

In order to evaluate the above gradients (2.54), (2.57) and (2.58) we represent the vectors $\boldsymbol{r}$ and $\boldsymbol{a}$ with respect to the linear coordinates (2.19) as

$$\boldsymbol{r}=x^i\boldsymbol{h}_i, \quad \boldsymbol{a}=a^i\boldsymbol{h}_i. \tag{2.59}$$

With the aid of the chain rule we can further write for the directional derivative of the function $\Phi=\Phi(\boldsymbol{r})$:

$$\begin{aligned}\frac{\mathrm{d}}{\mathrm{d}s}\Phi(\boldsymbol{r}+s\boldsymbol{a})\Big|_{s=0} &= \frac{\mathrm{d}}{\mathrm{d}s}\Phi\left[\left(x^i+sa^i\right)\boldsymbol{h}_i\right]\Big|_{s=0}\\ &= \frac{\partial\Phi}{\partial\left(x^i+sa^i\right)}\frac{\mathrm{d}\left(x^i+sa^i\right)}{\mathrm{d}s}\Big|_{s=0} = \frac{\partial\Phi}{\partial x^i}a^i\\ &= \left(\frac{\partial\Phi}{\partial x^i}\boldsymbol{h}^i\right)\cdot\left(a^j\boldsymbol{h}_j\right) = \left(\frac{\partial\Phi}{\partial x^i}\boldsymbol{h}^i\right)\cdot\boldsymbol{a}, \quad \forall\boldsymbol{a}\in\mathbb{E}^n.\end{aligned}$$

Comparing this result with (2.54) and bearing in mind that it holds for all vectors $\boldsymbol{a}$ we obtain

$$\operatorname{grad}\Phi = \frac{\partial \Phi}{\partial x^i}\boldsymbol{h}^i. \tag{2.60}$$

The representation (2.60) can be rewritten in terms of arbitrary curvilinear coordinates $\boldsymbol{r} = \boldsymbol{r}\left(\theta^1, \theta^2, \ldots, \theta^n\right)$ and the corresponding tangent vectors (2.28). Indeed, in view of (2.33) and (2.60)

$$\operatorname{grad}\Phi = \frac{\partial \Phi}{\partial x^i}\boldsymbol{h}^i = \frac{\partial \Phi}{\partial \theta^k}\frac{\partial \theta^k}{\partial x^i}\boldsymbol{h}^i = \frac{\partial \Phi}{\partial \theta^i}\boldsymbol{g}^i. \tag{2.61}$$

Comparison of the last result with (2.55) yields

$$\left.\frac{\mathrm{d}}{\mathrm{d}s}\Phi\left(\boldsymbol{r} + s\boldsymbol{g}_i\right)\right|_{s=0} = \frac{\partial \Phi}{\partial \theta^i}, \quad i = 1, 2, \ldots, n. \tag{2.62}$$

According to the definition (2.54) the gradient is independent of the choice of the coordinate system. This can also be seen from relation (2.61). Indeed, taking (2.33) into account we can write for an arbitrary coordinate system $\bar{\theta}^i = \bar{\theta}^i\left(\theta^1, \theta^2, \ldots, \theta^n\right)$ $(i = 1, 2, \ldots, n)$:

$$\operatorname{grad}\Phi = \frac{\partial \Phi}{\partial \theta^i}\boldsymbol{g}^i = \frac{\partial \Phi}{\partial \bar{\theta}^j}\frac{\partial \bar{\theta}^j}{\partial \theta^i}\boldsymbol{g}^i = \frac{\partial \Phi}{\partial \bar{\theta}^j}\bar{\boldsymbol{g}}^j. \tag{2.63}$$

Similarly to relation (2.61) one can express the gradients of the vector-valued function $\boldsymbol{x} = \boldsymbol{x}\left(\boldsymbol{r}\right)$ and the tensor-valued function $\mathbf{A} = \mathbf{A}\left(\boldsymbol{r}\right)$ by

$$\operatorname{grad}\boldsymbol{x} = \frac{\partial \boldsymbol{x}}{\partial \theta^i} \otimes \boldsymbol{g}^i, \quad \operatorname{grad}\mathbf{A} = \frac{\partial \mathbf{A}}{\partial \theta^i} \otimes \boldsymbol{g}^i. \tag{2.64}$$

Example 2.5. Deformation gradient and its representation in the case of simple shear. Let $\boldsymbol{x}$ and $\boldsymbol{X}$ be the position vectors of a material point in the current and reference configuration, respectively. The deformation gradient $\mathbf{F} \in \mathbf{Lin}^3$ is defined as the gradient of the function $\boldsymbol{x}\left(\boldsymbol{X}\right)$ as

$$\mathbf{F} = \operatorname{grad}\boldsymbol{x}. \tag{2.65}$$

For the Cartesian coordinates in $\mathbb{E}^3$ where $\boldsymbol{x} = x^i\boldsymbol{e}_i$ and $\boldsymbol{X} = X^i\boldsymbol{e}_i$ we can write by using $(2.64)_1$

$$\mathbf{F} = \frac{\partial \boldsymbol{x}}{\partial X^j} \otimes \boldsymbol{e}^j = \frac{\partial x^i}{\partial X^j}\boldsymbol{e}_i \otimes \boldsymbol{e}^j = \mathrm{F}^i_{\cdot j}\boldsymbol{e}_i \otimes \boldsymbol{e}^j, \tag{2.66}$$

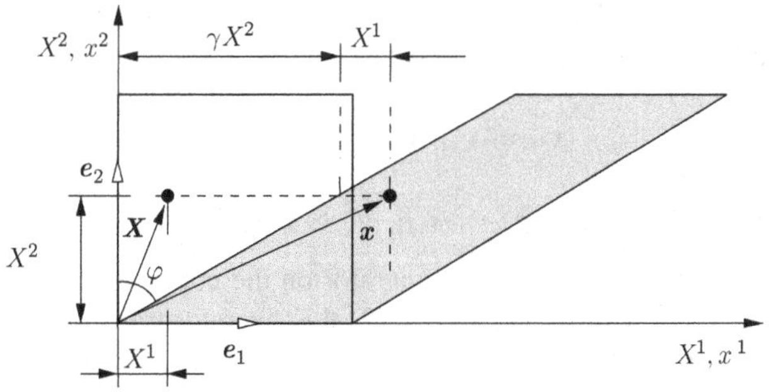

Fig. 2.3 Simple shear of a rectangular sheet

where the matrix $\left[\mathrm{F}^{i}_{\cdot j}\right]$ is given by

$$\left[\mathrm{F}^{i}_{\cdot j}\right] = \begin{bmatrix} \frac{\partial x^1}{\partial X^1} & \frac{\partial x^1}{\partial X^2} & \frac{\partial x^1}{\partial X^3} \\ \frac{\partial x^2}{\partial X^1} & \frac{\partial x^2}{\partial X^2} & \frac{\partial x^2}{\partial X^3} \\ \frac{\partial x^3}{\partial X^1} & \frac{\partial x^3}{\partial X^2} & \frac{\partial x^3}{\partial X^3} \end{bmatrix}. \tag{2.67}$$

In the case of simple shear it holds (see Fig. 2.3)

$$x^1 = X^1 + \gamma X^2, \quad x^2 = X^2, \quad x^3 = X^3, \tag{2.68}$$

where γ denotes the amount of shear. Insertion into (2.67) yields

$$\left[\mathrm{F}^{i}_{\cdot j}\right] = \begin{bmatrix} 1 & \gamma & 0 \\ 0 & 1 & 0 \\ 0 & 0 & 1 \end{bmatrix}. \tag{2.69}$$

Henceforth, the derivatives of the functions $\Phi = \Phi\left(\theta^1, \theta^2, \ldots, \theta^n\right)$, $\boldsymbol{x} = \boldsymbol{x}\left(\theta^1, \theta^2, \ldots, \theta^n\right)$ and $\mathbf{A} = \mathbf{A}\left(\theta^1, \theta^2, \ldots, \theta^n\right)$ with respect to curvilinear coordinates θ^i will be denoted shortly by

$$\Phi_{,i} = \frac{\partial \Phi}{\partial \theta^i}, \quad \boldsymbol{x}_{,i} = \frac{\partial \boldsymbol{x}}{\partial \theta^i}, \quad \mathbf{A}_{,i} = \frac{\partial \mathbf{A}}{\partial \theta^i}. \tag{2.70}$$

They obey the covariant transformation rule (2.32) with respect to the index i since

$$\frac{\partial \Phi}{\partial \theta^i} = \frac{\partial \Phi}{\partial \bar{\theta}^k}\frac{\partial \bar{\theta}^k}{\partial \theta^i}, \quad \frac{\partial \boldsymbol{x}}{\partial \theta^i} = \frac{\partial \boldsymbol{x}}{\partial \bar{\theta}^k}\frac{\partial \bar{\theta}^k}{\partial \theta^i}, \quad \frac{\partial \mathbf{A}}{\partial \theta^i} = \frac{\partial \mathbf{A}}{\partial \bar{\theta}^k}\frac{\partial \bar{\theta}^k}{\partial \theta^i} \tag{2.71}$$

and represent again a scalar, a vector and a second-order tensor, respectively. The latter ones can be represented with respect to a basis as

$$\boldsymbol{x}_{,i} = x^j|_i \, \boldsymbol{g}_j = x_j|_i \, \boldsymbol{g}^j,$$

$$\mathbf{A}_{,i} = \mathrm{A}^{kl}|_i \, \boldsymbol{g}_k \otimes \boldsymbol{g}_l = \mathrm{A}_{kl}|_i \, \boldsymbol{g}^k \otimes \boldsymbol{g}^l = \mathrm{A}^k_{\cdot l}|_i \, \boldsymbol{g}_k \otimes \boldsymbol{g}^l, \tag{2.72}$$

where $(\bullet)|_i$ denotes some differential operator on the components of the vector $\boldsymbol{x}$ or the tensor $\mathbf{A}$. In view of (2.71) and (2.72) this operator transforms with respect to the index i according to the covariant rule and is called covariant derivative. The covariant type of the derivative is accentuated by the lower position of the coordinate index.

On the basis of the covariant derivative we can also define the contravariant one. To this end, we formally apply the rule of component transformation $(1.95)_1$ as $(\bullet)|^i = g^{ij} \, (\bullet)|_j$. Accordingly,

$$x^j|^i = g^{ik} x^j|_k, \qquad x_j|^i = g^{ik} x_j|_k,$$

$$\mathrm{A}^{kl}|^i = g^{im} \mathrm{A}^{kl}|_m, \qquad \mathrm{A}_{kl}|^i = g^{im} \mathrm{A}_{kl}|_m, \qquad \mathrm{A}^k_{\cdot l}|^i = g^{im} \mathrm{A}^k_{\cdot l}|_m \,. \tag{2.73}$$

For scalar functions the covariant and the contravariant derivative are defined to be equal to the partial one so that:

$$\Phi|_i = \Phi|^i = \Phi_{,i} \,. \tag{2.74}$$

In view of (2.63)–(2.70), (2.72) and (2.74) the gradients of the functions $\Phi = \Phi\left(\theta^1, \theta^2, \ldots, \theta^n\right)$, $\boldsymbol{x} = \boldsymbol{x}\left(\theta^1, \theta^2, \ldots, \theta^n\right)$ and $\mathbf{A} = \mathbf{A}\left(\theta^1, \theta^2, \ldots, \theta^n\right)$ take the form

$$\mathrm{grad}\Phi = \Phi|_i \, \boldsymbol{g}^i = \Phi|^i \, \boldsymbol{g}_i,$$

$$\mathrm{grad}\boldsymbol{x} = x^j|_i \, \boldsymbol{g}_j \otimes \boldsymbol{g}^i = x_j|_i \, \boldsymbol{g}^j \otimes \boldsymbol{g}^i = x^j|^i \, \boldsymbol{g}_j \otimes \boldsymbol{g}_i = x_j|^i \, \boldsymbol{g}^j \otimes \boldsymbol{g}_i,$$

$$\begin{aligned} \mathrm{grad}\mathbf{A} &= \mathrm{A}^{kl}|_i \, \boldsymbol{g}_k \otimes \boldsymbol{g}_l \otimes \boldsymbol{g}^i = \mathrm{A}_{kl}|_i \, \boldsymbol{g}^k \otimes \boldsymbol{g}^l \otimes \boldsymbol{g}^i = \mathrm{A}^k_{\cdot l}|_i \, \boldsymbol{g}_k \otimes \boldsymbol{g}^l \otimes \boldsymbol{g}^i \\ &= \mathrm{A}^{kl}|^i \, \boldsymbol{g}_k \otimes \boldsymbol{g}_l \otimes \boldsymbol{g}_i = \mathrm{A}_{kl}|^i \, \boldsymbol{g}^k \otimes \boldsymbol{g}^l \otimes \boldsymbol{g}_i = \mathrm{A}^k_{\cdot l}|^i \, \boldsymbol{g}_k \otimes \boldsymbol{g}^l \otimes \boldsymbol{g}_i. \end{aligned} \tag{2.75}$$

2.5 Christoffel Symbols, Representation of the Covariant Derivative

In the previous section we have introduced the notion of the covariant derivative but have not so far discussed how it can be taken. Now, we are going to formulate a procedure constructing the differential operator of the covariant derivative. In other words, we would like to express the covariant derivative in terms of the vector or tensor components. To this end, the partial derivatives of the tangent vectors (2.28) with respect to the coordinates are first needed. Since these derivatives again represent vectors in $\mathbb{E}^n$, they can be expressed in terms of the tangent vectors $\boldsymbol{g}_i$ or dual vectors $\boldsymbol{g}^i$ $(i = 1, 2, \ldots, n)$ both forming bases of $\mathbb{E}^n$. Thus, one can write

$$\boldsymbol{g}_i,_j = \Gamma_{ijk}\boldsymbol{g}^k = \Gamma_{ij}^k \boldsymbol{g}_k, \quad i, j = 1, 2, \ldots, n, \tag{2.76}$$

where the components Γ_{ijk} and Γ_{ij}^k $(i, j, k = 1, 2, \ldots, n)$ are referred to as the Christoffel symbols of the first and second kind, respectively. In view of the relation $\boldsymbol{g}^k = g^{kl}\boldsymbol{g}_l$ $(k = 1, 2, \ldots, n)$ (1.21) these symbols are connected with each other by

$$\Gamma_{ij}^k = g^{kl}\Gamma_{ijl}, \quad i, j, k = 1, 2, \ldots, n. \tag{2.77}$$

Keeping the definition of tangent vectors (2.28) in mind we further obtain

$$\boldsymbol{g}_i,_j = \boldsymbol{r},_{ij} = \boldsymbol{r},_{ji} = \boldsymbol{g}_j,_i, \quad i, j = 1, 2, \ldots, n. \tag{2.78}$$

With the aid of (1.28) the Christoffel symbols can thus be expressed by

$$\Gamma_{ijk} = \Gamma_{jik} = \boldsymbol{g}_i,_j \cdot \boldsymbol{g}_k = \boldsymbol{g}_j,_i \cdot \boldsymbol{g}_k, \tag{2.79}$$

$$\Gamma_{ij}^k = \Gamma_{ji}^k = \boldsymbol{g}_i,_j \cdot \boldsymbol{g}^k = \boldsymbol{g}_j,_i \cdot \boldsymbol{g}^k, \quad i, j, k = 1, 2, \ldots, n. \tag{2.80}$$

For the dual basis $\boldsymbol{g}^i$ $(i = 1, 2, \ldots, n)$ one further gets by differentiating the identities $\boldsymbol{g}^i \cdot \boldsymbol{g}_j = \delta_j^i$ (1.15):

$$\begin{aligned}0 = \left(\delta_j^i\right),_k &= \left(\boldsymbol{g}^i \cdot \boldsymbol{g}_j\right),_k = \boldsymbol{g}^i,_k \cdot \boldsymbol{g}_j + \boldsymbol{g}^i \cdot \boldsymbol{g}_j,_k \\ &= \boldsymbol{g}^i,_k \cdot \boldsymbol{g}_j + \boldsymbol{g}^i \cdot \left(\Gamma_{jk}^l \boldsymbol{g}_l\right) = \boldsymbol{g}^i,_k \cdot \boldsymbol{g}_j + \Gamma_{jk}^i, \quad i, j, k = 1, 2, \ldots, n.\end{aligned}$$

Hence,

$$\Gamma_{jk}^i = \Gamma_{kj}^i = -\boldsymbol{g}^i,_k \cdot \boldsymbol{g}_j = -\boldsymbol{g}^i,_j \cdot \boldsymbol{g}_k, \quad i, j, k = 1, 2, \ldots, n \tag{2.81}$$

and consequently

$$\boldsymbol{g}^i,_k = -\Gamma_{jk}^i \boldsymbol{g}^j = -\Gamma_{kj}^i \boldsymbol{g}^j, \quad i, k = 1, 2, \ldots, n. \tag{2.82}$$

By means of the identities following from (2.79)

$$g_{ij},_k = \left(\boldsymbol{g}_i \cdot \boldsymbol{g}_j\right),_k = \boldsymbol{g}_i,_k \cdot \boldsymbol{g}_j + \boldsymbol{g}_i \cdot \boldsymbol{g}_j,_k = \Gamma_{ikj} + \Gamma_{jki}, \tag{2.83}$$

where $i, j, k = 1, 2, \ldots, n$ and in view of (2.77) we finally obtain

$$\Gamma_{ijk} = \frac{1}{2}\left(g_{ki},_j + g_{kj},_i - g_{ij},_k\right), \tag{2.84}$$

$$\Gamma_{ij}^k = \frac{1}{2} g^{kl}\left(g_{li},_j + g_{lj},_i - g_{ij},_l\right), \quad i,j,k = 1,2,\ldots,n. \tag{2.85}$$

It is seen from (2.84) and (2.85) that all Christoffel symbols identically vanish in the Cartesian coordinates (2.20). Indeed, in this case

$$g_{ij} = \boldsymbol{e}_i \cdot \boldsymbol{e}_j = \delta_{ij}, \quad i,j = 1,2,\ldots,n \tag{2.86}$$

and hence

$$\Gamma_{ijk} = \Gamma_{ij}^k = 0, \quad i,j,k = 1,2,\ldots,n. \tag{2.87}$$

Example 2.6. Christoffel symbols for cylindrical coordinates in $\mathbb{E}^3$ *(2.17).* By virtue of relation $(2.30)_1$ we realize that $g_{11},_3 = 2r$, while all other derivatives $g_{ik},_j$ $(i,j,k = 1,2,3)$ (2.83) are zero. Thus, Eq. (2.84) delivers

$$\Gamma_{131} = \Gamma_{311} = r, \quad \Gamma_{113} = -r, \tag{2.88}$$

while all other Christoffel symbols of the first kind Γ_{ijk} $(i,j,k = 1,2,3)$ are likewise zero. With the aid of (2.77) and $(2.30)_2$ we further obtain

$$\Gamma_{ij}^1 = g^{11}\Gamma_{ij1} = r^{-2}\Gamma_{ij1}, \quad \Gamma_{ij}^2 = g^{22}\Gamma_{ij2} = \Gamma_{ij2},$$

$$\Gamma_{ij}^3 = g^{33}\Gamma_{ij3} = \Gamma_{ij3}, \quad i,j = 1,2,3. \tag{2.89}$$

By virtue of (2.88) we can further write

$$\Gamma_{13}^1 = \Gamma_{31}^1 = \frac{1}{r}, \quad \Gamma_{11}^3 = -r, \tag{2.90}$$

while all remaining Christoffel symbols of the second kind Γ_{ij}^k $(i,j,k = 1,2,3)$ (2.85) vanish.

Now, we are in a position to express the covariant derivative in terms of the vector or tensor components by means of the Christoffel symbols. For the vector-valued function $\boldsymbol{x} = \boldsymbol{x}\left(\theta^1, \theta^2, \ldots, \theta^n\right)$ we can write using (2.76)

$$\begin{aligned}\boldsymbol{x},_j &= \left(x^i \boldsymbol{g}_i\right),_j = x^i,_j \boldsymbol{g}_i + x^i \boldsymbol{g}_i,_j \\ &= x^i,_j \boldsymbol{g}_i + x^i \Gamma_{ij}^k \boldsymbol{g}_k = \left(x^i,_j + x^k \Gamma_{kj}^i\right) \boldsymbol{g}_i,\end{aligned} \tag{2.91}$$

or alternatively using (2.82)

$$\boldsymbol{x},_j = \left(x_i \boldsymbol{g}^i\right),_j = x_i,_j \boldsymbol{g}^i + x_i \boldsymbol{g}^i,_j$$

$$= x_i,_j \boldsymbol{g}^i - x_i \Gamma^i_{kj} \boldsymbol{g}^k = \left(x_i,_j - x_k \Gamma^k_{ij}\right) \boldsymbol{g}^i. \tag{2.92}$$

Comparing these results with (2.72) yields

$$x^i|_j = x^i,_j + x^k \Gamma^i_{kj}, \quad x_i|_j = x_i,_j - x_k \Gamma^k_{ij}, \quad i, j = 1, 2, \ldots, n. \tag{2.93}$$

Similarly, we treat the tensor-valued function $\mathbf{A} = \mathbf{A}\left(\theta^1, \theta^2, \ldots, \theta^n\right)$:

$$\mathbf{A},_k = \left(\mathrm{A}^{ij} \boldsymbol{g}_i \otimes \boldsymbol{g}_j\right),_k$$

$$= \mathrm{A}^{ij},_k \boldsymbol{g}_i \otimes \boldsymbol{g}_j + \mathrm{A}^{ij} \boldsymbol{g}_i,_k \otimes \boldsymbol{g}_j + \mathrm{A}^{ij} \boldsymbol{g}_i \otimes \boldsymbol{g}_j,_k$$

$$= \mathrm{A}^{ij},_k \boldsymbol{g}_i \otimes \boldsymbol{g}_j + \mathrm{A}^{ij} \left(\Gamma^l_{ik} \boldsymbol{g}_l\right) \otimes \boldsymbol{g}_j + \mathrm{A}^{ij} \boldsymbol{g}_i \otimes \left(\Gamma^l_{jk} \boldsymbol{g}_l\right)$$

$$= \left(\mathrm{A}^{ij},_k + \mathrm{A}^{lj} \Gamma^i_{lk} + \mathrm{A}^{il} \Gamma^j_{lk}\right) \boldsymbol{g}_i \otimes \boldsymbol{g}_j. \tag{2.94}$$

Thus,

$$\mathrm{A}^{ij}|_k = \mathrm{A}^{ij},_k + \mathrm{A}^{lj} \Gamma^i_{lk} + \mathrm{A}^{il} \Gamma^j_{lk}, \quad i, j, k = 1, 2, \ldots, n. \tag{2.95}$$

By analogy, we further obtain

$$\mathrm{A}_{ij}|_k = \mathrm{A}_{ij},_k - \mathrm{A}_{lj} \Gamma^l_{ik} - \mathrm{A}_{il} \Gamma^l_{jk},$$

$$\mathrm{A}^i_{\cdot j}|_k = \mathrm{A}^i_{\cdot j},_k + \mathrm{A}^l_{\cdot j} \Gamma^i_{lk} - \mathrm{A}^i_{\cdot l} \Gamma^l_{jk}, \quad i, j, k = 1, 2, \ldots, n. \tag{2.96}$$

Similar expressions for the covariant derivative can also be formulated for tensors of higher orders.

From (2.87), (2.93), (2.95) and (2.96) it is seen that the covariant derivative taken in Cartesian coordinates (2.20) coincides with the partial derivative:

$$x^i|_j = x^i,_j, \quad x_i|_j = x_i,_j,$$

$$\mathrm{A}^{ij}|_k = \mathrm{A}^{ij},_k, \; \mathrm{A}_{ij}|_k = \mathrm{A}_{ij},_k, \; \mathrm{A}^i_{\cdot j}|_k = \mathrm{A}^i_{\cdot j},_k, \quad i, j, k = 1, 2, \ldots, n. \tag{2.97}$$

Formal application of the covariant derivative (2.93), (2.95) and (2.96) to the tangent vectors (2.28) and metric coefficients $(1.90)_{1,2}$ yields by virtue of (2.76), (2.77), (2.82) and (2.84) the following identities referred to as Ricci's Theorem:

$$\boldsymbol{g}_i|_j = \boldsymbol{g}_i,_j - \boldsymbol{g}_l \Gamma^l_{ij} = \boldsymbol{0}, \quad \boldsymbol{g}^i|_j = \boldsymbol{g}^i,_j + \boldsymbol{g}^l \Gamma^i_{lj} = \boldsymbol{0}, \tag{2.98}$$

$$g_{ij}|_k = g_{ij},_k - g_{lj}\Gamma^l_{ik} - g_{il}\Gamma^l_{jk} = g_{ij},_k - \Gamma_{ikj} - \Gamma_{jki} = 0, \quad (2.99)$$

$$g^{ij}|_k = g^{ij},_k + g^{lj}\Gamma^i_{lk} + g^{il}\Gamma^j_{lk} = g^{il}g^{jm}\left(-g_{lm},_k + \Gamma_{mkl} + \Gamma_{lkm}\right) = 0, \quad (2.100)$$

where $i, j, k = 1, 2, \ldots, n$. The latter two identities can alternatively be proved by taking (1.25) into account and using the product rules of differentiation for the covariant derivative which can be written as (Exercise 2.7)

$$\mathrm{A}_{ij}|_k = a_i|_k\, b_j + a_i b_j|_k \quad \text{for} \quad \mathrm{A}_{ij} = a_i b_j, \quad (2.101)$$

$$\mathrm{A}^{ij}|_k = a^i|_k\, b^j + a^i b^j|_k \quad \text{for} \quad \mathrm{A}^{ij} = a^i b^j, \quad (2.102)$$

$$\mathrm{A}^i_{j}|_k = a^i|_k\, b_j + a^i b_j|_k \quad \text{for} \quad \mathrm{A}^i_{j} = a^i b_j, \quad i, j, k = 1, 2, \ldots, n. \quad (2.103)$$

2.6 Applications in Three-Dimensional Space: Divergence and Curl

Divergence of a tensor field. One defines the divergence of a tensor field $\mathbf{S}(\boldsymbol{r})$ by

$$\mathrm{div}\mathbf{S} = \lim_{V \to 0} \frac{1}{V} \int_A \mathbf{S}\boldsymbol{n}\,\mathrm{d}A, \quad (2.104)$$

where the integration is carried out over a closed surface area A with the volume V and the outer unit normal vector $\boldsymbol{n}$ illustrated in Fig. 2.4.

For the integration we consider a curvilinear parallelepiped with the edges formed by the coordinate lines $\theta^1, \theta^2, \theta^3$ and $\theta^1 + \Delta\theta^1, \theta^2 + \Delta\theta^2, \theta^3 + \Delta\theta^3$ (Fig. 2.5). The infinitesimal surface elements of the parallelepiped can be defined in a vector form by

$$\mathrm{d}\boldsymbol{A}^{(i)} = \pm\left(d\theta^j \boldsymbol{g}_j\right) \times \left(d\theta^k \boldsymbol{g}_k\right) = \pm g\boldsymbol{g}^i d\theta^j d\theta^k, \quad i = 1, 2, 3, \quad (2.105)$$

where $g = [\boldsymbol{g}_1\boldsymbol{g}_2\boldsymbol{g}_3]$ (1.31) and i, j, k is an even permutation of 1,2,3. The corresponding infinitesimal volume element can thus be given by (no summation over i)

$$\mathrm{d}V = \mathrm{d}\boldsymbol{A}^{(i)} \cdot \left(d\theta^i \boldsymbol{g}_i\right) = \left[d\theta^1\boldsymbol{g}_1\; d\theta^2\boldsymbol{g}_2\; d\theta^3\boldsymbol{g}_3\right]$$

$$= [\boldsymbol{g}_1\boldsymbol{g}_2\boldsymbol{g}_3]\, d\theta^1 d\theta^2 d\theta^3 = g d\theta^1 d\theta^2 d\theta^3. \quad (2.106)$$

We also need the identities

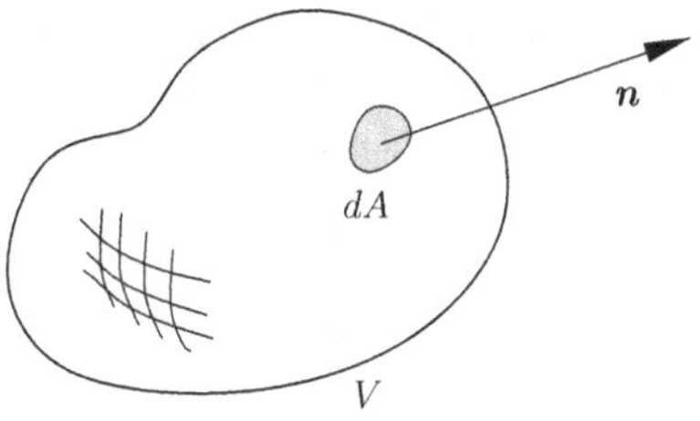

Fig. 2.4 Definition of the divergence: closed surface with the area A, volume V and the outer unit normal vector $\boldsymbol{n}$

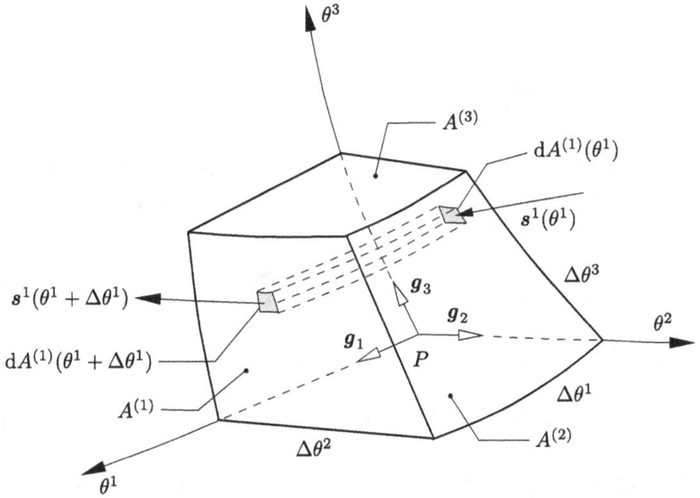

Fig. 2.5 Derivation of the divergence in three-dimensional space

$$
\begin{aligned}
g_{,k} = \left[\boldsymbol{g}_1\boldsymbol{g}_2\boldsymbol{g}_3\right]_{,k} &= \Gamma^l_{1k}\left[\boldsymbol{g}_l\boldsymbol{g}_2\boldsymbol{g}_3\right] + \Gamma^l_{2k}\left[\boldsymbol{g}_1\boldsymbol{g}_l\boldsymbol{g}_3\right] + \Gamma^l_{3k}\left[\boldsymbol{g}_1\boldsymbol{g}_2\boldsymbol{g}_l\right] \\
&= \Gamma^l_{lk}\left[\boldsymbol{g}_1\boldsymbol{g}_2\boldsymbol{g}_3\right] = \Gamma^l_{lk} g,
\end{aligned}
\tag{2.107}
$$

$$
\left(g\boldsymbol{g}^i\right)_{,i} = g_{,i}\,\boldsymbol{g}^i + g\boldsymbol{g}^i{}_{,i} = \Gamma^l_{li} g\boldsymbol{g}^i - \Gamma^i_{li} g\boldsymbol{g}^l = \boldsymbol{0},
\tag{2.108}
$$

following from (1.39), (2.76) and (2.82). With these results in hand, one can express the divergence (2.104) as follows

$$
\begin{aligned}
\operatorname{div}\mathbf{S} &= \lim_{V\to 0}\frac{1}{V}\int_A \mathbf{S}\boldsymbol{n}\,\mathrm{d}A \\
&= \lim_{V\to 0}\frac{1}{V}\sum_{i=1}^{3}\int_{A^{(i)}}\left[\mathbf{S}\left(\theta^i + \Delta\theta^i\right)\mathrm{d}\boldsymbol{A}^{(i)}\left(\theta^i + \Delta\theta^i\right) + \mathbf{S}\left(\theta^i\right)\mathrm{d}\boldsymbol{A}^{(i)}\left(\theta^i\right)\right].
\end{aligned}
$$

Keeping (2.105) and (2.106) in mind and using the abbreviation

$$s^i\left(\theta^i\right) = \mathbf{S}\left(\theta^i\right) g\left(\theta^i\right) \boldsymbol{g}^i\left(\theta^i\right), \quad i = 1,2,3 \tag{2.109}$$

we can thus write

$$\begin{aligned}
\operatorname{div}\mathbf{S} &= \lim_{V\to 0} \frac{1}{V} \sum_{i=1}^{3} \int_{\theta^k}^{\theta^k+\Delta\theta^k} \int_{\theta^j}^{\theta^j+\Delta\theta^j} \left[s^i\left(\theta^i + \Delta\theta^i\right) - s^i\left(\theta^i\right)\right] \mathrm{d}\theta^j \mathrm{d}\theta^k \\
&= \lim_{V\to 0} \frac{1}{V} \sum_{i=1}^{3} \int_{\theta^k}^{\theta^k+\Delta\theta^k} \int_{\theta^j}^{\theta^j+\Delta\theta^j} \int_{\theta^i}^{\theta^i+\Delta\theta^i} \frac{\partial s^i}{\partial\theta^i} \mathrm{d}\theta^i \mathrm{d}\theta^j \mathrm{d}\theta^k \\
&= \lim_{V\to 0} \frac{1}{V} \sum_{i=1}^{3} \int_V \frac{s^i{}_{,i}}{g} \mathrm{d}V,
\end{aligned} \tag{2.110}$$

where i, j, k is again an even permutation of 1,2,3. Assuming continuity of the integrand in (2.110) and applying (2.108) and (2.109) we obtain

$$\operatorname{div}\mathbf{S} = \frac{1}{g} s^i{}_{,i} = \frac{1}{g}\left[\mathbf{S} g \boldsymbol{g}^i\right]_{,i} = \frac{1}{g}\left[\mathbf{S}_{,i}\, g \boldsymbol{g}^i + \mathbf{S}\left(g\boldsymbol{g}^i\right)_{,i}\right] = \mathbf{S}_{,i}\, \boldsymbol{g}^i, \tag{2.111}$$

which finally yields by virtue of $(2.72)_2$

$$\operatorname{div}\mathbf{S} = \mathbf{S}_{,i}\, \boldsymbol{g}^i = \mathrm{S}^{i}_{j\cdot}|_i\, \boldsymbol{g}^j = \mathrm{S}^{ji}|_i\, \boldsymbol{g}_j. \tag{2.112}$$

Example 2.7. The momentum balance in Cartesian and cylindrical coordinates. Let us consider a material body or a part of it with a mass m, volume V and outer surface A. According to the Euler law of motion the vector sum of external volume forces $\boldsymbol{f}\mathrm{d}V$ and surface tractions $\boldsymbol{t}\mathrm{d}A$ results in the vector sum of inertia forces $\ddot{\boldsymbol{x}}\mathrm{d}m$, where $\boldsymbol{x}$ stands for the position vector of a material element $\mathrm{d}m$ and the superposed dot denotes the material time derivative. Hence,

$$\int_m \ddot{\boldsymbol{x}}\mathrm{d}m = \int_A \boldsymbol{t}\mathrm{d}A + \int_V \boldsymbol{f}\mathrm{d}V. \tag{2.113}$$

Applying the Cauchy theorem (1.77) to the first integral on the right hand side and using the identity $\mathrm{d}m = \rho\mathrm{d}V$ it further delivers

$$\int_V \rho\ddot{\boldsymbol{x}}\mathrm{d}V = \int_A \boldsymbol{\sigma}\boldsymbol{n}\mathrm{d}A + \int_V \boldsymbol{f}\mathrm{d}V, \tag{2.114}$$

where ρ denotes the density of the material. Dividing this equation by V and considering the limit case $V \to 0$ we obtain by virtue of (2.104)

$$\rho\ddot{\boldsymbol{x}} = \mathrm{div}\boldsymbol{\sigma} + \boldsymbol{f}. \tag{2.115}$$

This vector equation is referred to as the momentum balance.

Representing vector and tensor variables with respect to the tangent vectors $\boldsymbol{g}_i$ $(i = 1, 2, 3)$ of an arbitrary curvilinear coordinate system as

$$\ddot{\boldsymbol{x}} = a^i \boldsymbol{g}_i, \quad \boldsymbol{\sigma} = \sigma^{ij} \boldsymbol{g}_i \otimes \boldsymbol{g}_j, \quad \boldsymbol{f} = f^i \boldsymbol{g}_i$$

and expressing the divergence of the Cauchy stress tensor by (2.112) we obtain the component form of the momentum balance (2.115) by

$$\rho a^i = \sigma^{ij}|_j + f^i, \quad i = 1, 2, 3. \tag{2.116}$$

With the aid of (2.95) the covariant derivative of the Cauchy stress tensor can further be written by

$$\sigma^{ij}|_k = \sigma^{ij},_k + \sigma^{lj}\Gamma^i_{lk} + \sigma^{il}\Gamma^j_{lk}, \quad i, j, k = 1, 2, 3 \tag{2.117}$$

and thus,

$$\sigma^{ij}|_j = \sigma^{ij},_j + \sigma^{lj}\Gamma^i_{lj} + \sigma^{il}\Gamma^j_{lj}, \quad i = 1, 2, 3. \tag{2.118}$$

By virtue of the expressions for the Christoffel symbols (2.90) and keeping in mind the symmetry of the Cauchy stress tensors $\sigma^{ij} = \sigma^{ji}$ $(i \neq j = 1, 2, 3)$ we thus obtain for cylindrical coordinates:

$$\begin{aligned}
\sigma^{1j}|_j &= \sigma^{11},_\varphi + \sigma^{12},_z + \sigma^{13},_r + \frac{3\sigma^{31}}{r}, \\
\sigma^{2j}|_j &= \sigma^{21},_\varphi + \sigma^{22},_z + \sigma^{23},_r + \frac{\sigma^{32}}{r}, \\
\sigma^{3j}|_j &= \sigma^{31},_\varphi + \sigma^{32},_z + \sigma^{33},_r - r\sigma^{11} + \frac{\sigma^{33}}{r}.
\end{aligned} \tag{2.119}$$

The balance equations finally take the form

$$\begin{aligned}
\rho a^1 &= \sigma^{11},_\varphi + \sigma^{12},_z + \sigma^{13},_r + \frac{3\sigma^{31}}{r} + f^1, \\
\rho a^2 &= \sigma^{21},_\varphi + \sigma^{22},_z + \sigma^{23},_r + \frac{\sigma^{32}}{r} + f^2, \\
\rho a^3 &= \sigma^{31},_\varphi + \sigma^{32},_z + \sigma^{33},_r - r\sigma^{11} + \frac{\sigma^{33}}{r} + f^3.
\end{aligned} \tag{2.120}$$

In Cartesian coordinates, where $\boldsymbol{g}_i = \boldsymbol{e}_i$ $(i = 1, 2, 3)$, the covariant derivative coincides with the partial one, so that

$$\sigma^{ij}|_j = \sigma^{ij},_j = \sigma_{ij},_j\,. \tag{2.121}$$

Thus, the balance equations reduce to

$$\begin{aligned}\rho\ddot{x}_1 &= \sigma_{11},_1 + \sigma_{12},_2 + \sigma_{13},_3 + f_1,\\ \rho\ddot{x}_2 &= \sigma_{21},_1 + \sigma_{22},_2 + \sigma_{23},_3 + f_2,\\ \rho\ddot{x}_3 &= \sigma_{31},_1 + \sigma_{32},_2 + \sigma_{33},_3 + f_3,\end{aligned} \tag{2.122}$$

where $\ddot{x}_i = a_i$ $(i = 1, 2, 3)$.

Divergence and curl of a vector field. Now, we consider a differentiable vector field $\boldsymbol{t}\left(\theta^1, \theta^2, \theta^3\right)$. One defines the divergence and curl of $\boldsymbol{t}\left(\theta^1, \theta^2, \theta^3\right)$ respectively by

$$\mathrm{div}\boldsymbol{t} = \lim_{V\to 0}\frac{1}{V}\int_A (\boldsymbol{t}\cdot\boldsymbol{n})\,\mathrm{d}A, \tag{2.123}$$

$$\mathrm{curl}\boldsymbol{t} = \lim_{V\to 0}\frac{1}{V}\int_A (\boldsymbol{n}\times\boldsymbol{t})\,\mathrm{d}A = -\lim_{V\to 0}\frac{1}{V}\int_A (\boldsymbol{t}\times\boldsymbol{n})\,\mathrm{d}A, \tag{2.124}$$

where the integration is again carried out over a closed surface area A with the volume V and the outer unit normal vector $\boldsymbol{n}$ (see Fig. 2.4). Considering (1.66) and (2.104), the curl can also be represented by

$$\mathrm{curl}\boldsymbol{t} = -\lim_{V\to 0}\frac{1}{V}\int_A \hat{\boldsymbol{t}}\boldsymbol{n}\mathrm{d}A = -\mathrm{div}\hat{\boldsymbol{t}}. \tag{2.125}$$

Treating the vector field in the same manner as the tensor field we can write

$$\mathrm{div}\boldsymbol{t} = \boldsymbol{t},_i \cdot \boldsymbol{g}^i = t^i|_i \tag{2.126}$$

and in view of $(2.75)_2$ (see also Exercise 1.44)

$$\mathrm{div}\boldsymbol{t} = \mathrm{tr}\left(\mathrm{grad}\boldsymbol{t}\right). \tag{2.127}$$

The same procedure applied to the curl (2.124) leads to

$$\mathrm{curl}\boldsymbol{t} = \boldsymbol{g}^i \times \boldsymbol{t},_i\,. \tag{2.128}$$

By virtue of $(2.72)_1$ and (1.44) we further obtain (see also Exercise 2.8)

$$\mathrm{curl}\boldsymbol{t} = t_i|_j\, \boldsymbol{g}^j \times \boldsymbol{g}^i = e^{jik}\frac{1}{g}t_i|_j\, \boldsymbol{g}_k. \tag{2.129}$$

With respect to the Cartesian coordinates (2.20) with $\boldsymbol{g}_i = \boldsymbol{e}_i$ $(i = 1, 2, 3)$ the divergence (2.126) and curl (2.129) simplify to

$$\text{div}\boldsymbol{t} = t^i{}_{,i} = t^1{}_{,1} + t^2{}_{,2} + t^3{}_{,3} = t_{1,1} + t_{2,2} + t_{3,3}, \tag{2.130}$$

$$\begin{aligned}\text{curl}\boldsymbol{t} &= e^{jik} t_{i,j}\, \boldsymbol{e}_k \\ &= (t_{3,2} - t_{2,3})\, \boldsymbol{e}_1 + (t_{1,3} - t_{3,1})\, \boldsymbol{e}_2 + (t_{2,1} - t_{1,2})\, \boldsymbol{e}_3.\end{aligned} \tag{2.131}$$

Now, we are going to discuss some combined operations with a gradient, divergence, curl, tensor mapping and products of various types (see also Exercise 2.12).

1. Curl of a gradient:
$$\text{curl}\,\text{grad}\Phi = \boldsymbol{0}. \tag{2.132}$$
2. Divergence of a curl:
$$\text{div}\,\text{curl}\boldsymbol{t} = 0. \tag{2.133}$$
3. Divergence of a vector product:
$$\text{div}\,(\boldsymbol{u} \times \boldsymbol{v}) = \boldsymbol{v} \cdot \text{curl}\boldsymbol{u} - \boldsymbol{u} \cdot \text{curl}\boldsymbol{v}. \tag{2.134}$$
4. Gradient of a divergence:
$$\text{grad}\,\text{div}\boldsymbol{t} = \text{div}\,(\text{grad}\boldsymbol{t})^{\mathrm{T}}, \tag{2.135}$$
$$\text{grad}\,\text{div}\boldsymbol{t} = \text{curl}\,\text{curl}\boldsymbol{t} + \text{div}\,\text{grad}\boldsymbol{t} = \text{curl}\,\text{curl}\boldsymbol{t} + \Delta\boldsymbol{t}, \tag{2.136}$$
where the combined operator $\Delta\boldsymbol{t} = \text{div}\,\text{grad}\boldsymbol{t}$ is known as the Laplacian.
5. Skew-symmetric part of a gradient
$$\text{skew}\,(\text{grad}\boldsymbol{t}) = \frac{1}{2}\widehat{\text{curl}\boldsymbol{t}}. \tag{2.137}$$
6. Divergence of a (left) mapping
$$\text{div}\,(\boldsymbol{t}\mathbf{A}) = \mathbf{A} : \text{grad}\boldsymbol{t} + \boldsymbol{t} \cdot \text{div}\mathbf{A}. \tag{2.138}$$
7. Divergence of a product of a scalar-valued function and a vector-valued function
$$\text{div}\,(\Phi\boldsymbol{t}) = \boldsymbol{t} \cdot \text{grad}\Phi + \Phi\text{div}\boldsymbol{t}. \tag{2.139}$$
8. Divergence of a product of a scalar-valued function and a tensor-valued function
$$\text{div}\,(\Phi\mathbf{A}) = \mathbf{A}\text{grad}\Phi + \Phi\text{div}\mathbf{A}. \tag{2.140}$$

We prove, for example, identity (2.132). To this end, we apply $(2.75)_1$, (2.82) and (2.128). Thus, we write

$$\text{curl}\,\text{grad}\Phi = \boldsymbol{g}^j \times \left(\Phi|_i\, \boldsymbol{g}^i\right)_{,j} = \Phi_{,ij}\, \boldsymbol{g}^j \times \boldsymbol{g}^i + \Phi_{,i}\, \boldsymbol{g}^j \times \boldsymbol{g}^i{}_{,j}$$

$$= \Phi_{,ij}\, \boldsymbol{g}^j \times \boldsymbol{g}^i - \Phi_{,i}\, \Gamma^i_{kj} \boldsymbol{g}^j \times \boldsymbol{g}^k = \boldsymbol{0} \tag{2.141}$$

taking into account that $\Phi_{,ij} = \Phi_{,ji}$, $\Gamma^l_{ij} = \Gamma^l_{ji}$ and $\boldsymbol{g}^i \times \boldsymbol{g}^j = -\boldsymbol{g}^j \times \boldsymbol{g}^i$ $(i \neq j,\ i, j = 1, 2, 3)$.

Example 2.8. Balance of mechanical energy as an integral form of the momentum balance. Using the above identities we are now able to formulate the balance of mechanical energy on the basis of the momentum balance (2.115). To this end, we multiply this vector relation scalarly by the velocity vector $\boldsymbol{v} = \dot{\boldsymbol{x}}$

$$\boldsymbol{v} \cdot (\rho \ddot{\boldsymbol{x}}) = \boldsymbol{v} \cdot \text{div}\boldsymbol{\sigma} + \boldsymbol{v} \cdot \boldsymbol{f}.$$

Using (2.138) we can further write

$$\boldsymbol{v} \cdot (\rho \ddot{\boldsymbol{x}}) + \boldsymbol{\sigma} : \text{grad}\boldsymbol{v} = \text{div}\,(\boldsymbol{v}\boldsymbol{\sigma}) + \boldsymbol{v} \cdot \boldsymbol{f}.$$

Integrating this relation over the volume of the body V yields

$$\frac{\text{d}}{\text{d}t} \int_m \left(\frac{1}{2}\boldsymbol{v} \cdot \boldsymbol{v}\right) \text{d}m + \int_V \boldsymbol{\sigma} : \text{grad}\boldsymbol{v}\,\text{d}V = \int_V \text{div}\,(\boldsymbol{v}\boldsymbol{\sigma})\,\text{d}V + \int_V \boldsymbol{v} \cdot \boldsymbol{f}\,\text{d}V,$$

where again $\text{d}m = \rho \text{d}V$ and m denotes the mass of the body. Keeping in mind the definition of the divergence (2.104) and applying the Cauchy theorem (1.77) according to which the Cauchy stress vector is given by $\boldsymbol{t} = \boldsymbol{\sigma}\boldsymbol{n}$, we finally obtain the relation

$$\frac{\text{d}}{\text{d}t} \int_m \left(\frac{1}{2}\boldsymbol{v} \cdot \boldsymbol{v}\right) \text{d}m + \int_V \boldsymbol{\sigma} : \text{grad}\boldsymbol{v}\,\text{d}V = \int_A \boldsymbol{v} \cdot \boldsymbol{t}\,\text{d}A + \int_V \boldsymbol{v} \cdot \boldsymbol{f}\,\text{d}V \tag{2.142}$$

expressing the balance of mechanical energy. Indeed, the first and the second integrals on the left hand side of (2.142) represent the time rate of the kinetic energy and the stress power, respectively. The right hand side of (2.142) expresses the power of external forces i.e. external tractions $\boldsymbol{t}$ on the boundary of the body A and external volume forces $\boldsymbol{f}$ inside of it.

Example 2.9. Axial vector of the spin tensor. The spin tensor is defined as a skew-symmetric part of the velocity gradient by

$$\mathbf{w} = \text{skew}\,(\text{grad}\boldsymbol{v}). \tag{2.143}$$

By virtue of (1.158) we can represent it in terms of the axial vector

$$\mathbf{w} = \hat{\boldsymbol{w}}, \tag{2.144}$$

which in view of (2.137) takes the form:

$$\boldsymbol{w} = \frac{1}{2}\mathrm{curl}\boldsymbol{v}. \tag{2.145}$$

Example 2.10. Navier-Stokes equations for a linear-viscous fluid in Cartesian and cylindrical coordinates. A linear-viscous fluid (also called Newton fluid or Navier-Poisson fluid) is defined by a constitutive equation

$$\boldsymbol{\sigma} = -p\mathbf{I} + 2\eta\mathbf{d} + \lambda\,(\mathrm{tr}\mathbf{d})\,\mathbf{I}, \tag{2.146}$$

where

$$\mathbf{d} = \mathrm{sym}\,(\mathrm{grad}\boldsymbol{v}) = \frac{1}{2}\left[\mathrm{grad}\boldsymbol{v} + (\mathrm{grad}\boldsymbol{v})^{\mathrm{T}}\right] \tag{2.147}$$

denotes the rate of deformation tensor, p is the hydrostatic pressure while η and λ represent material constants referred to as shear viscosity and second viscosity coefficient, respectively. Inserting (2.147) into (2.146) and taking (2.127) into account delivers

$$\boldsymbol{\sigma} = -p\mathbf{I} + \eta\left[\mathrm{grad}\boldsymbol{v} + (\mathrm{grad}\boldsymbol{v})^{\mathrm{T}}\right] + \lambda\,(\mathrm{div}\boldsymbol{v})\,\mathbf{I}. \tag{2.148}$$

Substituting this expression into the momentum balance (2.115) and using (2.135) and (2.140) we obtain the relation

$$\rho\dot{\boldsymbol{v}} = -\mathrm{grad}p + \eta\mathrm{div}\,\mathrm{grad}\boldsymbol{v} + (\eta + \lambda)\,\mathrm{grad}\,\mathrm{div}\boldsymbol{v} + \boldsymbol{f} \tag{2.149}$$

referred to as the Navier-Stokes equation. By means of (2.136) it can be rewritten as

$$\rho\dot{\boldsymbol{v}} = -\mathrm{grad}p + (2\eta + \lambda)\,\mathrm{grad}\,\mathrm{div}\boldsymbol{v} - \eta\mathrm{curl}\,\mathrm{curl}\boldsymbol{v} + \boldsymbol{f}. \tag{2.150}$$

For an incompressible fluid characterized by the kinematic condition $\mathrm{tr}\mathbf{d} = \mathrm{div}\boldsymbol{v} = 0$, the latter two equations simplify to

$$\rho\dot{\boldsymbol{v}} = -\mathrm{grad}p + \eta\Delta\boldsymbol{v} + \boldsymbol{f}, \tag{2.151}$$

$$\rho\dot{\boldsymbol{v}} = -\mathrm{grad}p - \eta\mathrm{curl}\,\mathrm{curl}\boldsymbol{v} + \boldsymbol{f}. \tag{2.152}$$

With the aid of the identity $\Delta\boldsymbol{v} = \boldsymbol{v},_i|^i$ (see Exercise 2.14) we thus can write

$$\rho\dot{\boldsymbol{v}} = -\mathrm{grad}p + \eta\boldsymbol{v},_i|^i + \boldsymbol{f}. \tag{2.153}$$

In Cartesian coordinates this relation is thus written out as

$$\rho \dot{v}_i = -p_{,i} + \eta \left(v_{i,11} + v_{i,22} + v_{i,33} \right) + f_i, \qquad i = 1, 2, 3. \tag{2.154}$$

For arbitrary curvilinear coordinates we use the following representation for the vector Laplacian (see Exercise 2.16)

$$\Delta \boldsymbol{v} = g^{ij} \left(v^k{}_{,ij} + 2\Gamma^k_{li} v^l{}_{,j} - \Gamma^m_{ij} v^k{}_{,m} + \Gamma^k_{li,j} v^l + \Gamma^k_{mj} \Gamma^m_{li} v^l - \Gamma^m_{ij} \Gamma^k_{lm} v^l \right) \boldsymbol{g}_k. \tag{2.155}$$

For the cylindrical coordinates it takes by virtue of (2.30) and (2.90) the following form

$$\begin{aligned}
\Delta \boldsymbol{v} = & \left(r^{-2} v^1{}_{,11} + v^1{}_{,22} + v^1{}_{,33} + 3r^{-1} v^1{}_{,3} + 2r^{-3} v^3{}_{,1} \right) \boldsymbol{g}_1 \\
& + \left(r^{-2} v^2{}_{,11} + v^2{}_{,22} + v^2{}_{,33} + r^{-1} v^2{}_{,3} \right) \boldsymbol{g}_2 \\
& + \left(r^{-2} v^3{}_{,11} + v^3{}_{,22} + v^3{}_{,33} - 2r^{-1} v^1{}_{,1} + r^{-1} v^3{}_{,3} - r^{-2} v^3 \right) \boldsymbol{g}_3.
\end{aligned}$$

Inserting this result into (2.151) and using the representations $\dot{\boldsymbol{v}} = \dot{v}^i \boldsymbol{g}_i$ and $\boldsymbol{f} = f^i \boldsymbol{g}_i$ we finally obtain

$$\begin{aligned}
\rho \dot{v}^1 &= f^1 - \frac{\partial p}{\partial \varphi} + \eta \left(\frac{1}{r^2} \frac{\partial^2 v^1}{\partial \varphi^2} + \frac{\partial^2 v^1}{\partial z^2} + \frac{\partial^2 v^1}{\partial r^2} + \frac{3}{r} \frac{\partial v^1}{\partial r} + \frac{2}{r^3} \frac{\partial v^3}{\partial \varphi} \right), \\
\rho \dot{v}^2 &= f^2 - \frac{\partial p}{\partial z} + \eta \left(\frac{1}{r^2} \frac{\partial^2 v^2}{\partial \varphi^2} + \frac{\partial^2 v^2}{\partial z^2} + \frac{\partial^2 v^2}{\partial r^2} + \frac{1}{r} \frac{\partial v^2}{\partial r} \right), \\
\rho \dot{v}^3 &= f^3 - \frac{\partial p}{\partial r} + \eta \left(\frac{1}{r^2} \frac{\partial^2 v^3}{\partial \varphi^2} + \frac{\partial^2 v^3}{\partial z^2} + \frac{\partial^2 v^3}{\partial r^2} - \frac{2}{r} \frac{\partial v^1}{\partial \varphi} + \frac{1}{r} \frac{\partial v^3}{\partial r} - \frac{v^3}{r^2} \right).
\end{aligned} \tag{2.156}$$

Exercises

2.1. Evaluate tangent vectors, metric coefficients and the dual basis of spherical coordinates in $\mathbb{E}^3$ defined by (Fig. 2.6)

$$\boldsymbol{r}(\varphi, \phi, r) = r \sin\varphi \sin\phi \, \boldsymbol{e}_1 + r \cos\phi \, \boldsymbol{e}_2 + r \cos\varphi \sin\phi \, \boldsymbol{e}_3. \tag{2.157}$$

2.2. Evaluate the coefficients $\dfrac{\partial \bar{\theta}^i}{\partial \theta^k}$ (2.43) for the transformation of linear coordinates in the spherical ones and vice versa.

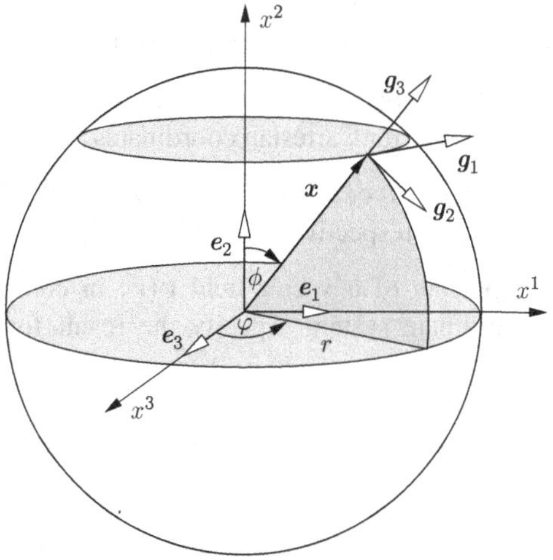

Fig. 2.6 Spherical coordinates in three-dimensional space

2.3. Evaluate gradients of the following functions of $\boldsymbol{r}$:

$$\text{(a)}\ \frac{1}{\|\boldsymbol{r}\|},\quad \text{(b)}\ \boldsymbol{r}\cdot\boldsymbol{w},\quad \text{(c)}\ \boldsymbol{r}\mathbf{A}\boldsymbol{r},\quad \text{(d)}\ \mathbf{A}\boldsymbol{r},\quad \text{(e)}\ \boldsymbol{w}\times\boldsymbol{r},$$

where $\boldsymbol{w}$ and $\mathbf{A}$ are some vector and tensor, respectively.

2.4. Evaluate the Christoffel symbols of the first and second kind for spherical coordinates (2.157).

2.5. Verify relations (2.96).

2.6. Prove identities (2.99) and (2.100) by using (1.91).

2.7. Prove the product rules of differentiation for the covariant derivative (2.101)–(2.103).

2.8. Verify relation (2.129) applying (2.112), (2.125) and using the results of Exercise 1.23.

2.9. Write out the balance equations (2.116) in spherical coordinates (2.157).

2.10. Evaluate tangent vectors, metric coefficients, the dual basis and Christoffel symbols for cylindrical surface coordinates defined by

$$\boldsymbol{r}\,(r,s,z) = r\cos\frac{s}{r}\boldsymbol{e}_1 + r\sin\frac{s}{r}\boldsymbol{e}_2 + z\boldsymbol{e}_3. \tag{2.158}$$

2.11. Write out the balance equations (2.116) in cylindrical surface coordinates (2.158).

2.12. Prove identities (2.133)–(2.140).

2.13. Write out the gradient, divergence and curl of a vector field $\boldsymbol{t}(\boldsymbol{r})$ in cylindrical and spherical coordinates (2.17) and (2.157), respectively.

2.14. Prove that the Laplacian of a vector-valued function $\boldsymbol{t}(\boldsymbol{r})$ can be given by $\Delta\boldsymbol{t} = \boldsymbol{t},_i|^i$. Specify this identity for Cartesian coordinates.

2.15. Write out the Laplacian $\Delta\Phi$ of a scalar field $\Phi(\boldsymbol{r})$ in cylindrical and spherical coordinates (2.17) and (2.157), respectively.

2.16. Write out the Laplacian of a vector field $\boldsymbol{t}(\boldsymbol{r})$ in component form in an arbitrary curvilinear coordinate system. Specify the result for spherical coordinates (2.157).

Chapter 3
Curves and Surfaces in Three-Dimensional Euclidean Space

3.1 Curves in Three-Dimensional Euclidean Space

A curve in three-dimensional space is defined by a vector function

$$\boldsymbol{r} = \boldsymbol{r}(t), \quad \boldsymbol{r} \in \mathbb{E}^3, \tag{3.1}$$

where the real variable t belongs to some interval: $t_1 \leq t \leq t_2$. Henceforth, we assume that the function $\boldsymbol{r}(t)$ is sufficiently differentiable and

$$\frac{\mathrm{d}\boldsymbol{r}}{\mathrm{d}t} \neq \boldsymbol{0} \tag{3.2}$$

over the whole definition domain. Specifying an arbitrary coordinate system (2.16) as

$$\theta^i = \theta^i(\boldsymbol{r}), \quad i = 1, 2, 3, \tag{3.3}$$

the curve (3.1) can alternatively be defined by

$$\theta^i = \theta^i(t), \quad i = 1, 2, 3. \tag{3.4}$$

Example 3.1. Straight line. A straight line can be defined by

$$\boldsymbol{r}(t) = \boldsymbol{a} + \boldsymbol{b}t, \quad \boldsymbol{a}, \boldsymbol{b} \in \mathbb{E}^3. \tag{3.5}$$

With respect to linear coordinates related to a basis $\mathcal{H} = \{\boldsymbol{h}_1, \boldsymbol{h}_2, \boldsymbol{h}_3\}$ it is equivalent to

$$r^i(t) = a^i + b^i t, \quad i = 1, 2, 3, \tag{3.6}$$

where $\boldsymbol{r} = r^i \boldsymbol{h}_i$, $\boldsymbol{a} = a^i \boldsymbol{h}_i$ and $\boldsymbol{b} = b^i \boldsymbol{h}_i$.

M. Itskov, *Tensor Algebra and Tensor Analysis for Engineers*, Mathematical Engineering, DOI 10.1007/978-3-642-30879-6_3, © Springer-Verlag Berlin Heidelberg 2013

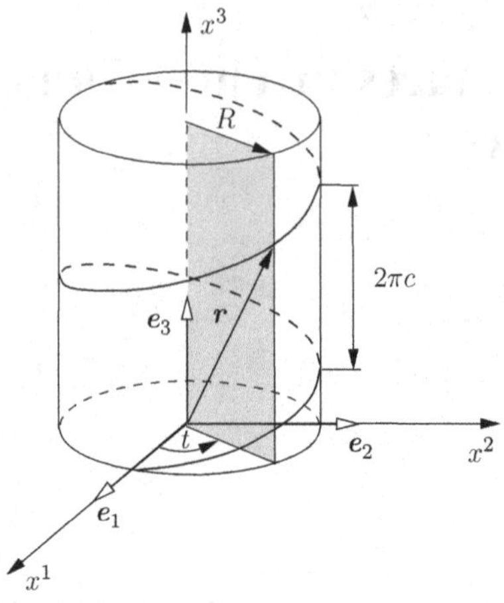

Fig. 3.1 Circular helix

Example 3.2. Circular helix. The circular helix (Fig. 3.1) is defined by

$$\boldsymbol{r}(t) = R\cos(t)\,\boldsymbol{e}_1 + R\sin(t)\,\boldsymbol{e}_2 + ct\boldsymbol{e}_3, \quad c \neq 0, \tag{3.7}$$

where $\boldsymbol{e}_i$ $(i = 1, 2, 3)$ form an orthonormal basis in $\mathbb{E}^3$. For the definition of the circular helix the cylindrical coordinates (2.17) appear to be very suitable. Indeed, alternatively to (3.7) we can write

$$r = R, \quad \varphi = t, \quad z = ct. \tag{3.8}$$

In the previous chapter we defined tangent vectors to the coordinate lines. By analogy one can also define a vector tangent to the curve (3.1) as

$$\boldsymbol{g}_t = \frac{\mathrm{d}\boldsymbol{r}}{\mathrm{d}t}. \tag{3.9}$$

It is advantageous to parametrize the curve (3.1) in terms of the so-called arc length. To this end, we first calculate the length of a curve segment between the points corresponding to parameters t_1 and t as

$$s(t) = \int\limits_{\boldsymbol{r}(t_1)}^{\boldsymbol{r}(t)} \sqrt{\mathrm{d}\boldsymbol{r} \cdot \mathrm{d}\boldsymbol{r}}. \tag{3.10}$$

With the aid of (3.9) we can write $\mathrm{d}\boldsymbol{r} = \boldsymbol{g}_t \mathrm{d}t$ and consequently

$$s(t) = \int\limits_{t_1}^{t} \sqrt{\boldsymbol{g}_t \cdot \boldsymbol{g}_t}\,\mathrm{d}t = \int\limits_{t_1}^{t} \|\boldsymbol{g}_t\|\,\mathrm{d}t = \int\limits_{t_1}^{t} \sqrt{g_{tt}(t)}\mathrm{d}t. \tag{3.11}$$

Using this equation and keeping in mind assumption (3.2) we have

$$\frac{\mathrm{d}s}{\mathrm{d}t} = \sqrt{g_{tt}(t)} \neq 0. \tag{3.12}$$

This implies that the function $s = s(t)$ is invertible and

$$t(s) = \int_{s(t_1)}^{s} \|\boldsymbol{g}_t\|^{-1} \mathrm{d}s = \int_{s(t_1)}^{s} \frac{\mathrm{d}s}{\sqrt{g_{tt}(t)}}. \tag{3.13}$$

Thus, the curve (3.1) can be redefined in terms of the arc length s as

$$\boldsymbol{r} = \boldsymbol{r}(t(s)) = \widehat{\boldsymbol{r}}(s). \tag{3.14}$$

In analogy with (3.9) one defines the vector tangent to the curve $\widehat{\boldsymbol{r}}(s)$ (3.14) as

$$\boldsymbol{a}_1 = \frac{\mathrm{d}\widehat{\boldsymbol{r}}}{\mathrm{d}s} = \frac{\mathrm{d}\boldsymbol{r}}{\mathrm{d}t}\frac{\mathrm{d}t}{\mathrm{d}s} = \frac{\boldsymbol{g}_t}{\|\boldsymbol{g}_t\|} \tag{3.15}$$

being a unit vector: $\|\boldsymbol{a}_1\| = 1$. Differentiation of this vector with respect to s further yields

$$\boldsymbol{a}_{1,s} = \frac{\mathrm{d}\boldsymbol{a}_1}{\mathrm{d}s} = \frac{\mathrm{d}^2\widehat{\boldsymbol{r}}}{\mathrm{d}s^2}. \tag{3.16}$$

It can be shown that the tangent vector $\boldsymbol{a}_1$ is orthogonal to $\boldsymbol{a}_{1,s}$ provided the latter one is not zero. Indeed, differentiating the identity $\boldsymbol{a}_1 \cdot \boldsymbol{a}_1 = 1$ with respect to s we have

$$\boldsymbol{a}_1 \cdot \boldsymbol{a}_{1,s} = 0. \tag{3.17}$$

The length of the vector $\boldsymbol{a}_{1,s}$

$$\varkappa(s) = \|\boldsymbol{a}_{1,s}\| \tag{3.18}$$

plays an important role in the theory of curves and is called curvature. The inverse value

$$\rho(s) = \frac{1}{\varkappa(s)} \tag{3.19}$$

is referred to as the radius of curvature of the curve at the point $\widehat{\boldsymbol{r}}(s)$. Henceforth, we focus on curves with non-zero curvature. The case of zero curvature corresponds to a straight line (see Exercise 3.1) and is trivial.

Next, we define the unit vector in the direction of $\boldsymbol{a}_{1,s}$

$$\boldsymbol{a}_2 = \frac{\boldsymbol{a}_{1,s}}{\|\boldsymbol{a}_{1,s}\|} = \frac{\boldsymbol{a}_{1,s}}{\varkappa(s)} \tag{3.20}$$

called the principal normal vector to the curve. The orthogonal vectors $\boldsymbol{a}_1$ and $\boldsymbol{a}_2$ can further be completed to an orthonormal basis in $\mathbb{E}^3$ by the vector

$$\boldsymbol{a}_3 = \boldsymbol{a}_1 \times \boldsymbol{a}_2 \tag{3.21}$$

called the unit binormal vector. The triplet of vectors $\boldsymbol{a}_1$, $\boldsymbol{a}_2$ and $\boldsymbol{a}_3$ is referred to as the moving trihedron of the curve.

In order to study the rotation of the trihedron along the curve we again consider the arc length s as a coordinate. In this case, we can write similarly to (2.76)

$$\boldsymbol{a}_{i,s} = \Gamma_{is}^{k} \boldsymbol{a}_k, \quad i = 1, 2, 3, \tag{3.22}$$

where $\Gamma_{is}^{k} = \boldsymbol{a}_{i,s} \cdot \boldsymbol{a}_k$ $(i, k = 1, 2, 3)$. From (3.17), (3.20) and (3.21) we immediately observe that $\Gamma_{1s}^{2} = \varkappa$ and $\Gamma_{1s}^{1} = \Gamma_{1s}^{3} = 0$. Further, differentiating the identities

$$\boldsymbol{a}_3 \cdot \boldsymbol{a}_3 = 1, \qquad \boldsymbol{a}_1 \cdot \boldsymbol{a}_3 = 0 \tag{3.23}$$

with respect to s yields

$$\boldsymbol{a}_3 \cdot \boldsymbol{a}_{3,s} = 0, \qquad \boldsymbol{a}_{1,s} \cdot \boldsymbol{a}_3 + \boldsymbol{a}_1 \cdot \boldsymbol{a}_{3,s} = 0. \tag{3.24}$$

Taking into account (3.20) this results in the following identity

$$\boldsymbol{a}_1 \cdot \boldsymbol{a}_{3,s} = -\boldsymbol{a}_{1,s} \cdot \boldsymbol{a}_3 = -\varkappa \boldsymbol{a}_2 \cdot \boldsymbol{a}_3 = 0. \tag{3.25}$$

Relations (3.24) and (3.25) suggest that

$$\boldsymbol{a}_{3,s} = -\tau(s)\, \boldsymbol{a}_2, \tag{3.26}$$

where the function

$$\tau(s) = -\boldsymbol{a}_{3,s} \cdot \boldsymbol{a}_2 \tag{3.27}$$

is called torsion of the curve at the point $\hat{\boldsymbol{r}}(s)$. Thus, $\Gamma_{3s}^{2} = -\tau$ and $\Gamma_{3s}^{1} = \Gamma_{3s}^{3} = 0$. The sign of the torsion (3.27) has a geometric meaning and remains unaffected by the change of the positive sense of the curve, i.e. by the transformation $s = -s'$ (see Exercise 3.2). Accordingly, one distinguishes right-handed curves with a positive torsion and left-handed curves with a negative torsion. In the case of zero torsion the curve is referred to as a plane curve.

Finally, differentiating the identities

$$\boldsymbol{a}_2 \cdot \boldsymbol{a}_1 = 0, \quad \boldsymbol{a}_2 \cdot \boldsymbol{a}_2 = 1, \quad \boldsymbol{a}_2 \cdot \boldsymbol{a}_3 = 0$$

with respect to s and using (3.20) and (3.27) we get

$$\boldsymbol{a}_{2,s} \cdot \boldsymbol{a}_1 = -\boldsymbol{a}_2 \cdot \boldsymbol{a}_{1,s} = -\varkappa \boldsymbol{a}_2 \cdot \boldsymbol{a}_2 = -\varkappa, \tag{3.28}$$

$$\boldsymbol{a}_2 \cdot \boldsymbol{a}_{2,s} = 0, \quad \boldsymbol{a}_{2,s} \cdot \boldsymbol{a}_3 = -\boldsymbol{a}_2 \cdot \boldsymbol{a}_{3,s} = \tau, \tag{3.29}$$

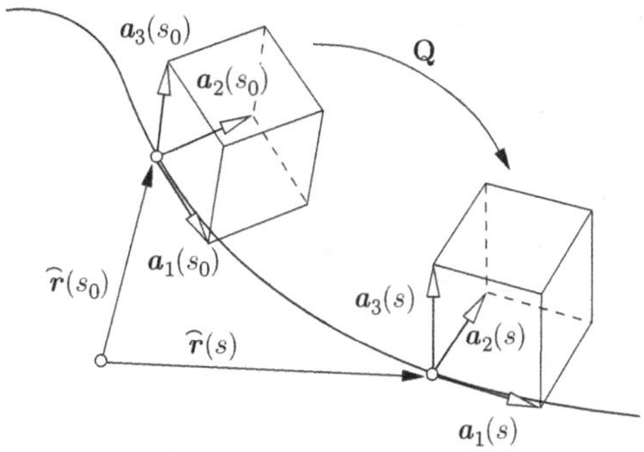

Fig. 3.2 Rotation of the moving trihedron

so that $\Gamma^1_{2s} = -\varkappa$, $\Gamma^2_{2s} = 0$ and $\Gamma^3_{2s} = \tau$. Summarizing the above results we can write

$$\left[\Gamma^j_{is}\right] = \begin{bmatrix} 0 & \varkappa & 0 \\ -\varkappa & 0 & \tau \\ 0 & -\tau & 0 \end{bmatrix} \tag{3.30}$$

and

$$\begin{aligned} \boldsymbol{a}_{1,s} &= \quad\quad \varkappa \boldsymbol{a}_2, \\ \boldsymbol{a}_{2,s} &= -\varkappa \boldsymbol{a}_1 \quad\quad + \tau \boldsymbol{a}_3, \\ \boldsymbol{a}_{3,s} &= \quad\quad -\tau \boldsymbol{a}_2. \end{aligned} \tag{3.31}$$

Relations (3.31) are known as the Frenet formulas.

A useful illustration of the Frenet formulas can be gained with the aid of a skew-symmetric tensor. To this end, we consider the rotation of the trihedron from some initial position at s_0 to the actual state at s. This rotation can be described by an orthogonal tensor $\mathbf{Q}(s)$ as (Fig. 3.2)

$$\boldsymbol{a}_i(s) = \mathbf{Q}(s)\,\boldsymbol{a}_i(s_0), \quad i = 1, 2, 3. \tag{3.32}$$

Differentiating this relation with respect to s yields

$$\boldsymbol{a}_{i,s}(s) = \mathbf{Q}_{,s}(s)\,\boldsymbol{a}_i(s_0), \quad i = 1, 2, 3. \tag{3.33}$$

Mapping both sides of (3.32) by $\mathbf{Q}^{\mathrm{T}}(s)$ and inserting the result into (3.33) we further obtain

$$\boldsymbol{a}_{i,s}(s) = \mathbf{Q}_{,s}(s)\,\mathbf{Q}^{\mathrm{T}}(s)\,\boldsymbol{a}_i(s), \quad i = 1, 2, 3. \tag{3.34}$$

Differentiating the identity (1.135) $\mathbf{Q}(s)\,\mathbf{Q}^{\mathrm{T}}(s) = \mathbf{I}$ with respect to s we have $\mathbf{Q}_{,s}(s)\,\mathbf{Q}^{\mathrm{T}}(s) + \mathbf{Q}(s)\,\mathbf{Q}^{\mathrm{T}}{}_{,s}(s) = \mathbf{0}$, which implies that the tensor $\mathbf{W}(s) = \mathbf{Q}_{,s}(s)\,\mathbf{Q}^{\mathrm{T}}(s)$ is skew-symmetric. Hence, Eq. (3.34) can be rewritten as (see also [3])

$$\boldsymbol{a}_{i,s}(s) = \mathbf{W}(s)\,\boldsymbol{a}_i(s), \quad \mathbf{W} \in \mathrm{Skew}^3, \quad i = 1, 2, 3, \tag{3.35}$$

where according to (3.31)

$$\mathbf{W}(s) = \tau(s)(\boldsymbol{a}_3 \otimes \boldsymbol{a}_2 - \boldsymbol{a}_2 \otimes \boldsymbol{a}_3) + \varkappa(s)(\boldsymbol{a}_2 \otimes \boldsymbol{a}_1 - \boldsymbol{a}_1 \otimes \boldsymbol{a}_2). \tag{3.36}$$

By virtue of (1.136) and (1.137) we further obtain

$$\mathbf{W} = \tau \hat{\boldsymbol{a}}_1 + \varkappa \hat{\boldsymbol{a}}_3 \tag{3.37}$$

and consequently

$$\boldsymbol{a}_{i},_{s} = \boldsymbol{d} \times \boldsymbol{a}_i = \hat{\boldsymbol{d}} \boldsymbol{a}_i, \quad i = 1, 2, 3, \tag{3.38}$$

where

$$\boldsymbol{d} = \tau \boldsymbol{a}_1 + \varkappa \boldsymbol{a}_3 \tag{3.39}$$

is referred to as the Darboux vector.

Example 3.3. Curvature, torsion, moving trihedron and Darboux vector for a circular helix. Inserting (3.7) into (3.9) delivers

$$\boldsymbol{g}_t = \frac{\mathrm{d}\boldsymbol{r}}{\mathrm{d}t} = -R\sin(t)\,\boldsymbol{e}_1 + R\cos(t)\,\boldsymbol{e}_2 + c\boldsymbol{e}_3, \tag{3.40}$$

so that

$$g_{tt} = \boldsymbol{g}_t \cdot \boldsymbol{g}_t = R^2 + c^2 = const. \tag{3.41}$$

Thus, using (3.13) we may set

$$t(s) = \frac{s}{\sqrt{R^2 + c^2}}. \tag{3.42}$$

Using this result, the circular helix (3.7) can be parametrized in terms of the arc length s by

$$\widehat{\boldsymbol{r}}(s) = R\cos\left(\frac{s}{\sqrt{R^2+c^2}}\right)\boldsymbol{e}_1 + R\sin\left(\frac{s}{\sqrt{R^2+c^2}}\right)\boldsymbol{e}_2 + \frac{cs}{\sqrt{R^2+c^2}}\boldsymbol{e}_3. \tag{3.43}$$

With the aid of (3.15) we further write

$$\begin{aligned}\boldsymbol{a}_1 = \frac{\mathrm{d}\widehat{\boldsymbol{r}}}{\mathrm{d}s} = \frac{1}{\sqrt{R^2+c^2}}\left[-R\sin\left(\frac{s}{\sqrt{R^2+c^2}}\right)\boldsymbol{e}_1\right.\\ \left. + R\cos\left(\frac{s}{\sqrt{R^2+c^2}}\right)\boldsymbol{e}_2 + c\boldsymbol{e}_3\right],\end{aligned} \tag{3.44}$$

$$\boldsymbol{a}_{1},_{s} = -\frac{R}{R^2+c^2}\left[\cos\left(\frac{s}{\sqrt{R^2+c^2}}\right)\boldsymbol{e}_1 + \sin\left(\frac{s}{\sqrt{R^2+c^2}}\right)\boldsymbol{e}_2\right]. \tag{3.45}$$

According to (3.18) the curvature of the helix is thus

$$\varkappa = \frac{R}{R^2 + c^2}. \tag{3.46}$$

By virtue of (3.20), (3.21) and (3.27) we have

$$\boldsymbol{a}_2 = \frac{\boldsymbol{a}_{1,s}}{\varkappa} = -\cos\left(\frac{s}{\sqrt{R^2+c^2}}\right)\boldsymbol{e}_1 - \sin\left(\frac{s}{\sqrt{R^2+c^2}}\right)\boldsymbol{e}_2, \tag{3.47}$$

$$\boldsymbol{a}_3 = \boldsymbol{a}_1 \times \boldsymbol{a}_2 = \frac{1}{\sqrt{R^2+c^2}}\left[c\sin\left(\frac{s}{\sqrt{R^2+c^2}}\right)\boldsymbol{e}_1 \right.$$
$$\left. -c\cos\left(\frac{s}{\sqrt{R^2+c^2}}\right)\boldsymbol{e}_2 + R\boldsymbol{e}_3\right]. \tag{3.48}$$

$$\boldsymbol{a}_{3,s} = \frac{c}{R^2+c^2}\left[\cos\left(\frac{s}{\sqrt{R^2+c^2}}\right)\boldsymbol{e}_1 + \sin\left(\frac{s}{\sqrt{R^2+c^2}}\right)\boldsymbol{e}_2\right], \tag{3.49}$$

$$\tau = \frac{c}{R^2+c^2}. \tag{3.50}$$

It is seen that the circular helix is right-handed for $c > 0$, left-handed for $c < 0$ and becomes a circle for $c = 0$. For the Darboux vector (3.39) we finally obtain

$$\boldsymbol{d} = \tau\boldsymbol{a}_1 + \varkappa\boldsymbol{a}_3 = \frac{1}{\sqrt{R^2+c^2}}\boldsymbol{e}_3. \tag{3.51}$$

3.2 Surfaces in Three-Dimensional Euclidean Space

A surface in three-dimensional Euclidean space is defined by a vector function

$$\boldsymbol{r} = \boldsymbol{r}\left(t^1, t^2\right), \quad \boldsymbol{r} \in \mathbb{E}^3, \tag{3.52}$$

of two real variables t^1 and t^2 referred to as Gauss coordinates. With the aid of the coordinate system (3.3) one can alternatively write

$$\theta^i = \theta^i\left(t^1, t^2\right), \quad i = 1, 2, 3. \tag{3.53}$$

In the following, we assume that the function $\boldsymbol{r}\left(t^1, t^2\right)$ is sufficiently differentiable with respect to both arguments and

$$\frac{\mathrm{d}\boldsymbol{r}}{\mathrm{d}t^\alpha} \neq \boldsymbol{0}, \quad \alpha = 1, 2 \tag{3.54}$$

over the whole definition domain.

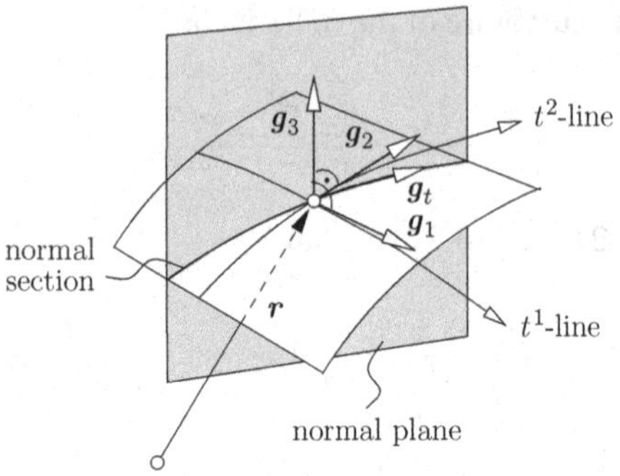

Fig. 3.3 Coordinate lines on the surface, normal section and tangent vectors

Example 3.4. Plane. Let us consider three linearly independent vectors $\boldsymbol{x}_i\,(i = 0, 1, 2)$ specifying three points in three-dimensional space. The plane going through these points can be defined by

$$\boldsymbol{r}\left(t^1, t^2\right) = \boldsymbol{x}_0 + t^1\left(\boldsymbol{x}_1 - \boldsymbol{x}_0\right) + t^2\left(\boldsymbol{x}_2 - \boldsymbol{x}_0\right). \tag{3.55}$$

Example 3.5. Cylinder. A cylinder of radius R with the axis parallel to $\boldsymbol{e}_3$ is defined by

$$\boldsymbol{r}\left(t^1, t^2\right) = R\cos t^1 \boldsymbol{e}_1 + R\sin t^1 \boldsymbol{e}_2 + t^2 \boldsymbol{e}_3, \tag{3.56}$$

where $\boldsymbol{e}_i\ (i = 1, 2, 3)$ again form an orthonormal basis in $\mathbb{E}^3$. With the aid of the cylindrical coordinates (2.17) we can alternatively write

$$\varphi = t^1, \quad z = t^2, \quad r = R. \tag{3.57}$$

Example 3.6. Sphere. A sphere of radius R with the center at $\boldsymbol{r} = \boldsymbol{0}$ is defined by

$$\boldsymbol{r}\left(t^1, t^2\right) = R\sin t^1 \sin t^2 \boldsymbol{e}_1 + R\cos t^2 \boldsymbol{e}_2 + R\cos t^1 \sin t^2 \boldsymbol{e}_3, \tag{3.58}$$

or by means of spherical coordinates (2.157) as

$$\varphi = t^1, \quad \phi = t^2, \quad r = R. \tag{3.59}$$

Using a parametric representation (see, e.g., [26])

$$t^1 = t^1(t), \quad t^2 = t^2(t) \tag{3.60}$$

one defines a curve on the surface (3.52). In particular, the curves $t^1 = const$ and $t^2 = const$ are called t^2 and t^1 coordinate lines, respectively (Fig. 3.3). The vector tangent to the curve (3.60) can be expressed by

$$\boldsymbol{g}_t = \frac{\mathrm{d}\boldsymbol{r}}{\mathrm{d}t} = \frac{\partial \boldsymbol{r}}{\partial t^1}\frac{\mathrm{d}t^1}{\mathrm{d}t} + \frac{\partial \boldsymbol{r}}{\partial t^2}\frac{\mathrm{d}t^2}{\mathrm{d}t} = \boldsymbol{g}_1\frac{\mathrm{d}t^1}{\mathrm{d}t} + \boldsymbol{g}_2\frac{\mathrm{d}t^2}{\mathrm{d}t}, \tag{3.61}$$

where

$$\boldsymbol{g}_\alpha = \frac{\partial \boldsymbol{r}}{\partial t^\alpha} = \boldsymbol{r}_{,\alpha}, \quad \alpha = 1, 2 \tag{3.62}$$

represent tangent vectors to the coordinate lines. For the length of an infinitesimal element of the curve (3.60) we thus write

$$(\mathrm{d}s)^2 = \mathrm{d}\boldsymbol{r} \cdot \mathrm{d}\boldsymbol{r} = (\boldsymbol{g}_t \mathrm{d}t) \cdot (\boldsymbol{g}_t \mathrm{d}t) = \left(\boldsymbol{g}_1 \mathrm{d}t^1 + \boldsymbol{g}_2 \mathrm{d}t^2\right) \cdot \left(\boldsymbol{g}_1 \mathrm{d}t^1 + \boldsymbol{g}_2 \mathrm{d}t^2\right). \tag{3.63}$$

With the aid of the abbreviation

$$g_{\alpha\beta} = g_{\beta\alpha} = \boldsymbol{g}_\alpha \cdot \boldsymbol{g}_\beta, \quad \alpha, \beta = 1, 2, \tag{3.64}$$

it delivers the quadratic form

$$(\mathrm{d}s)^2 = g_{11}\left(\mathrm{d}t^1\right)^2 + 2g_{12}\mathrm{d}t^1\mathrm{d}t^2 + g_{22}\left(\mathrm{d}t^2\right)^2 \tag{3.65}$$

referred to as the first fundamental form of the surface. The latter result can briefly be written as

$$(\mathrm{d}s)^2 = g_{\alpha\beta}\mathrm{d}t^\alpha \mathrm{d}t^\beta, \tag{3.66}$$

where and henceforth within this chapter the summation convention is implied for repeated Greek indices taking the values from 1 to 2. Similar to the metric coefficients $(1.90)_{1,2}$ in n-dimensional Euclidean space $g_{\alpha\beta}$ (3.64) describe the metric on a surface. Generally, the metric described by a differential quadratic form like (3.66) is referred to as Riemannian metric.

The tangent vectors (3.62) can be completed to a basis in $\mathbb{E}^3$ by the unit vector

$$\boldsymbol{g}_3 = \frac{\boldsymbol{g}_1 \times \boldsymbol{g}_2}{\|\boldsymbol{g}_1 \times \boldsymbol{g}_2\|} \tag{3.67}$$

called principal normal vector to the surface.

In the following, we focus on a special class of surface curves called normal sections. These are curves passing through a point of the surface $\boldsymbol{r}\left(t^1, t^2\right)$ and obtained by intersection of this surface with a plane involving the principal normal vector. Such a plane is referred to as the normal plane.

In order to study curvature properties of normal sections we first express the derivatives of the basis vectors $\boldsymbol{g}_i$ $(i = 1, 2, 3)$ with respect to the surface coordinates. Using the formalism of Christoffel symbols we can write

$$\boldsymbol{g}_{i,\alpha} = \frac{\partial \boldsymbol{g}_i}{\partial t^\alpha} = \Gamma_{i\alpha k}\boldsymbol{g}^k = \Gamma_{i\alpha}^k \boldsymbol{g}_k, \quad i = 1, 2, 3, \tag{3.68}$$

where

$$\Gamma_{i\alpha k} = \boldsymbol{g}_{i,\alpha} \cdot \boldsymbol{g}_k, \quad \Gamma_{i\alpha}^k = \boldsymbol{g}_{i,\alpha} \cdot \boldsymbol{g}^k, \quad i = 1, 2, 3, \quad \alpha = 1, 2. \tag{3.69}$$

Taking into account the identity $\boldsymbol{g}_3 = \boldsymbol{g}^3$ resulting from (3.67) we immediately observe that

$$\Gamma_{i\alpha 3} = \Gamma^3_{i\alpha}, \quad i = 1, 2, 3, \quad \alpha = 1, 2. \tag{3.70}$$

Differentiating the relations

$$\boldsymbol{g}_\alpha \cdot \boldsymbol{g}_3 = 0, \quad \boldsymbol{g}_3 \cdot \boldsymbol{g}_3 = 1 \tag{3.71}$$

with respect to the Gauss coordinates we further obtain

$$\boldsymbol{g}_{\alpha,\beta} \cdot \boldsymbol{g}_3 = -\boldsymbol{g}_\alpha \cdot \boldsymbol{g}_{3,\beta}, \quad \boldsymbol{g}_{3,\alpha} \cdot \boldsymbol{g}^3 = 0, \quad \alpha, \beta = 1, 2 \tag{3.72}$$

and consequently

$$\Gamma^3_{\alpha\beta} = -\Gamma_{3\beta\alpha}, \quad \Gamma^3_{3\alpha} = 0, \quad \alpha, \beta = 1, 2. \tag{3.73}$$

Using in (3.68) the abbreviation

$$b_{\alpha\beta} = b_{\beta\alpha} = \Gamma^3_{\alpha\beta} = -\Gamma_{3\alpha\beta} = \boldsymbol{g}_{\alpha,\beta} \cdot \boldsymbol{g}_3, \quad \alpha, \beta = 1, 2, \tag{3.74}$$

we arrive at the relations

$$\boldsymbol{g}_{\alpha,\beta} = \Gamma^\rho_{\alpha\beta} \boldsymbol{g}_\rho + b_{\alpha\beta} \boldsymbol{g}_3, \quad \alpha, \beta = 1, 2 \tag{3.75}$$

called the Gauss formulas.

Similarly to a coordinate system one can notionally define the covariant derivative also on the surface. To this end, relations (2.93), (2.95) and (2.96) are specified to the two-dimensional space in a straight forward manner as

$$f^\alpha|_\beta = f^\alpha{}_{,\beta} + f^\rho \Gamma^\alpha_{\rho\beta}, \quad f_\alpha|_\beta = f_{\alpha,\beta} - f_\rho \Gamma^\rho_{\alpha\beta}, \tag{3.76}$$

$$\mathrm{F}^{\alpha\beta}|_\gamma = \mathrm{F}^{\alpha\beta}{}_{,\gamma} + \mathrm{F}^{\rho\beta} \Gamma^\alpha_{\rho\gamma} + \mathrm{F}^{\alpha\rho} \Gamma^\beta_{\rho\gamma}, \quad \mathrm{F}_{\alpha\beta}|_\gamma = \mathrm{F}_{\alpha\beta,\gamma} - \mathrm{F}_{\rho\beta} \Gamma^\rho_{\alpha\gamma} - \mathrm{F}_{\alpha\rho} \Gamma^\rho_{\beta\gamma},$$

$$\mathrm{F}^\alpha_{\cdot\beta}|_\gamma = \mathrm{F}^\alpha_{\cdot\beta,\gamma} + \mathrm{F}^\rho_{\cdot\beta} \Gamma^\alpha_{\rho\gamma} - \mathrm{F}^\alpha_{\cdot\rho} \Gamma^\rho_{\beta\gamma}, \quad \alpha, \beta, \gamma = 1, 2. \tag{3.77}$$

Thereby, with the aid of $(3.76)_2$ the Gauss formulas (3.75) can alternatively be given by (cf. (2.98))

$$\boldsymbol{g}_\alpha|_\beta = b_{\alpha\beta} \boldsymbol{g}_3, \quad \alpha, \beta = 1, 2. \tag{3.78}$$

Further, we can write

$$b^\beta_\alpha = b_{\alpha\rho} g^{\rho\beta} = -\Gamma_{3\alpha\rho} g^{\rho\beta} = -\Gamma^\beta_{3\alpha}, \quad \alpha, \beta = 1, 2. \tag{3.79}$$

Inserting the latter relation into (3.68) and considering $(3.73)_2$, this yields the identities

$$\boldsymbol{g}_{3,\alpha} = \boldsymbol{g}_3|_\alpha = -b_\alpha^\rho \boldsymbol{g}_\rho, \quad \alpha = 1, 2 \tag{3.80}$$

referred to as the Weingarten formulas.

Now, we are in a position to express the curvature of a normal section. It is called normal curvature and denoted in the following by $\varkappa_n$. At first, we observe that the principal normals of the surface and of the normal section coincide in the sense that $\boldsymbol{a}_2 = \pm \boldsymbol{g}_3$. Using (3.13), (3.28), (3.61), $(3.72)_1$ and (3.74) and assuming for the moment that $\boldsymbol{a}_2 = \boldsymbol{g}_3$ we get

$$\varkappa_n = -\boldsymbol{a}_{2,s} \cdot \boldsymbol{a}_1 = -\boldsymbol{g}_{3,s} \cdot \frac{\boldsymbol{g}_t}{\|\boldsymbol{g}_t\|} = -\left(\boldsymbol{g}_{3,t} \frac{\mathrm{d}t}{\mathrm{d}s}\right) \cdot \frac{\boldsymbol{g}_t}{\|\boldsymbol{g}_t\|} = -\boldsymbol{g}_{3,t} \cdot \frac{\boldsymbol{g}_t}{\|\boldsymbol{g}_t\|^2}$$

$$= -\left(\boldsymbol{g}_{3,\alpha} \frac{\mathrm{d}t^\alpha}{\mathrm{d}t}\right) \cdot \left(\boldsymbol{g}_\beta \frac{\mathrm{d}t^\beta}{\mathrm{d}t}\right) \|\boldsymbol{g}_t\|^{-2} = b_{\alpha\beta} \frac{\mathrm{d}t^\alpha}{\mathrm{d}t} \frac{\mathrm{d}t^\beta}{\mathrm{d}t} \|\boldsymbol{g}_t\|^{-2}.$$

By virtue of (3.63) and (3.66) this leads to the following result

$$\varkappa_n = \frac{b_{\alpha\beta} \mathrm{d}t^\alpha \mathrm{d}t^\beta}{g_{\alpha\beta} \mathrm{d}t^\alpha \mathrm{d}t^\beta}, \tag{3.81}$$

where the quadratic form

$$b_{\alpha\beta} \mathrm{d}t^\alpha \mathrm{d}t^\beta = -\mathrm{d}\boldsymbol{r} \cdot \mathrm{d}\boldsymbol{g}_3 \tag{3.82}$$

is referred to as the second fundamental form of the surface. In the case $\boldsymbol{a}_2 = -\boldsymbol{g}_3$ the sign of the expression for $\varkappa_n$ (3.81) must be changed. Instead of that, we assume that the normal curvature can, in contrast to the curvature of space curves (3.18), be negative. However, the sign of $\varkappa_n$ (3.81) has no geometrical meaning. Indeed, it depends on the orientation of $\boldsymbol{g}_3$ with respect to $\boldsymbol{a}_2$ which is immaterial. For example, $\boldsymbol{g}_3$ changes the sign in coordinate transformations like $\bar{t}^1 = t^2$, $\bar{t}^2 = t^1$.

Of special interest is the dependence of the normal curvature $\varkappa_n$ on the direction of the normal section. For example, for the normal sections passing through the coordinate lines we have

$$\varkappa_n|_{t^2=const} = \frac{b_{11}}{g_{11}}, \quad \varkappa_n|_{t^1=const} = \frac{b_{22}}{g_{22}}. \tag{3.83}$$

In the following, we are going to find the directions of the maximal and minimal curvature. Necessary conditions for the extremum of the normal curvature (3.81) are given by

$$\frac{\partial \varkappa_n}{\partial t^\alpha} = 0, \quad \alpha = 1, 2. \tag{3.84}$$

Rewriting (3.81) as

$$\left(b_{\alpha\beta} - \varkappa_n g_{\alpha\beta}\right) \mathrm{d}t^\alpha \mathrm{d}t^\beta = 0 \tag{3.85}$$

and differentiating with respect to t^α we obtain

$$\left(b_{\alpha\beta} - \varkappa_n g_{\alpha\beta}\right) \mathrm{d}t^\beta = 0, \quad \alpha = 1, 2. \tag{3.86}$$

Multiplying both sides of this equation system by $g^{\alpha\rho}$ and summing up over α we have with the aid of (3.79)

$$\left(b_\beta^\rho - \varkappa_n \delta_\beta^\rho\right) \mathrm{d}t^\beta = 0, \quad \rho = 1, 2. \tag{3.87}$$

A nontrivial solution of this homogeneous equation system exists if and only if

$$\begin{vmatrix} b_1^1 - \varkappa_n & b_2^1 \\ b_1^2 & b_2^2 - \varkappa_n \end{vmatrix} = 0. \tag{3.88}$$

Writing out the above determinant we can also write

$$\varkappa_n^2 - b_\alpha^\alpha \varkappa_n + \left|b_\beta^\alpha\right| = 0. \tag{3.89}$$

The maximal and minimal curvatures $\varkappa_1$ and $\varkappa_2$ resulting from this quadratic equation are called the principal curvatures. One can show that directions of principal curvatures are mutually orthogonal (see Theorem 4.5, Sect. 4). These directions are called principal directions of normal curvature or curvature directions (see also [26]).

According to the Vieta theorem the product of principal curvatures can be expressed by

$$K = \varkappa_1 \varkappa_2 = \left|b_\beta^\alpha\right| = \frac{b}{g^2}, \tag{3.90}$$

where

$$b = \left|b_{\alpha\beta}\right| = \begin{vmatrix} b_{11} & b_{12} \\ b_{21} & b_{22} \end{vmatrix} = b_{11} b_{22} - \left(b_{12}\right)^2, \tag{3.91}$$

$$g^2 = \left[\mathbf{g}_1 \mathbf{g}_2 \mathbf{g}_3\right]^2 = \begin{vmatrix} g_{11} & g_{12} & 0 \\ g_{21} & g_{22} & 0 \\ 0 & 0 & 1 \end{vmatrix} = g_{11} g_{22} - \left(g_{12}\right)^2. \tag{3.92}$$

For the arithmetic mean of the principal curvatures we further obtain

$$H = \frac{1}{2}\left(\varkappa_1 + \varkappa_2\right) = \frac{1}{2} b_\alpha^\alpha. \tag{3.93}$$

The values K (3.90) and H (3.93) do not depend on the direction of the normal section and are called the Gaussian and mean curvatures, respectively. In terms of K and H the solutions of the quadratic equation (3.89) can simply be given by

$$\varkappa_{1,2} = H \pm \sqrt{H^2 - K}. \tag{3.94}$$

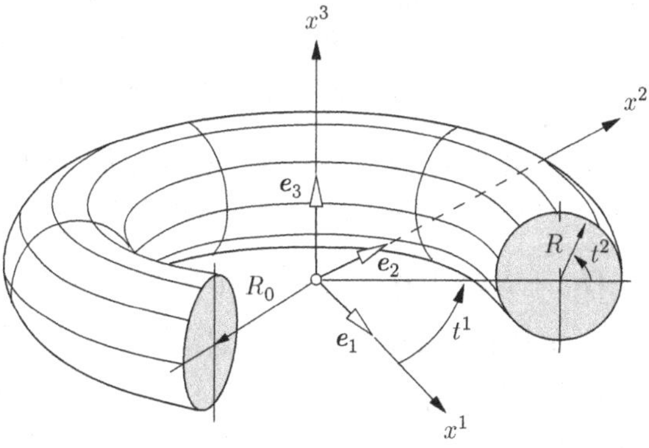

Fig. 3.4 Torus

One recognizes that the sign of the Gaussian curvature K (3.90) is defined by the sign of b (3.91). For positive b both $\varkappa_1$ and $\varkappa_2$ are positive or negative so that $\varkappa_n$ has the same sign for all directions of the normal sections at $\boldsymbol{r}\left(t^1, t^2\right)$. In other words, the orientation of $\boldsymbol{a}_2$ with respect to $\boldsymbol{g}_3$ remains constant. Such a point of the surface is called elliptic.

For $b < 0$, principal curvatures are of different signs so that different normal sections are characterized by different orientations of $\boldsymbol{a}_2$ with respect to $\boldsymbol{g}_3$. There are two directions of the normal sections with zero curvature. Such normal sections are referred to as asymptotic directions. The corresponding point of the surface is called hyperbolic or saddle point.

In the intermediate case $b = 0$, $\varkappa_n$ does not change sign. There is only one asymptotic direction which coincides with one of the principal directions (of $\varkappa_1$ or $\varkappa_2$). The corresponding point of the surface is called parabolic point.

Example 3.7. Torus. A torus is a surface obtained by rotating a circle about a coplanar axis (see Fig. 3.4). Additionally we assume that the rotation axis lies outside of the circle. Accordingly, the torus can be defined by

$$\boldsymbol{r}\left(t^1, t^2\right) = \left(R_0 + R\cos t^2\right)\cos t^1 \boldsymbol{e}_1$$
$$+ \left(R_0 + R\cos t^2\right)\sin t^1 \boldsymbol{e}_2 + R\sin t^2 \boldsymbol{e}_3, \tag{3.95}$$

where R is the radius of the circle and $R_0 > R$ is the distance between its center and the rotation axis. By means of (3.62) and (3.67) we obtain

$$\boldsymbol{g}_1 = -\left(R_0 + R\cos t^2\right)\sin t^1 \boldsymbol{e}_1 + \left(R_0 + R\cos t^2\right)\cos t^1 \boldsymbol{e}_2,$$
$$\boldsymbol{g}_2 = -R\cos t^1 \sin t^2 \boldsymbol{e}_1 - R\sin t^1 \sin t^2 \boldsymbol{e}_2 + R\cos t^2 \boldsymbol{e}_3,$$
$$\boldsymbol{g}_3 = \cos t^1 \cos t^2 \boldsymbol{e}_1 + \sin t^1 \cos t^2 \boldsymbol{e}_2 + \sin t^2 \boldsymbol{e}_3. \tag{3.96}$$

Thus, the coefficients (3.64) of the first fundamental form (3.65) are given by

$$g_{11} = \left(R_0 + R\cos t^2\right)^2, \quad g_{12} = 0, \quad g_{22} = R^2. \tag{3.97}$$

In order to express coefficients (3.74) of the second fundamental form (3.82) we first calculate derivatives of the tangent vectors $(3.96)_{1,2}$

$$\begin{aligned} g_{1,1} &= -\left(R_0 + R\cos t^2\right)\cos t^1 e_1 - \left(R_0 + R\cos t^2\right)\sin t^1 e_2, \\ g_{1,2} &= g_{2,1} = R\sin t^1 \sin t^2 e_1 - R\cos t^1 \sin t^2 e_2, \\ g_{2,2} &= -R\cos t^1 \cos t^2 e_1 - R\sin t^1 \cos t^2 e_2 - R\sin t^2 e_3. \end{aligned} \tag{3.98}$$

Inserting these expressions as well as $(3.96)_3$ into (3.74) we obtain

$$b_{11} = -\left(R_0 + R\cos t^2\right)\cos t^2, \quad b_{12} = b_{21} = 0, \quad b_{22} = -R. \tag{3.99}$$

In view of (3.79) and (3.97) $b_1^2 = b_2^1 = 0$. Thus, the solution of the equation system (3.88) delivers

$$\varkappa_1 = b_1^1 = \frac{b_{11}}{g_{11}} = -\frac{\cos t^2}{R_0 + R\cos t^2}, \quad \varkappa_2 = b_2^2 = \frac{b_{22}}{g_{22}} = -R^{-1}. \tag{3.100}$$

Comparing this result with (3.83) we see that the coordinate lines of the torus (3.95) coincide with the principal directions of the normal curvature. Hence, by (3.90)

$$K = \varkappa_1 \varkappa_2 = \frac{\cos t^2}{R\left(R_0 + R\cos t^2\right)}. \tag{3.101}$$

Thus, points of the torus for which $-\pi/2 < t^2 < \pi/2$ are elliptic while points for which $\pi/2 < t^2 < 3\pi/2$ are hyperbolic. Points of the coordinates lines $t^2 = -\pi/2$ and $t^2 = \pi/2$ are parabolic.

3.3 Application to Shell Theory

Geometry of the shell continuum. Let us consider a surface in the three-dimensional Euclidean space defined by (3.52) as

$$\boldsymbol{r} = \boldsymbol{r}\left(t^1, t^2\right), \quad \boldsymbol{r} \in \mathbb{E}^3 \tag{3.102}$$

and bounded by a closed curve C (Fig. 3.5). The shell continuum can then be described by a vector function

$$\boldsymbol{r}^* = \boldsymbol{r}^*\left(t^1, t^2, t^3\right) = \boldsymbol{r}\left(t^1, t^2\right) + \boldsymbol{g}_3 t^3, \tag{3.103}$$

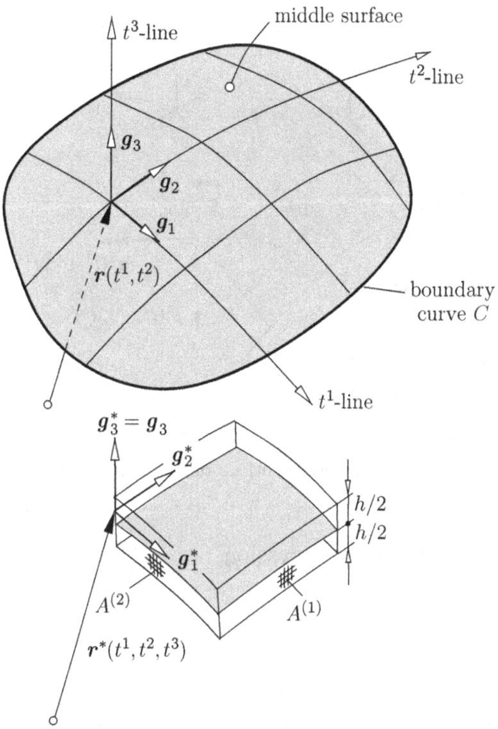

Fig. 3.5 Geometry of the shell continuum

where the unit vector $\boldsymbol{g}_3$ is defined by (3.62) and (3.67) while $-h/2 \leq t^3 \leq h/2$. The surface (3.102) is referred to as the middle surface of the shell. The thickness of the shell h is assumed to be small in comparison to its other dimensions as for example the minimal curvature radius of the middle surface.

Every fixed value of the thickness coordinate t^3 defines a surface $\boldsymbol{r}^*\left(t^1, t^2\right)$ whose geometrical variables are obtained according to (1.39), (3.62), (3.64), (3.79), (3.80), (3.90), (3.93) and (3.103) as follows.

$$\boldsymbol{g}_\alpha^* = \boldsymbol{r}^*{}_{,\alpha} = \boldsymbol{g}_\alpha + t^3 \boldsymbol{g}_{3,\alpha} = \left(\delta_\alpha^\rho - t^3 b_\alpha^\rho\right) \boldsymbol{g}_\rho, \quad \alpha = 1, 2, \tag{3.104}$$

$$\boldsymbol{g}_3^* = \frac{\boldsymbol{g}_1^* \times \boldsymbol{g}_2^*}{\left\|\boldsymbol{g}_1^* \times \boldsymbol{g}_2^*\right\|} = \boldsymbol{r}^*{}_{,3} = \boldsymbol{g}_3, \tag{3.105}$$

$$g_{\alpha\beta}^* = \boldsymbol{g}_\alpha^* \cdot \boldsymbol{g}_\beta^* = g_{\alpha\beta} - 2t^3 b_{\alpha\beta} + \left(t^3\right)^2 b_{\alpha\rho} b_\beta^\rho, \quad \alpha, \beta = 1, 2, \tag{3.106}$$

$$\begin{aligned} g^* &= \left[\boldsymbol{g}_1^* \boldsymbol{g}_2^* \boldsymbol{g}_3^*\right] = \left[\left(\delta_1^\rho - t^3 b_1^\rho\right) \boldsymbol{g}_\rho \left(\delta_2^\gamma - t^3 b_2^\gamma\right) \boldsymbol{g}_\gamma \boldsymbol{g}_3\right] \\ &= \left(\delta_1^\rho - t^3 b_1^\rho\right)\left(\delta_2^\gamma - t^3 b_2^\gamma\right) g e_{\rho\gamma 3} = g \left|\delta_\beta^\alpha - t^3 b_\beta^\alpha\right| \\ &= g\left[1 - 2t^3 H + \left(t^3\right)^2 K\right]. \end{aligned} \tag{3.107}$$

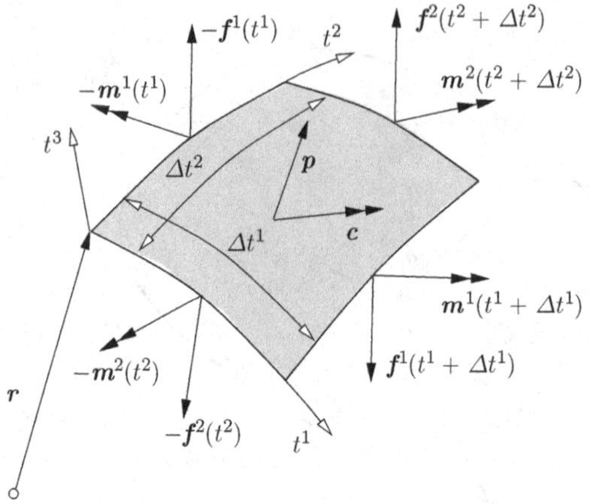

Fig. 3.6 Force variables related to the middle surface of the shell

The factor in brackets in the latter expression

$$\mu = \frac{g^*}{g} = 1 - 2t^3 H + \left(t^3\right)^2 K \tag{3.108}$$

is called the shell shifter.

Internal force variables. Let us consider an element of the shell continuum (see Fig. 3.6) bounded by the coordinate lines t^α and $t^\alpha + \Delta t^\alpha$ $(\alpha = 1, 2)$. One defines the force vector $\boldsymbol{f}^\alpha$ and the couple vector $\boldsymbol{m}^\alpha$ relative to the middle surface of the shell, respectively, by

$$\boldsymbol{f}^\alpha = \int_{-h/2}^{h/2} \mu \boldsymbol{\sigma} \boldsymbol{g}^{*\alpha} \mathrm{d}t^3, \quad \boldsymbol{m}^\alpha = \int_{-h/2}^{h/2} \mu \boldsymbol{r}^* \times \left(\boldsymbol{\sigma} \boldsymbol{g}^{*\alpha}\right) \mathrm{d}t^3, \quad \alpha = 1, 2, \tag{3.109}$$

where $\boldsymbol{\sigma}$ denotes the Cauchy stress tensor on the boundary surface $A^{(\alpha)}$ spanned on the coordinate lines t^3 and t^β $(\beta \neq \alpha)$. The unit normal to this boundary surface is given by

$$\boldsymbol{n}^{(\alpha)} = \frac{\boldsymbol{g}^{*\alpha}}{\left\|\boldsymbol{g}^{*\alpha}\right\|} = \frac{\boldsymbol{g}^{*\alpha}}{\sqrt{g^{*\alpha\alpha}}} = \frac{g^*}{\sqrt{g^*_{\beta\beta}}} \boldsymbol{g}^{*\alpha}, \quad \beta \neq \alpha = 1, 2, \tag{3.110}$$

where we keep in mind that $\boldsymbol{g}^{*\alpha} \cdot \boldsymbol{g}^*_\beta = \boldsymbol{g}^{*\alpha} \cdot \boldsymbol{g}_3 = 0$ and (see Exercise 3.8)

$$g^{*\alpha\alpha} = \frac{g^*_{\beta\beta}}{g^{*2}}, \quad \beta \neq \alpha = 1, 2. \tag{3.111}$$

Applying the Cauchy theorem (1.77) and bearing (3.108) in mind we obtain

$$\boldsymbol{f}^\alpha = \frac{1}{g}\int_{-h/2}^{h/2} \sqrt{g^*_{\beta\beta}}\,\boldsymbol{t}\,\mathrm{d}t^3, \quad \boldsymbol{m}^\alpha = \frac{1}{g}\int_{-h/2}^{h/2} \sqrt{g^*_{\beta\beta}}\left(\boldsymbol{r}^* \times \boldsymbol{t}\right)\mathrm{d}t^3, \tag{3.112}$$

where again $\beta \neq \alpha = 1,2$ and $\boldsymbol{t}$ denotes the Cauchy stress vector. The force and couple resulting on the whole boundary surface can thus be expressed respectively by

$$\int_{A^{(\alpha)}} \boldsymbol{t}\,\mathrm{d}A^{(\alpha)} = \int_{t^\beta}^{t^\beta+\Delta t^\beta}\int_{-h/2}^{h/2} \boldsymbol{t}\sqrt{g^*_{\beta\beta}}\,\mathrm{d}t^3\mathrm{d}t^\beta = \int_{t^\beta}^{t^\beta+\Delta t^\beta} g\boldsymbol{f}^\alpha\mathrm{d}t^\beta, \tag{3.113}$$

$$\int_{A^{(\alpha)}} \left(\boldsymbol{r}^* \times \boldsymbol{t}\right)\mathrm{d}A^{(\alpha)} = \int_{t^\beta}^{t^\beta+\Delta t^\beta}\int_{-h/2}^{h/2} \left(\boldsymbol{r}^* \times \boldsymbol{t}\right)\sqrt{g^*_{\beta\beta}}\,\mathrm{d}t^3\mathrm{d}t^\beta$$

$$= \int_{t^\beta}^{t^\beta+\Delta t^\beta} g\boldsymbol{m}^\alpha\mathrm{d}t^\beta, \quad \beta \neq \alpha = 1,2, \tag{3.114}$$

where we make use of the relation

$$\mathrm{d}A^{(\alpha)} = g^*\sqrt{g^{*\alpha\alpha}}\,\mathrm{d}t^\beta\mathrm{d}t^3 = \sqrt{g^*_{\beta\beta}}\,\mathrm{d}t^\beta\mathrm{d}t^3, \quad \beta \neq \alpha = 1,2 \tag{3.115}$$

following immediately from (2.105) and (3.111).

The force and couple vectors (3.109) are usually represented with respect to the basis related to the middle surface as (see also [1])

$$\boldsymbol{f}^\alpha = f^{\alpha\beta}\boldsymbol{g}_\beta + q^\alpha\boldsymbol{g}_3, \quad \boldsymbol{m}^\alpha = m^{\alpha\beta}\boldsymbol{g}_3 \times \boldsymbol{g}_\beta = g\,e_{3\beta\rho}m^{\alpha\beta}\boldsymbol{g}^\rho. \tag{3.116}$$

In shell theory, their components are denoted as follows.

$f^{\alpha\beta}$ Components of the stress resultant tensor,

q^α Components of the transverse shear stress vector,

$m^{\alpha\beta}$ Components of the moment tensor.

External force variables. One defines the load force vector and the load moment vector related to a unit area of the middle surface, respectively by

$$\boldsymbol{p} = p^i\boldsymbol{g}_i, \quad \boldsymbol{c} = c^\rho\boldsymbol{g}_3 \times \boldsymbol{g}_\rho. \tag{3.117}$$

The load moment vector $\boldsymbol{c}$ is thus assumed to be tangential to the middle surface. The resulting force and couple can be expressed respectively by

$$\int\limits_{t^2}^{t^2+\Delta t^2}\int\limits_{t^1}^{t^1+\Delta t^1} \boldsymbol{p}g\mathrm{d}t^1\mathrm{d}t^2, \quad \int\limits_{t^2}^{t^2+\Delta t^2}\int\limits_{t^1}^{t^1+\Delta t^1} \boldsymbol{c}g\mathrm{d}t^1\mathrm{d}t^2. \tag{3.118}$$

Equilibrium conditions. Taking (3.113) and $(3.118)_1$ into account the force equilibrium condition of the shell element can be expressed as

$$\sum_{\substack{\alpha,\beta=1\\ \alpha\neq\beta}}^{2} \int\limits_{t^\beta}^{t^\beta+\Delta t^\beta} \left[g\left(t^\alpha+\Delta t^\alpha\right)\boldsymbol{f}^\alpha\left(t^\alpha+\Delta t^\alpha\right)-g\left(t^\alpha\right)\boldsymbol{f}^\alpha\left(t^\alpha\right)\right]\mathrm{d}t^\beta$$

$$+\int\limits_{t^2}^{t^2+\Delta t^2}\int\limits_{t^1}^{t^1+\Delta t^1} \boldsymbol{p}g\mathrm{d}t^1\mathrm{d}t^2=\boldsymbol{0}. \tag{3.119}$$

Rewriting the first integral in (3.119) we further obtain

$$\int\limits_{t^2}^{t^2+\Delta t^2}\int\limits_{t^1}^{t^1+\Delta t^1} \left[\left(g\boldsymbol{f}^\alpha\right)_{,\alpha}+g\boldsymbol{p}\right]\mathrm{d}t^1\mathrm{d}t^2=\boldsymbol{0}. \tag{3.120}$$

Since the latter condition holds for all shell elements we infer that

$$\left(g\boldsymbol{f}^\alpha\right)_{,\alpha}+g\boldsymbol{p}=\boldsymbol{0}, \tag{3.121}$$

which leads by virtue of (2.107) and $(3.73)_2$ to

$$\boldsymbol{f}^\alpha|_\alpha+\boldsymbol{p}=\boldsymbol{0}, \tag{3.122}$$

where the covariant derivative is formally applied to the vectors $\boldsymbol{f}^\alpha$ according to $(3.76)_1$.

In a similar fashion we can treat the moment equilibrium. In this case, we obtain instead of (3.121) the following condition

$$\left[g\left(\boldsymbol{m}^\alpha+\boldsymbol{r}\times\boldsymbol{f}^\alpha\right)\right]_{,\alpha}+g\boldsymbol{r}\times\boldsymbol{p}+g\boldsymbol{c}=\boldsymbol{0}. \tag{3.123}$$

With the aid of (3.62) and keeping (3.122) in mind, it finally delivers

$$\boldsymbol{m}^\alpha|_\alpha+\boldsymbol{g}_\alpha\times\boldsymbol{f}^\alpha+\boldsymbol{c}=\boldsymbol{0}. \tag{3.124}$$

In order to rewrite the equilibrium conditions (3.122) and (3.124) in component form we further utilize representations (3.116), (3.117) and apply the product rule of differentiation for the covariant derivative (see, e.g., (2.101)–(2.103)). By virtue of (3.78) and (3.80) it delivers

$$\left(f^{\alpha\rho}|_{\alpha} - b_{\alpha}^{\rho} q^{\alpha} + p^{\rho}\right) \boldsymbol{g}_{\rho} + \left(f^{\alpha\beta} b_{\alpha\beta} + q^{\alpha}|_{\alpha} + p^{3}\right) \boldsymbol{g}_{3} = \boldsymbol{0}, \tag{3.125}$$

$$\left(m^{\alpha\rho}|_{\alpha} - q^{\rho} + c^{\rho}\right) \boldsymbol{g}_{3} \times \boldsymbol{g}_{\rho} + g\, e_{\alpha\beta 3} \tilde{f}^{\alpha\beta} \boldsymbol{g}_{3} = \boldsymbol{0} \tag{3.126}$$

with a new variable

$$\tilde{f}^{\alpha\beta} = f^{\alpha\beta} + b_{\gamma}^{\beta} m^{\gamma\alpha}, \quad \alpha, \beta = 1, 2 \tag{3.127}$$

called pseudo-stress resultant. Keeping in mind that the vectors $\boldsymbol{g}_i$ $(i = 1, 2, 3)$ are linearly independent we thus obtain the following scalar force equilibrium conditions

$$f^{\alpha\rho}|_{\alpha} - b_{\alpha}^{\rho} q^{\alpha} + p^{\rho} = 0, \quad \rho = 1, 2, \tag{3.128}$$

$$b_{\alpha\beta} f^{\alpha\beta} + q^{\alpha}|_{\alpha} + p^{3} = 0 \tag{3.129}$$

and moment equilibrium conditions

$$m^{\alpha\rho}|_{\alpha} - q^{\rho} + c^{\rho} = 0, \quad \rho = 1, 2, \tag{3.130}$$

$$\tilde{f}^{\alpha\beta} = \tilde{f}^{\beta\alpha}, \quad \alpha, \beta = 1, 2, \quad \alpha \neq \beta. \tag{3.131}$$

With the aid of (3.127) one can finally eliminate the components of the stress resultant tensor $f^{\alpha\beta}$ from (3.128) and (3.129). This leads to the following equation system

$$\tilde{f}^{\alpha\rho}|_{\alpha} - \left(b_{\gamma}^{\rho} m^{\gamma\alpha}\right)|_{\alpha} - b_{\alpha}^{\rho} q^{\alpha} + p^{\rho} = 0, \quad \rho = 1, 2, \tag{3.132}$$

$$b_{\alpha\beta} \tilde{f}^{\alpha\beta} - b_{\alpha\beta} b_{\gamma}^{\beta} m^{\gamma\alpha} + q^{\alpha}|_{\alpha} + p^{3} = 0, \tag{3.133}$$

$$m^{\alpha\rho}|_{\alpha} - q^{\rho} + c^{\rho} = 0, \quad \rho = 1, 2, \tag{3.134}$$

where the latter relation is repeated from (3.130) for completeness.

Example 3.8. Equilibrium equations of plate theory. In this case, the middle surface of the shell is a plane (3.55) for which

$$b_{\alpha\beta} = b_{\beta}^{\alpha} = 0, \quad \alpha, \beta = 1, 2. \tag{3.135}$$

Thus, the equilibrium equations (3.132)–(3.134) simplify to

$$f^{\alpha\rho}{}_{,\alpha} + p^{\rho} = 0, \quad \rho = 1, 2, \tag{3.136}$$

$$q^{\alpha}{}_{,\alpha} + p^3 = 0, \tag{3.137}$$

$$m^{\alpha\rho}{}_{,\alpha} - q^{\rho} + c^{\rho} = 0, \quad \rho = 1, 2, \tag{3.138}$$

where in view of (3.127) and (3.131) $f^{\alpha\beta} = f^{\beta\alpha}$ $(\alpha \neq \beta = 1, 2)$.

Example 3.9. Equilibrium equations of membrane theory. The membrane theory assumes that the shell is moment free so that

$$m^{\alpha\beta} = 0, \quad c^{\beta} = 0, \quad \alpha, \beta = 1, 2. \tag{3.139}$$

In this case, the equilibrium equations (3.132)–(3.134) reduce to

$$f^{\alpha\rho}|_{\alpha} + p^{\rho} = 0, \quad \rho = 1, 2, \tag{3.140}$$

$$b_{\alpha\beta} f^{\alpha\beta} + p^3 = 0, \tag{3.141}$$

$$q^{\rho} = 0, \quad \rho = 1, 2, \tag{3.142}$$

where again $f^{\alpha\beta} = f^{\beta\alpha}$ $(\alpha \neq \beta = 1, 2)$.

Exercises

3.1. Show that a curve $\boldsymbol{r}(s)$ is a straight line if $\varkappa(s) \equiv 0$ for any s.

3.2. Show that the curves $\boldsymbol{r}(s)$ and $\boldsymbol{r}'(s) = \boldsymbol{r}(-s)$ have the same curvature and torsion.

3.3. Show that a curve $\boldsymbol{r}(s)$ characterized by zero torsion $\tau(s) \equiv 0$ for any s lies in a plane.

3.4. Evaluate the Christoffel symbols of the second kind, the coefficients of the first and second fundamental forms, the Gaussian and mean curvatures for the cylinder (3.56).

3.5. Evaluate the Christoffel symbols of the second kind, the coefficients of the first and second fundamental forms, the Gaussian and mean curvatures for the sphere (3.58).

3.6. For the so-called hyperbolic paraboloidal surface defined by

$$\boldsymbol{r}\left(t^1, t^2\right) = t^1 \boldsymbol{e}_1 + t^2 \boldsymbol{e}_2 + \frac{t^1 t^2}{c} \boldsymbol{e}_3, \quad c > 0, \tag{3.143}$$

evaluate the tangent vectors to the coordinate lines, the coefficients of the first and second fundamental forms, the Gaussian and mean curvatures.

3.7. For a cone of revolution defined by

$$\boldsymbol{r}\left(t^1,t^2\right) = ct^2 \cos t^1 \boldsymbol{e}_1 + ct^2 \sin t^1 \boldsymbol{e}_2 + t^2 \boldsymbol{e}_3, \quad c \neq 0, \tag{3.144}$$

evaluate the vectors tangent to the coordinate lines, the coefficients of the first and second fundamental forms, the Gaussian and mean curvatures.

3.8. Verify relation (3.111).

3.9. Write out equilibrium equations (3.140) and (3.141) of the membrane theory for a cylindrical shell and a spherical shell.

Chapter 4
Eigenvalue Problem and Spectral Decomposition of Second-Order Tensors

4.1 Complexification

So far we have considered solely real vectors and real vector spaces. For the purposes of this chapter an introduction of complex vectors is, however, necessary. Indeed, in the following we will see that the existence of a solution of an eigenvalue problem even for real second-order tensors can be guaranteed only within a complex vector space. In order to define the complex vector space let us consider ordered pairs $\langle \boldsymbol{x}, \boldsymbol{y} \rangle$ of real vectors $\boldsymbol{x}$ and $\boldsymbol{y} \in \mathbb{E}^n$. The sum of two such pairs is defined by [15]

$$\langle \boldsymbol{x}_1, \boldsymbol{y}_1 \rangle + \langle \boldsymbol{x}_2, \boldsymbol{y}_2 \rangle = \langle \boldsymbol{x}_1 + \boldsymbol{x}_2, \boldsymbol{y}_1 + \boldsymbol{y}_2 \rangle . \tag{4.1}$$

Further, we define the product of a pair $\langle \boldsymbol{x}, \boldsymbol{y} \rangle$ by a complex number $\alpha + \mathrm{i}\beta$ by

$$(\alpha + \mathrm{i}\beta) \langle \boldsymbol{x}, \boldsymbol{y} \rangle = \langle \alpha \boldsymbol{x} - \beta \boldsymbol{y}, \beta \boldsymbol{x} + \alpha \boldsymbol{y} \rangle , \tag{4.2}$$

where $\alpha, \beta \in \mathbb{R}$ and $\mathrm{i} = \sqrt{-1}$. These formulas can easily be recovered assuming that

$$\langle \boldsymbol{x}, \boldsymbol{y} \rangle = \boldsymbol{x} + \mathrm{i}\boldsymbol{y} . \tag{4.3}$$

The definitions (4.1) and (4.2) enriched by the zero pair $\langle \boldsymbol{0}, \boldsymbol{0} \rangle$ are sufficient to ensure that the axioms (A.1–A.4) and (B.1–B.4) of Chap. 1 are valid. Thus, the set of all pairs $\boldsymbol{z} = \langle \boldsymbol{x}, \boldsymbol{y} \rangle$ characterized by the above properties forms a vector space referred to as complex vector space. Every basis $\mathcal{G} = \{\boldsymbol{g}_1, \boldsymbol{g}_2, \ldots, \boldsymbol{g}_n\}$ of the underlying Euclidean space $\mathbb{E}^n$ represents simultaneously a basis of the corresponding complexified space. Indeed, for every complex vector within this space

$$\boldsymbol{z} = \boldsymbol{x} + \mathrm{i}\boldsymbol{y} , \tag{4.4}$$

M. Itskov, *Tensor Algebra and Tensor Analysis for Engineers*, Mathematical Engineering,
DOI 10.1007/978-3-642-30879-6_4,

where $\boldsymbol{x}, \boldsymbol{y} \in \mathbb{E}^n$ and consequently

$$\boldsymbol{x} = x^i \boldsymbol{g}_i, \quad \boldsymbol{y} = y^i \boldsymbol{g}_i, \tag{4.5}$$

we can write

$$\boldsymbol{z} = \left(x^i + \mathrm{i} y^i\right) \boldsymbol{g}_i. \tag{4.6}$$

Thus, the dimension of the complexified space coincides with the dimension of the original real vector space. Using this fact we will denote the complex vector space based on $\mathbb{E}^n$ by $\mathbb{C}^n$. Clearly, $\mathbb{E}^n$ represents a subspace of $\mathbb{C}^n$.

For every vector $\boldsymbol{z} \in \mathbb{C}^n$ given by (4.4) one defines a complex conjugate counterpart by

$$\bar{\boldsymbol{z}} = \boldsymbol{x} - \mathrm{i} \boldsymbol{y}. \tag{4.7}$$

Of special interest is the scalar product of two complex vectors, say $\boldsymbol{z}_1 = \boldsymbol{x}_1 + \mathrm{i}\boldsymbol{y}_1$ and $\boldsymbol{z}_2 = \boldsymbol{x}_2 + \mathrm{i}\boldsymbol{y}_2$, which we define by (see also [4])

$$(\boldsymbol{x}_1 + \mathrm{i}\boldsymbol{y}_1) \cdot (\boldsymbol{x}_2 + \mathrm{i}\boldsymbol{y}_2) = \boldsymbol{x}_1 \cdot \boldsymbol{x}_2 - \boldsymbol{y}_1 \cdot \boldsymbol{y}_2 + \mathrm{i}\left(\boldsymbol{x}_1 \cdot \boldsymbol{y}_2 + \boldsymbol{y}_1 \cdot \boldsymbol{x}_2\right). \tag{4.8}$$

This scalar product is commutative (C.1), distributive (C.2) and linear in each factor (C.3). Thus, it differs from the classical scalar product of complex vectors given in terms of the complex conjugate (see, e.g., [15]). As a result, the axiom (C.4) does not generally hold. For instance, one can easily imagine a non-zero complex vector (for example $\boldsymbol{e}_1 + \mathrm{i}\boldsymbol{e}_2$) whose scalar product with itself is zero. For complex vectors with the scalar product (4.8) the notions of length, orthogonality or parallelity can hardly be interpreted geometrically.

However, for complex vectors the axiom (C.4) can be reformulated by

$$\boldsymbol{z} \cdot \bar{\boldsymbol{z}} \geq 0, \quad \boldsymbol{z} \cdot \bar{\boldsymbol{z}} = 0 \quad \text{if and only if} \quad \boldsymbol{z} = \boldsymbol{0}. \tag{4.9}$$

Indeed, using (4.4), (4.7) and (4.8) we obtain $\boldsymbol{z} \cdot \bar{\boldsymbol{z}} = \boldsymbol{x} \cdot \boldsymbol{x} + \boldsymbol{y} \cdot \boldsymbol{y}$. Bearing in mind that the vectors $\boldsymbol{x}$ and $\boldsymbol{y}$ belong to the Euclidean space this immediately implies (4.9).

As we learned in Chap. 1, the Euclidean space $\mathbb{E}^n$ is characterized by the existence of an orthonormal basis (1.8). This can now be postulated for the complex vector space $\mathbb{C}^n$ as well, because $\mathbb{C}^n$ includes $\mathbb{E}^n$ by the very definition. Also Theorem 1.6 remains valid since it has been proved without making use of the property (C.4). Thus, we may state that for every basis in $\mathbb{C}^n$ there exists a unique dual basis.

The last step of the complexification is a generalization of a linear mapping on complex vectors. This can be achieved by setting for every tensor $\mathbf{A} \in \mathbf{Lin}^n$

$$\mathbf{A}\left(\boldsymbol{x} + \mathrm{i}\boldsymbol{y}\right) = \mathbf{A}\boldsymbol{x} + \mathrm{i}\left(\mathbf{A}\boldsymbol{y}\right). \tag{4.10}$$

4.2 Eigenvalue Problem, Eigenvalues and Eigenvectors

Let $\mathbf{A} \in \mathbf{Lin}^n$ be a second-order tensor. The equation

$$\mathbf{A}\boldsymbol{a} = \lambda \boldsymbol{a}, \quad \boldsymbol{a} \neq \boldsymbol{0} \tag{4.11}$$

is referred to as the eigenvalue problem of the tensor $\mathbf{A}$. The non-zero vector $\boldsymbol{a} \in \mathbb{C}^n$ satisfying this equation is called an eigenvector of $\mathbf{A}$; $\lambda \in \mathbb{C}$ is called an eigenvalue of $\mathbf{A}$. It is clear that any product of an eigenvector with any (real or complex) scalar is again an eigenvector.

The eigenvalue problem (4.11) and the corresponding eigenvector $\boldsymbol{a}$ can be regarded as the right eigenvalue problem and the right eigenvector, respectively. In contrast, one can define the left eigenvalue problem by

$$\boldsymbol{b}\mathbf{A} = \lambda \boldsymbol{b}, \quad \boldsymbol{b} \neq \boldsymbol{0}, \tag{4.12}$$

where $\boldsymbol{b} \in \mathbb{C}^n$ is the left eigenvector. In view of (1.115), every right eigenvector of $\mathbf{A}$ represents the left eigenvector of $\mathbf{A}^{\mathrm{T}}$ and vice versa. In the following, unless indicated otherwise, we will mean the right eigenvalue problem and the right eigenvector.

Mapping (4.11) by $\mathbf{A}$ several times we obtain

$$\mathbf{A}^k \boldsymbol{a} = \lambda^k \boldsymbol{a}, \quad k = 1, 2, \ldots \tag{4.13}$$

This leads to the following (spectral mapping) theorem.

Theorem 4.1. *Let λ be an eigenvalue of the tensor $\mathbf{A}$ and let $g(\mathbf{A}) = \sum_{k=0}^{m} a_k \mathbf{A}^k$ be a polynomial of $\mathbf{A}$. Then $g(\lambda) = \sum_{k=0}^{m} a_k \lambda^k$ is the eigenvalue of $g(\mathbf{A})$.*

Proof. Let $\boldsymbol{a}$ be an eigenvector of $\mathbf{A}$ associated with λ. Then, in view of (4.13)

$$g(\mathbf{A})\boldsymbol{a} = \sum_{k=0}^{m} a_k \mathbf{A}^k \boldsymbol{a} = \sum_{k=0}^{m} a_k \lambda^k \boldsymbol{a} = \left(\sum_{k=0}^{m} a_k \lambda^k \right) \boldsymbol{a} = g(\lambda)\boldsymbol{a}.$$

In order to find the eigenvalues of the tensor $\mathbf{A}$ we consider the following representations:

$$\mathbf{A} = \mathrm{A}^{i}_{\cdot j} \boldsymbol{g}_i \otimes \boldsymbol{g}^j, \quad \boldsymbol{a} = a^i \boldsymbol{g}_i, \quad \boldsymbol{b} = b_i \boldsymbol{g}^i, \tag{4.14}$$

where $\mathcal{G} = \{\boldsymbol{g}_1, \boldsymbol{g}_2, \ldots, \boldsymbol{g}_n\}$ and $\mathcal{G}' = \{\boldsymbol{g}^1, \boldsymbol{g}^2, \ldots, \boldsymbol{g}^n\}$ are two arbitrary mutually dual bases in $\mathbb{E}^n$ and consequently also in $\mathbb{C}^n$. Note that we prefer here the mixed variant representation of the tensor $\mathbf{A}$. Inserting (4.14) into (4.11) and (4.12) further yields

$$\mathrm{A}^{i}_{\cdot j} a^j \boldsymbol{g}_i = \lambda a^i \boldsymbol{g}_i, \quad \mathrm{A}^{i}_{\cdot j} b_i \boldsymbol{g}^j = \lambda b_j \boldsymbol{g}^j,$$

and therefore

$$\left(\mathrm{A}^{i}_{\cdot j} a^{j} - \lambda a^{i}\right) \boldsymbol{g}_{i} = \boldsymbol{0}, \quad \left(\mathrm{A}^{i}_{\cdot j} b_{i} - \lambda b_{j}\right) \boldsymbol{g}^{j} = \boldsymbol{0}. \tag{4.15}$$

Since both the vectors $\boldsymbol{g}_i$ and $\boldsymbol{g}^i$ $(i = 1, 2, \ldots, n)$ are linearly independent the associated scalar coefficients in (4.15) must be zero. This results in the following two linear homogeneous equation systems

$$\left(\mathrm{A}^{i}_{\cdot j} - \lambda \delta^{i}_{j}\right) a^{j} = 0, \quad \left(\mathrm{A}^{j}_{\cdot i} - \lambda \delta^{j}_{i}\right) b_{j} = 0, \quad i = 1, 2, \ldots, n \tag{4.16}$$

with respect to the components of the right eigenvector $\boldsymbol{a}$ and the left eigenvector $\boldsymbol{b}$, respectively. A non-trivial solution of these equation systems exists if and only if

$$\left|\mathrm{A}^{i}_{\cdot j} - \lambda \delta^{i}_{j}\right| = 0, \tag{4.17}$$

or equivalently

$$\begin{vmatrix} \mathrm{A}^{1}_{\cdot 1} - \lambda & \mathrm{A}^{1}_{\cdot 2} & \ldots & \mathrm{A}^{1}_{\cdot n} \\ \mathrm{A}^{2}_{\cdot 1} & \mathrm{A}^{2}_{\cdot 2} - \lambda & \ldots & \mathrm{A}^{2}_{\cdot n} \\ \vdots & \vdots & \ddots & \vdots \\ \mathrm{A}^{n}_{\cdot 1} & \mathrm{A}^{n}_{\cdot 2} & \ldots & \mathrm{A}^{n}_{\cdot n} - \lambda \end{vmatrix} = 0, \tag{4.18}$$

where $|\bullet|$ denotes the determinant of a matrix. Equation (4.17) is called the characteristic equation of the tensor $\mathbf{A}$. Writing out the determinant on the left hand side of this equation one obtains a polynomial of degree n with respect to the powers of λ

$$p_{\mathbf{A}}(\lambda) = (-1)^{n} \lambda^{n} + (-1)^{n-1} \lambda^{n-1} \mathrm{I}^{(1)}_{\mathbf{A}} + \ldots$$
$$+ (-1)^{n-k} \lambda^{n-k} \mathrm{I}^{(k)}_{\mathbf{A}} + \ldots + \mathrm{I}^{(n)}_{\mathbf{A}}, \tag{4.19}$$

referred to as the characteristic polynomial of the tensor $\mathbf{A}$. Thereby, it can easily be seen that

$$\mathrm{I}^{(1)}_{\mathbf{A}} = \mathrm{A}^{i}_{\cdot i} = \mathrm{tr}\mathbf{A}, \quad \mathrm{I}^{(n)}_{\mathbf{A}} = \left|\mathrm{A}^{i}_{\cdot j}\right|. \tag{4.20}$$

The characteristic equation (4.17) can briefly be written as

$$p_{\mathbf{A}}(\lambda) = 0. \tag{4.21}$$

According to the fundamental theorem of algebra, a polynomial of degree n has n complex roots which may be multiple. These roots are the eigenvalues λ_i $(i = 1, 2, \ldots, n)$ of the tensor $\mathbf{A}$.

Factorizing the characteristic polynomial (4.19) yields

$$p_{\mathbf{A}}(\lambda) = \prod_{i=1}^{n} (\lambda_i - \lambda). \tag{4.22}$$

Collecting multiple eigenvalues the polynomial (4.22) can further be rewritten as

$$p_{\mathbf{A}}(\lambda) = \prod_{i=1}^{s} (\lambda_i - \lambda)^{r_i}, \tag{4.23}$$

where s $(1 \le s \le n)$ denotes the number of distinct eigenvalues, while r_i is referred to as an algebraic multiplicity of the eigenvalue λ_i $(i = 1, 2, \ldots, s)$. It should formally be distinguished from the so-called geometric multiplicity t_i, which represents the number of linearly independent eigenvectors associated with this eigenvalue.

Example 4.1. Eigenvalues and eigenvectors of the deformation gradient in the case of simple shear. In simple shear, the deformation gradient can be given by $\mathbf{F} = \mathrm{F}^{i}_{\cdot j} \boldsymbol{e}_i \otimes \boldsymbol{e}^j$, where the matrix $\left[\mathrm{F}^{i}_{\cdot j}\right]$ is represented by (2.69). The characteristic equation (4.17) for the tensor $\mathbf{F}$ takes thus the form

$$\begin{vmatrix} 1-\lambda & \gamma & 0 \\ 0 & 1-\lambda & 0 \\ 0 & 0 & 1-\lambda \end{vmatrix} = 0.$$

Writing out this determinant we obtain

$$(1-\lambda)^3 = 0,$$

which yields one triple eigenvalue

$$\lambda_1 = \lambda_2 = \lambda_3 = 1.$$

The associated (right) eigenvectors $\boldsymbol{a} = a^i \boldsymbol{e}_i$ can be obtained from the equation system $(4.16)_1$ i.e.

$$\left(\mathrm{F}^{i}_{\cdot j} - \lambda \delta^{i}_{j}\right) a^j = 0, \quad i = 1, 2, 3.$$

In view of (2.69) it reduces to the only non-trivial equation

$$a^2 \gamma = 0.$$

Hence, all eigenvectors of $\mathbf{F}$ can be given by $\boldsymbol{a} = a^1 \boldsymbol{e}_1 + a^3 \boldsymbol{e}_3$. They are linear combinations of the only two linearly independent eigenvectors $\boldsymbol{e}_1$ and $\boldsymbol{e}_3$. Accordingly, the geometric and algebraic multiplicities of the eigenvalue 1 are $t_1 = 2$ and $r_1 = 3$, respectively.

4.3 Characteristic Polynomial

By the very definition of the eigenvalue problem (4.11) the eigenvalues are independent of the choice of the basis. This is also the case for the coefficients $\mathrm{I}_{\mathbf{A}}^{(i)}$ $(i = 1, 2, \ldots, n)$ of the characteristic polynomial (4.19) because they uniquely define the eigenvalues and vice versa. These coefficients are called principal invariants of $\mathbf{A}$. Writing out (4.22) and comparing with (4.19) one obtains the following relations between the principal invariants and eigenvalues

$$\begin{aligned}
\mathrm{I}_{\mathbf{A}}^{(1)} &= \lambda_1 + \lambda_2 + \ldots + \lambda_n, \\
\mathrm{I}_{\mathbf{A}}^{(2)} &= \lambda_1\lambda_2 + \lambda_1\lambda_3 + \ldots + \lambda_{n-1}\lambda_n, \\
&\vdots \\
\mathrm{I}_{\mathbf{A}}^{(k)} &= \sum_{o_1<o_2<\ldots<o_k}^{n} \lambda_{o_1}\lambda_{o_2}\ldots\lambda_{o_k}, \\
&\vdots \\
\mathrm{I}_{\mathbf{A}}^{(n)} &= \lambda_1\lambda_2\ldots\lambda_n,
\end{aligned} \tag{4.24}$$

referred to as the Vieta theorem. The principal invariants can also be expressed in terms of the so-called principal traces $\mathrm{tr}\mathbf{A}^k$ $(k = 1, 2, \ldots, n)$. Indeed, by use of (4.13), $(4.20)_1$ and $(4.24)_1$ we first write

$$\mathrm{tr}\mathbf{A}^k = \lambda_1^k + \lambda_2^k + \ldots + \lambda_n^k, \quad k = 1, 2, \ldots, n. \tag{4.25}$$

Then, we apply Newton's identities (also referred to as the Newton-Girard formulas) relating coefficients of a polynomial to its roots represented by the sum of the powers (see e.g. [10]) in the form of the right hand side of (4.25). Taking (4.25) into account, Newton's identities can thus be written as

$$\begin{aligned}
\mathrm{I}_{\mathbf{A}}^{(1)} &= \mathrm{tr}\mathbf{A}, \\
\mathrm{I}_{\mathbf{A}}^{(2)} &= \frac{1}{2}\left(\mathrm{I}_{\mathbf{A}}^{(1)}\mathrm{tr}\mathbf{A} - \mathrm{tr}\mathbf{A}^2\right), \\
\mathrm{I}_{\mathbf{A}}^{(3)} &= \frac{1}{3}\left(\mathrm{I}_{\mathbf{A}}^{(2)}\mathrm{tr}\mathbf{A} - \mathrm{I}_{\mathbf{A}}^{(1)}\mathrm{tr}\mathbf{A}^2 + \mathrm{tr}\mathbf{A}^3\right), \\
&\vdots
\end{aligned}$$

$$\mathrm{I}_{\mathbf{A}}^{(k)} = \frac{1}{k}\left(\mathrm{I}_{\mathbf{A}}^{(k-1)}\mathrm{tr}\mathbf{A} - \mathrm{I}_{\mathbf{A}}^{(k-2)}\mathrm{tr}\mathbf{A}^2 + \ldots + (-1)^{k-1}\,\mathrm{tr}\mathbf{A}^k\right)$$
$$= \frac{1}{k}\sum_{i=1}^{k}(-1)^{i-1}\,\mathrm{I}_{\mathbf{A}}^{(k-i)}\mathrm{tr}\mathbf{A}^i,$$
$$\vdots$$
$$\mathrm{I}_{\mathbf{A}}^{(n)} = \det\mathbf{A}, \tag{4.26}$$

where we set $\mathrm{I}_{\mathbf{A}}^{(0)} = 1$ and

$$\det\mathbf{A} = \left|\mathrm{A}^{i}_{\cdot j}\right| = \left|\mathrm{A}_{j\cdot}^{\;i}\right| \tag{4.27}$$

is called the determinant of the tensor $\mathbf{A}$.

Example 4.2. Three-dimensional space. For illustration, we consider a second-order tensor $\mathbf{A}$ in three-dimensional space. In this case, the characteristic polynomial (4.19) takes the form

$$p_{\mathbf{A}}(\lambda) = -\lambda^3 + \mathrm{I}_{\mathbf{A}}\lambda^2 - \mathrm{II}_{\mathbf{A}}\lambda + \mathrm{III}_{\mathbf{A}}, \tag{4.28}$$

where

$$\mathrm{I}_{\mathbf{A}} = \mathrm{I}_{\mathbf{A}}^{(1)} = \mathrm{tr}\mathbf{A},$$
$$\mathrm{II}_{\mathbf{A}} = \mathrm{I}_{\mathbf{A}}^{(2)} = \frac{1}{2}\left[(\mathrm{tr}\mathbf{A})^2 - \mathrm{tr}\mathbf{A}^2\right],$$
$$\mathrm{III}_{\mathbf{A}} = \mathrm{I}_{\mathbf{A}}^{(3)} = \frac{1}{3}\left[\mathrm{tr}\mathbf{A}^3 - \frac{3}{2}\mathrm{tr}\mathbf{A}^2\mathrm{tr}\mathbf{A} + \frac{1}{2}(\mathrm{tr}\mathbf{A})^3\right] = \det\mathbf{A} \tag{4.29}$$

are the principal invariants (4.26) of the tensor $\mathbf{A}$. They can alternatively be expressed by the Vieta theorem (4.24) in terms of the eigenvalues as follows

$$\mathrm{I}_{\mathbf{A}} = \lambda_1 + \lambda_2 + \lambda_3, \quad \mathrm{II}_{\mathbf{A}} = \lambda_1\lambda_2 + \lambda_2\lambda_3 + \lambda_3\lambda_1, \quad \mathrm{III}_{\mathbf{A}} = \lambda_1\lambda_2\lambda_3. \tag{4.30}$$

The roots of the cubic polynomial (4.28) can be obtained in a closed form by means of the Cardano formula (see, e.g. [5]) as

$$\lambda_k = \frac{1}{3}\left\{\mathrm{I}_{\mathbf{A}} + 2\sqrt{\mathrm{I}_{\mathbf{A}}^2 - 3\mathrm{II}_{\mathbf{A}}}\cos\frac{1}{3}\left[\vartheta + 2\pi(k-1)\right]\right\}, \quad k = 1,2,3, \tag{4.31}$$

where

$$\vartheta = \arccos\left[\frac{2\mathrm{I}_{\mathbf{A}}^3 - 9\mathrm{I}_{\mathbf{A}}\mathrm{II}_{\mathbf{A}} + 27\mathrm{III}_{\mathbf{A}}}{2\left(\mathrm{I}_{\mathbf{A}}^2 - 3\mathrm{II}_{\mathbf{A}}\right)^{3/2}}\right], \quad \mathrm{I}_{\mathbf{A}}^2 - 3\mathrm{II}_{\mathbf{A}} \neq 0. \tag{4.32}$$

In the case $\mathrm{I}_{\mathbf{A}}^2 - 3\mathrm{II}_{\mathbf{A}} = 0$, the eigenvalues of $\mathbf{A}$ take another form

$$\lambda_k = \frac{1}{3}\mathrm{I}_{\mathbf{A}} + \frac{1}{3}\left(27\mathrm{III}_{\mathbf{A}} - \mathrm{I}_{\mathbf{A}}^3\right)^{1/3}\left[\cos\left(\tfrac{2}{3}\pi k\right) + \mathrm{i}\sin\left(\tfrac{2}{3}\pi k\right)\right], \tag{4.33}$$

where $k = 1, 2, 3$.

4.4 Spectral Decomposition and Eigenprojections

The spectral decomposition is a powerful tool for the tensor analysis and tensor algebra. It enables to gain a deeper insight into the properties of second-order tensors and to represent various useful tensor operations in a relatively simple form. In the spectral decomposition, eigenvectors represent one of the most important ingredients.

Theorem 4.2. *The eigenvectors of a second-order tensor corresponding to pairwise distinct eigenvalues are linearly independent.*

Proof. Suppose that these eigenvectors are linearly dependent. Among all possible nontrivial linear relations connecting them we can choose one involving the minimal number, say r, of eigenvectors $\boldsymbol{a}_i \neq \boldsymbol{0}$ $(i = 1, 2, \ldots, r)$. Obviously, $1 < r \leq n$. Thus,

$$\sum_{i=1}^{r} \alpha_i \boldsymbol{a}_i = \boldsymbol{0}, \tag{4.34}$$

where all α_i $(i = 1, 2, \ldots, r)$ are non-zero. We can also write

$$\mathbf{A}\boldsymbol{a}_i = \lambda_i \boldsymbol{a}_i, \quad i = 1, 2, \ldots, r, \tag{4.35}$$

where $\lambda_i \neq \lambda_j$, $(i \neq j = 1, 2, \ldots, r)$. Mapping both sides of (4.34) by $\mathbf{A}$ and taking (4.35) into account we obtain

$$\sum_{i=1}^{r} \alpha_i \mathbf{A}\boldsymbol{a}_i = \sum_{i=1}^{r} \alpha_i \lambda_i \boldsymbol{a}_i = \boldsymbol{0}. \tag{4.36}$$

Multiplying (4.34) by λ_r and subtracting from (4.36) yield

$$\boldsymbol{0} = \sum_{i=1}^{r} \alpha_i \left(\lambda_i - \lambda_r\right) \boldsymbol{a}_i = \sum_{i=1}^{r-1} \alpha_i \left(\lambda_i - \lambda_r\right) \boldsymbol{a}_i.$$

In the latter linear combination none of the coefficients is zero. Thus, we have a linear relation involving only $r - 1$ eigenvectors. This contradicts, however, the earlier assumption that r is the smallest number of eigenvectors satisfying such a relation.

Theorem 4.3. *Let $\boldsymbol{b}_i$ be a left and $\boldsymbol{a}_j$ a right eigenvector associated with distinct eigenvalues $\lambda_i \neq \lambda_j$ of a tensor* $\mathbf{A}$. *Then,*

$$\boldsymbol{b}_i \cdot \boldsymbol{a}_j = 0. \tag{4.37}$$

Proof. With the aid of (1.78) and taking (4.11) into account we can write

$$\boldsymbol{b}_i \mathbf{A} \boldsymbol{a}_j = \boldsymbol{b}_i \cdot \left(\mathbf{A} \boldsymbol{a}_j\right) = \boldsymbol{b}_i \cdot \left(\lambda_j \boldsymbol{a}_j\right) = \lambda_j \boldsymbol{b}_i \cdot \boldsymbol{a}_j.$$

On the other hand, in view of (4.12)

$$\boldsymbol{b}_i \mathbf{A} \boldsymbol{a}_j = \left(\boldsymbol{b}_i \mathbf{A}\right) \cdot \boldsymbol{a}_j = \left(\boldsymbol{b}_i \lambda_i\right) \cdot \boldsymbol{a}_j = \lambda_i \boldsymbol{b}_i \cdot \boldsymbol{a}_j.$$

Subtracting one equation from another one we obtain

$$\left(\lambda_i - \lambda_j\right) \boldsymbol{b}_i \cdot \boldsymbol{a}_j = 0.$$

Since $\lambda_i \neq \lambda_j$ this immediately implies (4.37).

Now, we proceed with the spectral decomposition of a second-order tensor $\mathbf{A}$. First, we consider the case of n simple eigenvalues. Solving the equation systems (4.16) one obtains for every simple eigenvalue λ_i the components of the right eigenvector $\boldsymbol{a}_i$ and the components of the left eigenvector $\boldsymbol{b}_i$ $(i = 1, 2, \ldots, n)$. n right eigenvectors on the one hand and n left eigenvectors on the other hand are linearly independent and form bases of $\mathbb{C}^n$. Obviously, $\boldsymbol{b}_i \cdot \boldsymbol{a}_i \neq 0\, (i = 1, 2, \ldots, n)$ because otherwise it would contradict (4.37) (see Exercise 4.5). Normalizing the eigenvectors we can thus write

$$\boldsymbol{b}_i \cdot \boldsymbol{a}_j = \delta_{ij}, \quad i, j = 1, 2, \ldots, n. \tag{4.38}$$

Accordingly, the bases $\boldsymbol{a}_i$ and $\boldsymbol{b}_i$ are dual to each other such that $\boldsymbol{a}^i = \boldsymbol{b}_i$ and $\boldsymbol{b}^i = \boldsymbol{a}_i\, (i = 1, 2, \ldots, n)$. Now, representing $\mathbf{A}$ with respect to the basis $\boldsymbol{a}_i \otimes \boldsymbol{b}_j$ $(i, j = 1, 2, \ldots, n)$ as $\mathbf{A} = \mathrm{A}^{ij} \boldsymbol{a}_i \otimes \boldsymbol{b}_j$ we obtain with the aid of (1.88), (4.11) and (4.38)

$$\mathrm{A}^{ij} = \boldsymbol{a}^i \mathbf{A} \boldsymbol{b}^j = \boldsymbol{b}_i \mathbf{A} \boldsymbol{a}_j = \boldsymbol{b}_i \cdot \left(\mathbf{A} \boldsymbol{a}_j\right) = \boldsymbol{b}_i \cdot \left(\lambda_j \boldsymbol{a}_j\right) = \lambda_j \delta_{ij},$$

where $i, j = 1, 2, \ldots, n$. Thus,

$$\mathbf{A} = \sum_{i=1}^{n} \lambda_i \boldsymbol{a}_i \otimes \boldsymbol{b}_i. \tag{4.39}$$

Next, we consider second-order tensors with multiple eigenvalues. We assume, however, that the algebraic multiplicity r_i of every eigenvalue λ_i coincides with its geometric multiplicity t_i. In this case we again have n linearly independent right eigenvectors forming a basis of $\mathbb{C}^n$ (Exercise 4.4). We will denote these eigenvectors by $\boldsymbol{a}_i^{(k)}$ $(i = 1, 2, \ldots, s; k = 1, 2, \ldots, r_i)$ where s is the number of pairwise distinct eigenvalues. Constructing the basis $\boldsymbol{b}_j^{(l)}$ dual to $\boldsymbol{a}_i^{(k)}$ such that

$$\boldsymbol{a}_i^{(k)} \cdot \boldsymbol{b}_j^{(l)} = \delta_{ij}\delta^{kl}, \; i, j = 1, 2, \ldots, s; \; k = 1, 2, \ldots, r_i; \; l = 1, 2, \ldots, r_j \quad (4.40)$$

we can write similarly to (4.39)

$$\mathbf{A} = \sum_{i=1}^{s} \lambda_i \sum_{k=1}^{r_i} \boldsymbol{a}_i^{(k)} \otimes \boldsymbol{b}_i^{(k)}. \quad (4.41)$$

The representations of the form (4.39) or (4.41) are called spectral decomposition in diagonal form or, briefly, spectral decomposition. Note that not every second-order tensor $\mathbf{A} \in \mathbf{Lin}^n$ permits the spectral decomposition. The tensors which can be represented by (4.39) or (4.41) are referred to as diagonalizable tensors. For instance, we will show in the next sections that symmetric, skew-symmetric and orthogonal tensors are always diagonalizable. If, however, the algebraic multiplicity of at least one eigenvalue exceeds its geometric multiplicity, the spectral representation is not possible. Such eigenvalues (for which $r_i > t_i$) are called defective eigenvalues. A tensor that has one or more defective eigenvalues is called defective tensor. In Sect. 4.2 we have seen, for example, that the deformation gradient $\mathbf{F}$ represents in the case of simple shear a defective tensor since its triple eigenvalue 1 is defective. Clearly, a simple eigenvalue ($r_i = 1$) cannot be defective. For this reason, a tensor whose all eigenvalues are simple is diagonalizable.

Now, we look again at the spectral decompositions (4.39) and (4.41). With the aid of the abbreviation

$$\mathbf{P}_i = \sum_{k=1}^{r_i} \boldsymbol{a}_i^{(k)} \otimes \boldsymbol{b}_i^{(k)}, \quad i = 1, 2, \ldots, s \quad (4.42)$$

they can be given in a unified form by

$$\mathbf{A} = \sum_{i=1}^{s} \lambda_i \mathbf{P}_i. \quad (4.43)$$

The generally complex tensors $\mathbf{P}_i$ $(i = 1, 2, \ldots, s)$ defined by (4.42) are called eigenprojections. It follows from (4.40) and (4.42) that (Exercise 4.6)

$$\mathbf{P}_i \mathbf{P}_j = \delta_{ij} \mathbf{P}_i, \quad i, j = 1, 2, \ldots, s \quad (4.44)$$

and consequently

$$\mathbf{P}_i\mathbf{A} = \mathbf{A}\mathbf{P}_i = \lambda_i \mathbf{P}_i, \quad i = 1, 2, \ldots, s. \tag{4.45}$$

Bearing in mind that the eigenvectors $\boldsymbol{a}_i^{(k)}$ $(i = 1, 2, \ldots, s; k = 1, 2, \ldots, r_i)$ form a basis of $\mathbb{C}^n$ and taking (4.40) into account we also obtain (Exercise 4.7)

$$\sum_{i=1}^{s} \mathbf{P}_i = \mathbf{I}. \tag{4.46}$$

Due to these properties of eigenprojections (4.42) the spectral representation (4.43) is very suitable for calculating tensor powers, polynomials and other tensor functions defined in terms of power series. Indeed, in view of (4.44) powers of **A** can be expressed by

$$\mathbf{A}^k = \sum_{i=1}^{s} \lambda_i^k \mathbf{P}_i, \quad k = 0, 1, 2, \ldots \tag{4.47}$$

For a tensor polynomial it further yields

$$g(\mathbf{A}) = \sum_{i=1}^{s} g(\lambda_i) \mathbf{P}_i. \tag{4.48}$$

For example, the exponential tensor function (1.114) can thus be represented by

$$\exp(\mathbf{A}) = \sum_{i=1}^{s} \exp(\lambda_i) \mathbf{P}_i. \tag{4.49}$$

For an invertible second-order tensor we can also write

$$\mathbf{A}^{-1} = \sum_{i=1}^{s} \lambda_i^{-1} \mathbf{P}_i, \quad \mathbf{A} \in \mathbf{Inv}^n, \tag{4.50}$$

which implies that $\lambda_i \neq 0\,(i = 1, 2, \ldots, s)$. The latter property generally characterizes all (not necessarily diagonalizable) invertible tensors (see Exercise 4.9).

With the aid of (4.44) and (4.46) the eigenprojections can be obtained without solving the eigenvalue problem in the general form (4.11). To this end, we first consider s polynomial functions $p_i(\lambda)$ $(i = 1, 2, \ldots, s)$ satisfying the following conditions

$$p_i(\lambda_j) = \delta_{ij}, \quad i, j = 1, 2, \ldots, s. \tag{4.51}$$

Thus, by use of (4.48) we obtain

$$p_i(\mathbf{A}) = \sum_{j=1}^{s} p_i(\lambda_j)\,\mathbf{P}_j = \sum_{j=1}^{s} \delta_{ij}\mathbf{P}_j = \mathbf{P}_i, \quad i = 1, 2, \ldots, s. \tag{4.52}$$

Using Lagrange's interpolation formula (see, e.g., [5]) and assuming that $s \neq 1$ one can represent the functions $p_i(\lambda)$ (4.51) by the following polynomials of degree $s-1$:

$$p_i(\lambda) = \prod_{\substack{j=1\\ j\neq i}}^{s} \frac{\lambda - \lambda_j}{\lambda_i - \lambda_j}, \quad i = 1, 2, \ldots, s > 1. \tag{4.53}$$

Considering these expressions in (4.52) we obtain the so-called Sylvester formula as

$$\mathbf{P}_i = \prod_{\substack{j=1\\ j\neq i}}^{s} \frac{\mathbf{A} - \lambda_j \mathbf{I}}{\lambda_i - \lambda_j}, \quad i = 1, 2, \ldots, s > 1. \tag{4.54}$$

Note that according to (4.46), $\mathbf{P}_1 = \mathbf{I}$ in the the case of $s = 1$. With this result in hand the above representation can be generalized by

$$\mathbf{P}_i = \delta_{1s}\mathbf{I} + \prod_{\substack{j=1\\ j\neq i}}^{s} \frac{\mathbf{A} - \lambda_j \mathbf{I}}{\lambda_i - \lambda_j}, \quad i = 1, 2, \ldots, s. \tag{4.55}$$

Writing out the product on the right hand side of (4.55) also delivers (see, e.g., [48])

$$\mathbf{P}_i = \frac{1}{D_i} \sum_{p=0}^{s-1} \iota_{i\,s-p-1}\mathbf{A}^p, \quad i = 1, 2, \ldots, s, \tag{4.56}$$

where $\iota_{i0} = 1$,

$$\iota_{ip} = (-1)^p \sum_{1 \leq o_1 \leq \cdots \leq o_p \leq s} \lambda_{o_1} \cdots \lambda_{o_p} \left(1 - \delta_{io_1}\right) \cdots \left(1 - \delta_{io_p}\right),$$

$$D_i = \delta_{1s} + \prod_{\substack{j=1\\ j\neq i}}^{s} \left(\lambda_i - \lambda_j\right), \quad p = 1, 2, \ldots, s-1, \quad i = 1, 2, \ldots, s. \tag{4.57}$$

4.5 Spectral Decomposition of Symmetric Second-Order Tensors

We begin with some useful theorems concerning eigenvalues and eigenvectors of symmetric tensors.

Theorem 4.4. *The eigenvalues of a symmetric second-order tensor* $\mathbf{M} \in \mathbf{Sym}^n$ *are real, the eigenvectors belong to* $\mathbb{E}^n$.

Proof. Let λ be an eigenvalue of $\mathbf{M}$ and $\boldsymbol{a}$ a corresponding eigenvector such that according to (4.11)

$$\mathbf{M}\boldsymbol{a} = \lambda \boldsymbol{a}.$$

The complex conjugate counterpart of this equation is

$$\overline{\mathbf{M}}\,\overline{\boldsymbol{a}} = \overline{\lambda}\,\overline{\boldsymbol{a}}.$$

Taking into account that $\mathbf{M}$ is real and symmetric such that $\overline{\mathbf{M}} = \mathbf{M}$ and $\mathbf{M}^{\mathrm{T}} = \mathbf{M}$ we obtain in view of (1.115)

$$\overline{\boldsymbol{a}}\,\mathbf{M} = \overline{\lambda}\,\overline{\boldsymbol{a}}.$$

Hence, one can write

$$\begin{aligned} 0 = \overline{\boldsymbol{a}}\mathbf{M}\boldsymbol{a} - \overline{\boldsymbol{a}}\mathbf{M}\boldsymbol{a} &= \overline{\boldsymbol{a}} \cdot (\mathbf{M}\boldsymbol{a}) - (\overline{\boldsymbol{a}}\mathbf{M}) \cdot \boldsymbol{a} \\ &= \lambda\,(\overline{\boldsymbol{a}} \cdot \boldsymbol{a}) - \overline{\lambda}\,(\overline{\boldsymbol{a}} \cdot \boldsymbol{a}) = \left(\lambda - \overline{\lambda}\right)(\overline{\boldsymbol{a}} \cdot \boldsymbol{a}). \end{aligned}$$

Bearing in mind that $\boldsymbol{a} \neq \boldsymbol{0}$ and taking (4.9) into account we conclude that $\overline{\boldsymbol{a}} \cdot \boldsymbol{a} > 0$. Hence, $\overline{\lambda} = \lambda$. The components of $\boldsymbol{a}$ with respect to a basis $\mathcal{G} = \{\boldsymbol{g}_1, \boldsymbol{g}_2, \ldots, \boldsymbol{g}_n\}$ in $\mathbb{E}^n$ are real since they represent a solution of the linear equation system $(4.16)_1$ with real coefficients. Therefore, $\boldsymbol{a} \in \mathbb{E}^n$.

Theorem 4.5. *Eigenvectors of a symmetric second-order tensor corresponding to distinct eigenvalues are mutually orthogonal.*

Proof. According to Theorem 4.3 scalar product of a right and a left eigenvector associated with distinct eigenvalues is zero. However, for a symmetric tensor every right eigenvector represents the left eigenvector associated with the same eigenvalue and vice versa. Taking also into account that the eigenvectors are real we infer that right (left) eigenvectors associated with distinct eigenvalues are mutually orthogonal.

Theorem 4.6. *Let* λ_i *be an eigenvalue of a symmetric second order tensor* $\mathbf{M}$. *Then, the algebraic and geometric multiplicity of* λ_i *coincide.*

Proof. Let $\boldsymbol{a}_k \in \mathbb{E}^n$ $(k = 1, 2, \ldots, t_i)$ be all linearly independent eigenvectors associated with λ_i, while t_i and r_i denote its geometric and algebraic multiplicity, respectively. Every linear combination of $\boldsymbol{a}_k$ with not all zero coefficients is again an eigenvector associated with λ_i. Indeed,

$$\mathbf{M}\sum_{k=1}^{t_i}\alpha_k\boldsymbol{a}_k=\sum_{k=1}^{t_i}\alpha_k\left(\mathbf{M}\boldsymbol{a}_k\right)=\sum_{k=1}^{t_i}\alpha_k\lambda_i\boldsymbol{a}_k=\lambda_i\sum_{k=1}^{t_i}\alpha_k\boldsymbol{a}_k. \tag{4.58}$$

According to Theorem 1.4 the set of vectors $\boldsymbol{a}_k$ $(k=1,2,\ldots,t_i)$ can be completed to a basis of $\mathbb{E}^n$. With the aid of the Gram-Schmidt procedure described in Chap. 1 (Sect. 1.4) this basis can be transformed to an orthonormal basis $\boldsymbol{e}_l$ $(l=1,2,\ldots,n)$. Since the vectors $\boldsymbol{e}_j$ $(j=1,2,\ldots,t_i)$ are linear combinations of $\boldsymbol{a}_k$ $(k=1,2,\ldots,t_i)$ they likewise represent eigenvectors of $\mathbf{M}$ associated with λ_i. Further, we represent the tensor $\mathbf{M}$ with respect to the basis $\boldsymbol{e}_l\otimes\boldsymbol{e}_m$ $(l,m=1,2,\ldots,n)$. In view of the identities $\mathbf{M}\boldsymbol{e}_k=\boldsymbol{e}_k\mathbf{M}=\lambda_i\boldsymbol{e}_k$ $(k=1,2,\ldots,t_i)$ and keeping in mind the symmetry of $\mathbf{M}$ we can write using (1.88)

$$\mathbf{M}=\lambda_i\sum_{k=1}^{t_i}\boldsymbol{e}_k\otimes\boldsymbol{e}_k+\sum_{l,m=t_i+1}^{n}\mathrm{M}'_{lm}\boldsymbol{e}_l\otimes\boldsymbol{e}_m. \tag{4.59}$$

Thus, the characteristic polynomial of $\mathbf{M}$ can be given as

$$p_{\mathbf{M}}(\lambda)=\left|\mathrm{M}'_{lm}-\lambda\delta_{lm}\right|\left(\lambda_i-\lambda\right)^{t_i}, \tag{4.60}$$

which implies that $r_i\geq t_i$.

Now, we consider the vector space $\mathbb{E}^{n-t_i}$ of all linear combinations of the vectors $\boldsymbol{e}_l$ $(l=t_i+1,\ldots,n)$. The tensor

$$\mathbf{M}'=\sum_{l,m=t_i+1}^{n}\mathrm{M}'_{lm}\boldsymbol{e}_l\otimes\boldsymbol{e}_m$$

represents a linear mapping of this space into itself. The eigenvectors of $\mathbf{M}'$ are linear combinations of $\boldsymbol{e}_l$ $(l=t_i+1,\ldots,n)$ and therefore are linearly independent of $\boldsymbol{e}_k$ $(k=1,2,\ldots,t_i)$. Consequently, λ_i is not an eigenvalue of $\mathbf{M}'$. Otherwise, the eigenvector corresponding to this eigenvalue λ_i would be linearly independent of $\boldsymbol{e}_k$ $(k=1,2,\ldots,t_i)$ which contradicts the previous assumption. Thus, all the roots of the characteristic polynomial of this tensor

$$p_{\mathbf{M}'}(\lambda)=\left|\mathrm{M}'_{lm}-\lambda\delta_{lm}\right|$$

differ from λ_i. In view of (4.60) this implies that $r_i=t_i$.

As a result of this theorem and in view of (4.41) and (4.43), the spectral decomposition of a symmetric second-order tensor can be given by

$$\mathbf{M}=\sum_{i=1}^{s}\lambda_i\sum_{k=1}^{r_i}\boldsymbol{a}_i^{(k)}\otimes\boldsymbol{a}_i^{(k)}=\sum_{i=1}^{s}\lambda_i\mathbf{P}_i,\quad\mathbf{M}\in\mathbf{Sym}^n, \tag{4.61}$$

in terms of the real symmetric eigenprojections

$$\mathbf{P}_i = \sum_{k=1}^{r_i} \boldsymbol{a}_i^{(k)} \otimes \boldsymbol{a}_i^{(k)}, \tag{4.62}$$

where the eigenvectors $\boldsymbol{a}_i^{(k)}$ form an orthonormal basis in $\mathbb{E}^n$ so that

$$\boldsymbol{a}_i^{(k)} \cdot \boldsymbol{a}_j^{(l)} = \delta_{ij}\delta^{kl}, \tag{4.63}$$

where $i, j = 1, 2, \ldots, s$; $k = 1, 2, \ldots, r_i$; $l = 1, 2, \ldots, r_j$.

Of particular interest in continuum mechanics are the so-called positive-definite second-order tensors. They are defined by the following condition

$$\boldsymbol{x}\mathbf{A}\boldsymbol{x} > 0, \quad \forall \boldsymbol{x} \in \mathbb{E}^n, \quad \boldsymbol{x} \neq \boldsymbol{0}. \tag{4.64}$$

For a symmetric tensor $\mathbf{M}$ the above condition implies that all its eigenvalues are positive. Indeed, let $\boldsymbol{a}_i$ be a unit eigenvector associated with the eigenvalue λ_i $(i = 1, 2, \ldots, n)$. In view of (4.64) one can thus write

$$\lambda_i = \boldsymbol{a}_i \mathbf{M} \boldsymbol{a}_i > 0, \quad i = 1, 2, \ldots, n. \tag{4.65}$$

This allows to define powers of a symmetric positive-definite tensor with a real exponent as follows

$$\mathbf{M}^\alpha = \sum_{i=1}^{s} \lambda_i^\alpha \mathbf{P}_i, \quad \alpha \in \mathbb{R}. \tag{4.66}$$

4.6 Spectral Decomposition of Orthogonal and Skew-Symmetric Second-Order Tensors

We begin with the orthogonal tensors $\mathbf{Q} \in \mathbf{Orth}^n$ defined by the condition (1.135). For every eigenvector $\boldsymbol{a}$ and the corresponding eigenvalue λ we can write

$$\mathbf{Q}\boldsymbol{a} = \lambda \boldsymbol{a}, \quad \mathbf{Q}\overline{\boldsymbol{a}} = \overline{\lambda}\,\overline{\boldsymbol{a}}, \tag{4.67}$$

because $\mathbf{Q}$ is by definition a real tensor such that $\overline{\mathbf{Q}} = \mathbf{Q}$. Mapping both sides of these vector equations by $\mathbf{Q}^{\mathrm{T}}$ and taking (1.115) into account we have

$$\boldsymbol{a}\mathbf{Q} = \lambda^{-1}\boldsymbol{a}, \quad \overline{\boldsymbol{a}}\mathbf{Q} = \overline{\lambda}^{-1}\overline{\boldsymbol{a}}. \tag{4.68}$$

Thus, every right eigenvector of an orthogonal tensor represents its left eigenvector associated with the inverse eigenvalue.

Now, we consider the product $\overline{\boldsymbol{a}}\mathbf{Q}\boldsymbol{a}$. With the aid of $(4.67)_1$ and $(4.68)_2$ we obtain

$$\overline{\boldsymbol{a}}\mathbf{Q}\boldsymbol{a} = \lambda \left(\overline{\boldsymbol{a}} \cdot \boldsymbol{a}\right) = \overline{\lambda}^{-1} \left(\overline{\boldsymbol{a}} \cdot \boldsymbol{a}\right). \tag{4.69}$$

Since, however, $\overline{\boldsymbol{a}} \cdot \boldsymbol{a} = \boldsymbol{a} \cdot \overline{\boldsymbol{a}} > 0$ according to (4.8) and (4.9) we infer that

$$\lambda\overline{\lambda} = 1. \tag{4.70}$$

Thus, all eigenvalues of an orthogonal tensor have absolute value 1 so that we can write

$$\lambda = e^{\omega \mathrm{i}} = \cos\omega + \mathrm{i}\sin\omega. \tag{4.71}$$

By virtue of (4.70) one can further rewrite (4.68) as

$$\boldsymbol{a}\mathbf{Q} = \overline{\lambda}\boldsymbol{a}, \quad \overline{\boldsymbol{a}}\mathbf{Q} = \lambda\overline{\boldsymbol{a}}. \tag{4.72}$$

If further $\lambda \neq \lambda^{-1} = \overline{\lambda}$ or, in other words, λ is neither $+1$ nor -1, Theorem 4.3 immediately implies the relations

$$\boldsymbol{a} \cdot \boldsymbol{a} = 0, \quad \overline{\boldsymbol{a}} \cdot \overline{\boldsymbol{a}} = 0, \quad \lambda \neq \lambda^{-1} \tag{4.73}$$

indicating that $\boldsymbol{a}$ and consequently $\overline{\boldsymbol{a}}$ are complex (definitely not real) vectors. Using the representation

$$\boldsymbol{a} = \frac{1}{\sqrt{2}}\left(\boldsymbol{p} + \mathrm{i}\boldsymbol{q}\right), \quad \boldsymbol{p}, \boldsymbol{q} \in \mathbb{E}^n \tag{4.74}$$

and applying (4.8) one can write

$$\|\boldsymbol{p}\| = \|\boldsymbol{q}\| = 1, \quad \boldsymbol{p} \cdot \boldsymbol{q} = 0, \tag{4.75}$$

so that $\overline{\boldsymbol{a}} \cdot \boldsymbol{a} = 1/2\left(\boldsymbol{p} \cdot \boldsymbol{p} + \boldsymbol{q} \cdot \boldsymbol{q}\right) = 1$.

Summarizing these results we conclude that every complex (definitely not real) eigenvalue λ of an orthogonal tensor comes in pair with its complex conjugate counterpart $\overline{\lambda} = \lambda^{-1}$. If $\boldsymbol{a}$ is a right eigenvector associated with λ, then $\overline{\boldsymbol{a}}$ is its left eigenvector. For $\overline{\lambda}$, $\boldsymbol{a}$ is, vice versa, the left eigenvector and $\overline{\boldsymbol{a}}$ the right one.

Next, we show that the algebraic and geometric multiplicities of every eigenvalue of an orthogonal tensor $\mathbf{Q}$ coincide. Let $\tilde{\boldsymbol{a}}_k$ $(k = 1, 2, \ldots, t_i)$ be all linearly independent right eigenvectors associated with an eigenvalue λ_i. According to Theorem 1.4 these vectors can be completed to a basis of $\mathbb{C}^n$. With the aid of the Gram-Schmidt procedure (see Exercise 4.17) a linear combination of this basis can be constructed in such a way that $\boldsymbol{a}_k \cdot \overline{\boldsymbol{a}}_l = \delta_{kl}$ $(k, l = 1, 2, \ldots, n)$. Since the vectors $\boldsymbol{a}_k$ $(k = 1, 2, \ldots, t_i)$ are linear combinations of $\tilde{\boldsymbol{a}}_k$ $(k = 1, 2, \ldots, t_i)$ they likewise represent eigenvectors of $\mathbf{Q}$ associated with λ_i. Thus, representing $\mathbf{Q}$ with respect to the basis $\boldsymbol{a}_k \otimes \overline{\boldsymbol{a}}_l$ $(k, l = 1, 2, \ldots, n)$ we can write

$$\mathbf{Q} = \lambda_i \sum_{k=1}^{t_i} \boldsymbol{a}_k \otimes \overline{\boldsymbol{a}}_k + \sum_{l,m=t_i+1}^{n} \mathrm{Q}'_{lm} \boldsymbol{a}_l \otimes \overline{\boldsymbol{a}}_m.$$

Comparing this representation with (4.59) and using the same reasoning as applied for the proof of Theorem 4.6 we infer that λ_i cannot be an eigenvalue of $\mathbf{Q}' = \sum_{l,m=t_i+1}^{n} \mathrm{Q}'_{lm} \boldsymbol{a}_l \otimes \overline{\boldsymbol{a}}_m$. This means that the algebraic multiplicity r_i of λ_i coincides with its geometric multiplicity t_i. Thus, every orthogonal tensor $\mathbf{Q} \in \mathbf{Orth}^n$ is characterized by exactly n linearly independent eigenvectors forming a basis of $\mathbb{C}^n$. Using this fact the spectral decomposition of $\mathbf{Q}$ can be given by

$$\mathbf{Q} = \sum_{k=1}^{r_{+1}} \boldsymbol{a}_{+1}^{(k)} \otimes \boldsymbol{a}_{+1}^{(k)} - \sum_{l=1}^{r_{-1}} \boldsymbol{a}_{-1}^{(l)} \otimes \boldsymbol{a}_{-1}^{(l)} + \sum_{i=1}^{s} \left\{ \lambda_i \sum_{k=1}^{r_i} \boldsymbol{a}_i^{(k)} \otimes \overline{\boldsymbol{a}}_i^{(k)} + \overline{\lambda}_i \sum_{k=1}^{r_i} \overline{\boldsymbol{a}}_i^{(k)} \otimes \boldsymbol{a}_i^{(k)} \right\}, \tag{4.76}$$

where r_{+1} and r_{-1} denote the algebraic multiplicities of real eigenvalues $+1$ and -1, respectively, while $\boldsymbol{a}_{+1}^{(k)}$ $(k = 1, 2, \ldots, r_{+1})$ and $\boldsymbol{a}_{-1}^{(l)}$ $(l = 1, 2, \ldots, r_{-1})$ are the corresponding orthonormal real eigenvectors. s is the number of complex conjugate pairs of eigenvalues $\lambda_i = \cos\omega_i \pm \mathrm{i}\sin\omega_i$ with distinct arguments ω_i and the multiplicities r_i. The associated eigenvectors $\boldsymbol{a}_i^{(k)}$ and $\overline{\boldsymbol{a}}_i^{(k)}$ obey the following relations (see also Exercise 4.18)

$$\boldsymbol{a}_i^{(k)} \cdot \boldsymbol{a}_{+1}^{(o)} = 0, \quad \boldsymbol{a}_i^{(k)} \cdot \boldsymbol{a}_{-1}^{(p)} = 0, \quad \boldsymbol{a}_i^{(k)} \cdot \overline{\boldsymbol{a}}_j^{(l)} = \delta_{ij}\delta^{kl}, \quad \boldsymbol{a}_i^{(k)} \cdot \boldsymbol{a}_i^{(m)} = 0, \tag{4.77}$$

where $i, j = 1, 2, \ldots, s$; $k, m = 1, 2, \ldots, r_i$; $l = 1, 2, \ldots, r_j$; $o = 1, 2, \ldots, r_{+1}$; $p = 1, 2, \ldots, r_{-1}$. Using the representations (4.74) and (4.71) the spectral decomposition (4.76) can alternatively be written as

$$\mathbf{Q} = \sum_{k=1}^{r_{+1}} \boldsymbol{a}_{+1}^{(k)} \otimes \boldsymbol{a}_{+1}^{(k)} + \sum_{i=1}^{s} \cos\omega_i \sum_{k=1}^{r_i} \left(\boldsymbol{p}_i^{(k)} \otimes \boldsymbol{p}_i^{(k)} + \boldsymbol{q}_i^{(k)} \otimes \boldsymbol{q}_i^{(k)} \right) - \sum_{l=1}^{r_{-1}} \boldsymbol{a}_{-1}^{(l)} \otimes \boldsymbol{a}_{-1}^{(l)} + \sum_{i=1}^{s} \sin\omega_i \sum_{k=1}^{r_i} \left(\boldsymbol{p}_i^{(k)} \otimes \boldsymbol{q}_i^{(k)} - \boldsymbol{q}_i^{(k)} \otimes \boldsymbol{p}_i^{(k)} \right). \tag{4.78}$$

Now, we turn our attention to skew-symmetric tensors $\mathbf{W} \in \mathbf{Skew}^n$ as defined in (1.155). Instead of (4.68) and (4.69) we have in this case

$$\boldsymbol{a}\mathbf{W} = -\lambda \boldsymbol{a}, \quad \overline{\boldsymbol{a}}\,\mathbf{W} = -\overline{\lambda}\,\overline{\boldsymbol{a}}, \tag{4.79}$$

$$\overline{\boldsymbol{a}}\mathbf{W}\boldsymbol{a} = \lambda \left(\overline{\boldsymbol{a}} \cdot \boldsymbol{a} \right) = -\overline{\lambda} \left(\overline{\boldsymbol{a}} \cdot \boldsymbol{a} \right) \tag{4.80}$$

and consequently

$$\lambda = -\overline{\lambda}. \tag{4.81}$$

Thus, the eigenvalues of $\mathbf{W}$ are either zero or imaginary. The latter ones come in pairs with the complex conjugate like in the case of orthogonal tensors. Similarly to (4.76) and (4.78) we thus obtain

$$\begin{aligned}\mathbf{W} &= \sum_{i=1}^{s} \omega_i \mathrm{i} \sum_{k=1}^{r_i} \left(\boldsymbol{a}_i^{(k)} \otimes \overline{\boldsymbol{a}}_i^{(k)} - \overline{\boldsymbol{a}}_i^{(k)} \otimes \boldsymbol{a}_i^{(k)} \right) \\ &= \sum_{i=1}^{s} \omega_i \sum_{k=1}^{r_i} \left(\boldsymbol{p}_i^{(k)} \otimes \boldsymbol{q}_i^{(k)} - \boldsymbol{q}_i^{(k)} \otimes \boldsymbol{p}_i^{(k)} \right),\end{aligned} \tag{4.82}$$

where s denotes the number of pairwise distinct imaginary eigenvalues $\omega_i \mathrm{i}$ while the associated eigenvectors $\boldsymbol{a}_i^{(k)}$ and $\overline{\boldsymbol{a}}_i^{(k)}$ are subject to the restrictions $(4.77)_{3,4}$.

Orthogonal tensors in three-dimensional space. In the three-dimen-sional case $\mathbf{Q} \in \mathbf{Orth}^3$, at least one of the eigenvalues is real, since complex eigenvalues of orthogonal tensors appear in pairs with the complex conjugate. Hence, we can write

$$\lambda_1 = \pm 1, \quad \lambda_2 = \mathrm{e}^{\mathrm{i}\omega} = \cos\omega + \mathrm{i}\sin\omega, \quad \lambda_3 = \mathrm{e}^{-\mathrm{i}\omega} = \cos\omega - \mathrm{i}\sin\omega. \tag{4.83}$$

In the case $\sin\omega = 0$ all three eigenvalues become real. The principal invariants (4.30) take thus the form

$$\begin{aligned}\mathrm{I}_\mathbf{Q} &= \lambda_1 + 2\cos\omega = \pm 1 + 2\cos\omega, \\ \mathrm{II}_\mathbf{Q} &= 2\lambda_1 \cos\omega + 1 = \lambda_1 \mathrm{I}_\mathbf{Q} = \pm \mathrm{I}_\mathbf{Q}, \\ \mathrm{III}_\mathbf{Q} &= \lambda_1 = \pm 1.\end{aligned} \tag{4.84}$$

The spectral representation (4.76) takes the form

$$\mathbf{Q} = \pm \boldsymbol{a}_1 \otimes \boldsymbol{a}_1 + (\cos\omega + \mathrm{i}\sin\omega)\, \boldsymbol{a} \otimes \overline{\boldsymbol{a}} + (\cos\omega - \mathrm{i}\sin\omega)\, \overline{\boldsymbol{a}} \otimes \boldsymbol{a}, \tag{4.85}$$

where $\boldsymbol{a}_1 \in \mathbb{E}^3$ and $\boldsymbol{a} \in \mathbb{C}^3$ is given by (4.74) and (4.75). Taking into account that by (4.77)

$$\boldsymbol{a}_1 \cdot \boldsymbol{a} = \boldsymbol{a}_1 \cdot \boldsymbol{p} = \boldsymbol{a}_1 \cdot \boldsymbol{q} = 0 \tag{4.86}$$

we can set

$$\boldsymbol{a}_1 = \boldsymbol{q} \times \boldsymbol{p}. \tag{4.87}$$

Substituting (4.74) into (4.85) we also obtain

$$\mathbf{Q} = \pm \boldsymbol{a}_1 \otimes \boldsymbol{a}_1 + \cos\omega\, (\boldsymbol{p} \otimes \boldsymbol{p} + \boldsymbol{q} \otimes \boldsymbol{q}) + \sin\omega\, (\boldsymbol{p} \otimes \boldsymbol{q} - \boldsymbol{q} \otimes \boldsymbol{p}). \tag{4.88}$$

By means of the vector identity (1.136) and considering (1.66), (1.92) and (4.87) it finally leads to

$$\mathbf{Q} = \cos\omega\mathbf{I} + \sin\omega\hat{\boldsymbol{a}}_1 + (\pm 1 - \cos\omega)\, \boldsymbol{a}_1 \otimes \boldsymbol{a}_1. \tag{4.89}$$

Comparing this representation with (1.73) we observe that any orthogonal tensor $\mathbf{Q} \in \mathbf{Orth}^3$ describes a rotation in three-dimensional space if $\mathrm{III}_\mathbf{Q} = \lambda_1 = 1$. The eigenvector $\boldsymbol{a}_1$ corresponding to the eigenvalue 1 specifies the rotation axis. In this case, $\mathbf{Q}$ is referred to as a proper orthogonal tensor.

Skew-symmetric tensors in three-dimensional space. For a skew-symmetric tensor $\mathbf{W} \in \mathbf{Skew}^3$ we can write in view of (4.81)

$$\lambda_1 = 0, \quad \lambda_2 = \omega\mathrm{i}, \quad \lambda_3 = -\omega\mathrm{i}. \tag{4.90}$$

Similarly to (4.84) we further obtain (see Exercise 4.19)

$$\mathrm{I}_\mathbf{W} = 0, \quad \mathrm{II}_\mathbf{W} = \frac{1}{2}\|\mathbf{W}\|^2 = \omega^2, \quad \mathrm{III}_\mathbf{W} = 0. \tag{4.91}$$

The spectral representation (4.82) takes the form

$$\mathbf{W} = \omega\mathrm{i}\,(\boldsymbol{a} \otimes \bar{\boldsymbol{a}} - \bar{\boldsymbol{a}} \otimes \boldsymbol{a}) = \omega\,(\boldsymbol{p} \otimes \boldsymbol{q} - \boldsymbol{q} \otimes \boldsymbol{p}), \tag{4.92}$$

where $\boldsymbol{a}$, $\boldsymbol{p}$ and $\boldsymbol{q}$ are again related by (4.74) and (4.75). With the aid of the abbreviation

$$\boldsymbol{w} = \omega\boldsymbol{a}_1 = \omega\boldsymbol{q} \times \boldsymbol{p} \tag{4.93}$$

and bearing (1.169) in mind we finally arrive at the representation (1.158)

$$\mathbf{W} = \hat{\boldsymbol{w}}. \tag{4.94}$$

Thus, the axial vector $\boldsymbol{w}$ (4.93) of the skew-symmetric tensor $\mathbf{W}$ (4.92) in three-dimensional space represents its eigenvector corresponding to the zero eigenvalue in accordance with (1.160).

4.7 Cayley-Hamilton Theorem

Theorem 4.7. *Let $p_\mathbf{A}(\lambda)$ be the characteristic polynomial of a second-order tensor $\mathbf{A} \in \mathbf{Lin}^n$. Then,*

$$p_\mathbf{A}(\mathbf{A}) = \sum_{k=0}^{n} (-1)^{n-k}\, \mathrm{I}_\mathbf{A}^{(k)} \mathbf{A}^{n-k} = \mathbf{0}. \tag{4.95}$$

Proof. As a proof (see, e.g., [12]) we show that

$$p_{\mathbf{A}}(\mathbf{A})\,\boldsymbol{x} = \boldsymbol{0}, \quad \forall \boldsymbol{x} \in \mathbb{E}^n. \tag{4.96}$$

For $\boldsymbol{x} = \boldsymbol{0}$ it is trivial, so we suppose that $\boldsymbol{x} \neq \boldsymbol{0}$. Consider the vectors

$$\boldsymbol{y}_i = \mathbf{A}^{i-1}\boldsymbol{x}, \quad i = 1, 2, \ldots. \tag{4.97}$$

Obviously, there is an integer number k such that the vectors $\boldsymbol{y}_1, \boldsymbol{y}_2, \ldots, \boldsymbol{y}_k$ are linearly independent, but

$$a_1\boldsymbol{y}_1 + a_2\boldsymbol{y}_2 + \ldots + a_k\boldsymbol{y}_k + \mathbf{A}^k\boldsymbol{x} = \boldsymbol{0}. \tag{4.98}$$

Note that $1 \leq k \leq n$. If $k \neq n$ we can complete the vectors $\boldsymbol{y}_i$ $(i = 1, 2, \ldots, k)$ to a basis $\boldsymbol{y}_i$ $(i = 1, 2, \ldots, n)$ of $\mathbb{E}^n$. Let $\mathbf{A} = \mathrm{A}^{i}_{\cdot j}\boldsymbol{y}_i \otimes \boldsymbol{y}^j$, where the vectors $\boldsymbol{y}^i$ form the basis dual to $\boldsymbol{y}_i$ $(i = 1, 2, \ldots, n)$. By virtue of (4.97) and (4.98) we can write

$$\mathbf{A}\boldsymbol{y}_i = \begin{cases} \boldsymbol{y}_{i+1} & \text{if } i < k, \\ \mathbf{A}^k\boldsymbol{x} = -\sum\limits_{j=1}^{k} a_j\boldsymbol{y}_j & \text{if } i = k. \end{cases} \tag{4.99}$$

The components of $\mathbf{A}$ can thus be given by

$$\left[\mathrm{A}^{i}_{\cdot j}\right] = \left[\boldsymbol{y}^i\mathbf{A}\boldsymbol{y}_j\right] = \left[\begin{array}{cc} \begin{array}{ccccc} 0 & 0 & \ldots & 0 & -a_1 \\ 1 & 0 & \ldots & 0 & -a_2 \\ \vdots & \vdots & \ddots & \vdots & \vdots \\ 0 & 0 & \ldots & 1 & -a_k \end{array} & \mathrm{A}' \\ \hline 0 & \mathrm{A}'' \end{array}\right], \tag{4.100}$$

where A' and A'' denote some submatrices. Therefore, the characteristic polynomial of $\mathbf{A}$ takes the form

$$p_{\mathbf{A}}(\lambda) = p_{\mathrm{A}''}(\lambda) \begin{vmatrix} -\lambda & 0 \ldots 0 & -a_1 \\ 1 & -\lambda \ldots 0 & -a_2 \\ \vdots & \vdots \ddots \vdots & \vdots \\ 0 & 0 \ldots 1 & -a_k - \lambda \end{vmatrix}, \tag{4.101}$$

where $p_{\mathrm{A}''}(\lambda) = \det\left(\mathrm{A}'' - \lambda\mathrm{I}\right)$. By means of the Laplace expansion rule (see, e.g., [5]) we expand the determinant in (4.101) along the last column, which yields

$$p_{\mathbf{A}}(\lambda) = p_{\mathrm{A}''}(\lambda)(-1)^k\left(a_1 + a_2\lambda + \ldots + a_k\lambda^{k-1} + \lambda^k\right). \tag{4.102}$$

Bearing (4.97) and (4.98) in mind we finally prove (4.96) by

$$p_{\mathbf{A}}(\mathbf{A})\,\boldsymbol{x} = (-1)^k\, p_{\mathbf{A}''}(\mathbf{A})\left(a_1\mathbf{I} + a_2\mathbf{A} + \ldots + a_k\mathbf{A}^{k-1} + \mathbf{A}^k\right)\boldsymbol{x}$$

$$= (-1)^k\, p_{\mathbf{A}''}(\mathbf{A})\left(a_1\boldsymbol{x} + a_2\mathbf{A}\boldsymbol{x} + \ldots + a_k\mathbf{A}^{k-1}\boldsymbol{x} + \mathbf{A}^k\boldsymbol{x}\right)$$

$$= (-1)^k\, p_{\mathbf{A}''}(\mathbf{A})\left(a_1\boldsymbol{y}_1 + a_2\boldsymbol{y}_2 + \ldots + a_k\boldsymbol{y}_k + \mathbf{A}^k\boldsymbol{x}\right) = \boldsymbol{0}.$$

Exercises

4.1. Evaluate eigenvalues and eigenvectors of the right Cauchy-Green tensor $\mathbf{C} = \mathbf{F}^{\mathrm{T}}\mathbf{F}$ in the case of simple shear, where $\mathbf{F}$ is defined by (2.69).

4.2. Let $\boldsymbol{g}_i$ $(i = 1, 2, 3)$ be linearly independent vectors in $\mathbb{E}^3$. Prove that for any second order tensor $\mathbf{A} \in \mathbf{Lin}^3$

$$\det\mathbf{A} = \frac{[\mathbf{A}\boldsymbol{g}_1\,\mathbf{A}\boldsymbol{g}_2\,\mathbf{A}\boldsymbol{g}_3]}{[\boldsymbol{g}_1\boldsymbol{g}_2\boldsymbol{g}_3]}. \tag{4.103}$$

4.3. Prove identity $(4.29)_3$ using Newton's identities (4.26).

4.4. Prove that eigenvectors $\boldsymbol{a}_i^{(k)}$ $(i = 1, 2, \ldots, s; k = 1, 2, \ldots, t_i)$ of a second order tensor $\mathbf{A} \in \mathbf{Lin}^n$ are linearly independent and form a basis of $\mathbb{C}^n$ if for every eigenvalue the algebraic and geometric multiplicities coincide so that $r_i = t_i$ $(i = 1, 2, \ldots, s)$.

4.5. Generalize the proof of Exercise 1.8 for complex vectors in $\mathbb{C}^n$.

4.6. Prove identity (4.44) using (4.40) and (4.42).

4.7. Prove identity (4.46) taking (4.40) and (4.42) into account and using the results of Exercise 4.4.

4.8. Prove the identity $\det[\exp(\mathbf{A})] = \exp(\mathrm{tr}\mathbf{A})$.

4.9. Prove that a second-order tensor is invertible if and only if all its eigenvalues are non-zero.

4.10. Let λ_i be an eigenvalue of a tensor $\mathbf{A} \in \mathbf{Inv}^n$. Show that λ_i^{-1} represents then the eigenvalue of $\mathbf{A}^{-1}$.

4.11. Show that the tensor $\mathbf{MN}$ is diagonalizable if $\mathbf{M}, \mathbf{N} \in \mathbf{Sym}^n$ and at least one of the tensors $\mathbf{M}$ or $\mathbf{N}$ is positive-definite.

4.12. Verify the Sylvester formula for $s = 3$ by inserting (4.43) and (4.46) into (4.55).

4.13. Represent eigenprojections of the right Cauchy-Green tensor in the case of simple shear using the results of Exercise 4.1 by (4.42) and alternatively by the Sylvester formula (4.55). Compare both representations.

4.14. Calculate eigenvalues and eigenprojections of the tensor $\mathbf{A} = \mathrm{A}^i_{\cdot j} \boldsymbol{e}_i \otimes \boldsymbol{e}^j$, where

$$\left[\mathrm{A}^i_{\cdot j}\right] = \begin{bmatrix} -2 & 2 & 2 \\ 2 & 1 & 4 \\ 2 & 4 & 1 \end{bmatrix}.$$

Apply the Cardano formula (4.31) and Sylvester formula (4.55).

4.15. Calculate the exponential of the tensor **A** given in Exercise 4.14 using the spectral representation in terms of eigenprojections (4.43).

4.16. Calculate eigenvectors of the tensor **A** defined in Exercise 4.14. Express eigenprojections by (4.42) and compare the results with those obtained by the Sylvester formula (Exercise 4.14).

4.17. Let $\boldsymbol{c}_i$ $(i = 1, 2, \ldots, m) \in \mathbb{C}^n$ be a set of linearly independent complex vectors. Using the (Gram-Schmidt) procedure described in Chap. 1 (Sect. 1.4), construct linear combinations of these vectors, say $\boldsymbol{a}_i$ $(i = 1, 2, \ldots, m)$, again linearly independent, in such a way that $\boldsymbol{a}_i \cdot \overline{\boldsymbol{a}}_j = \delta_{ij}$ $(i, j = 1, 2, \ldots, m)$.

4.18. Let $\boldsymbol{a}_i^{(k)}$ $(k = 1, 2, \ldots, t_i)$ be all linearly independent right eigenvectors of an orthogonal tensor associated with a complex (definitely not real) eigenvalue λ_i. Show that $\boldsymbol{a}_i^{(k)} \cdot \boldsymbol{a}_i^{(l)} = 0$ $(k, l = 1, 2, \ldots, t_i)$.

4.19. Evaluate principal invariants of a skew-symmetric tensor in three-dimensional space using (4.29).

4.20. Evaluate eigenvalues, eigenvectors and eigenprojections of the tensor describing the rotation by the angle α about the axis $\boldsymbol{e}_3$ (see Exercise 1.24).

4.21. Verify the Cayley-Hamilton theorem for the tensor **A** defined in Exercise 4.14.

4.22. Verify the Cayley-Hamilton theorem for the deformation gradient in the case of simple shear (2.69).

Chapter 5
Fourth-Order Tensors

5.1 Fourth-Order Tensors as a Linear Mapping

Fourth-order tensors play an important role in continuum mechanics where they appear as elasticity and compliance tensors. In this section we define fourth-order tensors and learn some basic operations with them. To this end, we consider a set $\mathcal{L}\mathrm{in}^n$ of all linear mappings of one second-order tensor into another one within $\mathbf{Lin}^n$. Such mappings are denoted by a colon as

$$\mathbf{Y} = \mathcal{A} : \mathbf{X}, \quad \mathcal{A} \in \mathcal{L}\mathrm{in}^n, \ \mathbf{Y} \in \mathbf{Lin}^n, \ \forall \mathbf{X} \in \mathbf{Lin}^n. \tag{5.1}$$

The elements of $\mathcal{L}\mathrm{in}^n$ are called fourth-order tensors.

Example 5.1. Elasticity and compliance tensors. A constitutive law of a linearly elastic material establishes a linear relationship between the Cauchy stress tensor $\boldsymbol{\sigma}$ and Cauchy strain tensor $\boldsymbol{\epsilon}$. Since these tensors are of the second-order a linear relation between them can be expressed by fourth-order tensors like

$$\boldsymbol{\sigma} = \mathcal{C} : \boldsymbol{\epsilon} \quad \text{or} \quad \boldsymbol{\epsilon} = \mathcal{H} : \boldsymbol{\sigma}. \tag{5.2}$$

The fourth-order tensors $\mathcal{C}$ and $\mathcal{H}$ describe properties of the elastic material and are called the elasticity and compliance tensor, respectively.

Linearity of the mapping (5.1) implies that

$$\mathcal{A} : (\mathbf{X} + \mathbf{Y}) = \mathcal{A} : \mathbf{X} + \mathcal{A} : \mathbf{Y}, \tag{5.3}$$

$$\mathcal{A} : (\alpha \mathbf{X}) = \alpha (\mathcal{A} : \mathbf{X}), \quad \forall \mathbf{X}, \mathbf{Y} \in \mathbf{Lin}^n, \ \forall \alpha \in \mathbb{R}, \ \mathcal{A} \in \mathcal{L}\mathrm{in}^n. \tag{5.4}$$

Similarly to second-order tensors one defines the product of a fourth-order tensor with a scalar

$$(\alpha \mathcal{A}) : \mathbf{X} = \alpha (\mathcal{A} : \mathbf{X}) = \mathcal{A} : (\alpha \mathbf{X}) \tag{5.5}$$

M. Itskov, *Tensor Algebra and Tensor Analysis for Engineers*, Mathematical Engineering, DOI 10.1007/978-3-642-30879-6_5, © Springer-Verlag Berlin Heidelberg 2013

and the sum of two fourth-order tensors by

$$(\mathcal{A} + \mathcal{B}) : \mathbf{X} = \mathcal{A} : \mathbf{X} + \mathcal{B} : \mathbf{X}, \quad \forall \mathbf{X} \in \mathbf{Lin}^n. \tag{5.6}$$

Further, we define the zero-tensor $\mathcal{O}$ of the fourth-order by

$$\mathcal{O} : \mathbf{X} = \mathbf{0}, \quad \forall \mathbf{X} \in \mathbf{Lin}^n. \tag{5.7}$$

Thus, summarizing the properties of fourth-order tensors one can write similarly to second-order tensors

$$\mathcal{A} + \mathcal{B} = \mathcal{B} + \mathcal{A}, \quad \text{(addition is commutative)}, \tag{5.8}$$

$$\mathcal{A} + (\mathcal{B} + \mathcal{C}) = (\mathcal{A} + \mathcal{B}) + \mathcal{C}, \quad \text{(addition is associative)}, \tag{5.9}$$

$$\mathcal{O} + \mathcal{A} = \mathcal{A}, \tag{5.10}$$

$$\mathcal{A} + (-\mathcal{A}) = \mathcal{O}, \tag{5.11}$$

$$\alpha (\beta \mathcal{A}) = (\alpha\beta) \mathcal{A}, \quad \text{(multiplication by scalars is associative)}, \tag{5.12}$$

$$1\mathcal{A} = \mathcal{A}, \tag{5.13}$$

$$\alpha (\mathcal{A} + \mathcal{B}) = \alpha\mathcal{A} + \alpha\mathcal{B}, \quad \text{(multiplication by scalars is distributive with respect to tensor addition)}, \tag{5.14}$$

$$(\alpha + \beta) \mathcal{A} = \alpha\mathcal{A} + \beta\mathcal{A}, \quad \text{(multiplication by scalars is distributive with respect to scalar addition)}, \quad \forall \mathcal{A}, \mathcal{B}, \mathcal{C} \in \mathcal{L}\mathrm{in}^n, \ \forall \alpha, \beta \in \mathbb{R}. \tag{5.15}$$

Thus, the set of fourth-order tensors $\mathcal{L}\mathrm{in}^n$ forms a vector space.

On the basis of the "right" mapping (5.1) and the scalar product of two second-order tensors (1.143) we can also define the "left" mapping by

$$(\mathbf{Y} : \mathcal{A}) : \mathbf{X} = \mathbf{Y} : (\mathcal{A} : \mathbf{X}), \quad \mathbf{Y} \in \mathbf{Lin}^n, \quad \forall \mathbf{X} \in \mathbf{Lin}^n. \tag{5.16}$$

5.2 Tensor Products, Representation of Fourth-Order Tensors with Respect to a Basis

For the construction of fourth-order tensors from second-order ones we introduce two tensor products as follows

$$\mathbf{A} \otimes \mathbf{B} : \mathbf{X} = \mathbf{AXB}, \quad \mathbf{A} \odot \mathbf{B} : \mathbf{X} = \mathbf{A} (\mathbf{B} : \mathbf{X}), \quad \forall \mathbf{X} \in \mathbf{Lin}^n, \tag{5.17}$$

where $\mathbf{A}, \mathbf{B} \in \mathbf{Lin}^n$. Note, that the tensor product "$\otimes$" $(5.17)_1$ applied to second-order tensors differs from the tensor product of vectors (1.80). One can easily show that the mappings described by (5.17) are linear and therefore represent fourth-order tensors. Indeed, we have, for example, for the tensor product "$\otimes$" $(5.17)_1$

$$\begin{aligned}\mathbf{A}\otimes\mathbf{B}:(\mathbf{X}+\mathbf{Y}) &= \mathbf{A}(\mathbf{X}+\mathbf{Y})\mathbf{B}\\ &= \mathbf{AXB}+\mathbf{AYB}=\mathbf{A}\otimes\mathbf{B}:\mathbf{X}+\mathbf{A}\otimes\mathbf{B}:\mathbf{Y},\end{aligned}\tag{5.18}$$

$$\begin{aligned}\mathbf{A}\otimes\mathbf{B}:(\alpha\mathbf{X}) &= \mathbf{A}(\alpha\mathbf{X})\mathbf{B}=\alpha(\mathbf{AXB})\\ &= \alpha(\mathbf{A}\otimes\mathbf{B}:\mathbf{X}), \quad \forall\mathbf{X},\mathbf{Y}\in\mathbf{Lin}^n,\ \forall\alpha\in\mathbb{R}.\end{aligned}\tag{5.19}$$

With definitions (5.17) in hand one can easily prove the following identities

$$\mathbf{A}\otimes(\mathbf{B}+\mathbf{C})=\mathbf{A}\otimes\mathbf{B}+\mathbf{A}\otimes\mathbf{C},\quad (\mathbf{B}+\mathbf{C})\otimes\mathbf{A}=\mathbf{B}\otimes\mathbf{A}+\mathbf{C}\otimes\mathbf{A},\tag{5.20}$$

$$\mathbf{A}\odot(\mathbf{B}+\mathbf{C})=\mathbf{A}\odot\mathbf{B}+\mathbf{A}\odot\mathbf{C},\quad (\mathbf{B}+\mathbf{C})\odot\mathbf{A}=\mathbf{B}\odot\mathbf{A}+\mathbf{C}\odot\mathbf{A}.\tag{5.21}$$

For the left mapping (5.16) the tensor products (5.17) yield

$$\mathbf{Y}:\mathbf{A}\otimes\mathbf{B}=\mathbf{A}^{\mathrm{T}}\mathbf{Y}\mathbf{B}^{\mathrm{T}},\quad \mathbf{Y}:\mathbf{A}\odot\mathbf{B}=(\mathbf{Y}:\mathbf{A})\,\mathbf{B}.\tag{5.22}$$

As fourth-order tensors represent vectors they can be given with respect to a basis in $\mathcal{L}\mathrm{in}^n$.

Theorem 5.1. *Let* $\mathcal{F}=\{\mathbf{F}_1,\mathbf{F}_2,\ldots,\mathbf{F}_{n^2}\}$ *and* $\mathcal{G}=\{\mathbf{G}_1,\mathbf{G}_2,\ldots,\mathbf{G}_{n^2}\}$ *be two arbitrary (not necessarily distinct) bases of* $\mathbf{Lin}^n$. *Then, fourth-order tensors* $\mathbf{F}_i\odot\mathbf{G}_j\ \left(i,j=1,2,\ldots,n^2\right)$ *form a basis of* $\mathcal{L}in^n$. *The dimension of* $\mathcal{L}in^n$ *is thus* n^4.

Proof. See the proof of Theorem 1.7.

A basis in $\mathcal{L}\mathrm{in}^n$ can be represented in another way as by the tensors $\mathbf{F}_i\odot\mathbf{G}_j\ \left(i,j=1,2,\ldots,n^2\right)$. To this end, we prove the following identity

$$(\boldsymbol{a}\otimes\boldsymbol{d})\odot(\boldsymbol{b}\otimes\boldsymbol{c})=\boldsymbol{a}\otimes\boldsymbol{b}\otimes\boldsymbol{c}\otimes\boldsymbol{d},\tag{5.23}$$

where we set

$$(\boldsymbol{a}\otimes\boldsymbol{b})\otimes(\boldsymbol{c}\otimes\boldsymbol{d})=\boldsymbol{a}\otimes\boldsymbol{b}\otimes\boldsymbol{c}\otimes\boldsymbol{d}.\tag{5.24}$$

Indeed, let $\mathbf{X}\in\mathbf{Lin}^n$ be an arbitrary second-order tensor. Then, in view of (1.142) and $(5.17)_2$

$$(\boldsymbol{a}\otimes\boldsymbol{d})\odot(\boldsymbol{b}\otimes\boldsymbol{c}):\mathbf{X}=(\boldsymbol{b}\mathbf{X}\boldsymbol{c})(\boldsymbol{a}\otimes\boldsymbol{d}).\tag{5.25}$$

For the right hand side of (5.23) we obtain the same result using $(5.17)_1$ and (5.24)

$$\boldsymbol{a}\otimes\boldsymbol{b}\otimes\boldsymbol{c}\otimes\boldsymbol{d}:\mathbf{X}=(\boldsymbol{a}\otimes\boldsymbol{b})\otimes(\boldsymbol{c}\otimes\boldsymbol{d}):\mathbf{X}=(\boldsymbol{b}\mathbf{X}\boldsymbol{c})(\boldsymbol{a}\otimes\boldsymbol{d}).\tag{5.26}$$

For the left mapping (5.16) it thus holds

$$\mathbf{Y} : \boldsymbol{a} \otimes \boldsymbol{b} \otimes \boldsymbol{c} \otimes \boldsymbol{d} = (\boldsymbol{a}\mathbf{Y}\boldsymbol{d})(\boldsymbol{b} \otimes \boldsymbol{c}). \tag{5.27}$$

Now, we are in a position to prove the following theorem.

Theorem 5.2. *Let* $\mathcal{E} = \{\boldsymbol{e}_1, \boldsymbol{e}_2, \ldots, \boldsymbol{e}_n\}$, $\mathcal{F} = \{\boldsymbol{f}_1, \boldsymbol{f}_2, \ldots, \boldsymbol{f}_n\}$, $\mathcal{G} = \{\boldsymbol{g}_1, \boldsymbol{g}_2, \ldots, \boldsymbol{g}_n\}$ *and finally* $\mathcal{H} = \{\boldsymbol{h}_1, \boldsymbol{h}_2, \ldots, \boldsymbol{h}_n\}$ *be four arbitrary (not necessarily distinct) bases of* $\mathbb{E}^n$*. Then, fourth-order tensors* $\boldsymbol{e}_i \otimes \boldsymbol{f}_j \otimes \boldsymbol{g}_k \otimes \boldsymbol{h}_l$ $(i, j, k, l = 1, 2, \ldots, n)$ *represent a basis of* $\mathcal{L}\mathrm{in}^n$.

Proof. In view of (5.23)

$$\boldsymbol{e}_i \otimes \boldsymbol{f}_j \otimes \boldsymbol{g}_k \otimes \boldsymbol{h}_l = (\boldsymbol{e}_i \otimes \boldsymbol{h}_l) \odot \left(\boldsymbol{f}_j \otimes \boldsymbol{g}_k\right).$$

According to Theorem 1.7 the second-order tensors $\boldsymbol{e}_i \otimes \boldsymbol{h}_l$ $(i, l = 1, 2, \ldots, n)$ on the one hand and $\boldsymbol{f}_j \otimes \boldsymbol{g}_k$ $(j, k = 1, 2, \ldots, n)$ on the other hand form bases of Lin^n. According to Theorem 5.1 the fourth-order tensors $(\boldsymbol{e}_i \otimes \boldsymbol{h}_l) \odot \left(\boldsymbol{f}_j \otimes \boldsymbol{g}_k\right)$ and consequently $\boldsymbol{e}_i \otimes \boldsymbol{f}_j \otimes \boldsymbol{g}_k \otimes \boldsymbol{h}_l$ $(i, j, k, l = 1, 2, \ldots, n)$ represent thus a basis of $\mathcal{L}\mathrm{in}^n$.

As a result of this Theorem any fourth-order tensor can be represented by

$$\begin{aligned}
\mathcal{A} &= \mathcal{A}^{ijkl}\boldsymbol{g}_i \otimes \boldsymbol{g}_j \otimes \boldsymbol{g}_k \otimes \boldsymbol{g}_l = \mathcal{A}_{ijkl}\boldsymbol{g}^i \otimes \boldsymbol{g}^j \otimes \boldsymbol{g}^k \otimes \boldsymbol{g}^l \\
&= \mathcal{A}^{ij}_{\cdot\cdot kl}\boldsymbol{g}_i \otimes \boldsymbol{g}_j \otimes \boldsymbol{g}^k \otimes \boldsymbol{g}^l = \ldots
\end{aligned} \tag{5.28}$$

The components of $\mathcal{A}$ appearing in (5.28) can be expressed by

$$\begin{aligned}
&\mathcal{A}^{ijkl} = \boldsymbol{g}^i \otimes \boldsymbol{g}^l : \mathcal{A} : \boldsymbol{g}^j \otimes \boldsymbol{g}^k, \quad \mathcal{A}_{ijkl} = \boldsymbol{g}_i \otimes \boldsymbol{g}_l : \mathcal{A} : \boldsymbol{g}_j \otimes \boldsymbol{g}_k, \\
&\mathcal{A}^{ij}_{\cdot\cdot kl} = \boldsymbol{g}^i \otimes \boldsymbol{g}_l : \mathcal{A} : \boldsymbol{g}^j \otimes \boldsymbol{g}_k, \quad i, j, k, l = 1, 2, \ldots, n.
\end{aligned} \tag{5.29}$$

By virtue of (1.109), $(5.17)_1$ and $(5.22)_1$ the right and left mappings with a second-order tensor (5.1) and (5.16) can thus be represented by

$$\begin{aligned}
\mathcal{A} : \mathbf{X} &= \left(\mathcal{A}^{ijkl}\boldsymbol{g}_i \otimes \boldsymbol{g}_j \otimes \boldsymbol{g}_k \otimes \boldsymbol{g}_l\right) : \left(\mathrm{X}_{qp}\boldsymbol{g}^q \otimes \boldsymbol{g}^p\right) = \mathcal{A}^{ijkl}\mathrm{X}_{jk}\boldsymbol{g}_i \otimes \boldsymbol{g}_l, \\
\mathbf{X} : \mathcal{A} &= \left(\mathrm{X}_{qp}\boldsymbol{g}^q \otimes \boldsymbol{g}^p\right) : \left(\mathcal{A}^{ijkl}\boldsymbol{g}_i \otimes \boldsymbol{g}_j \otimes \boldsymbol{g}_k \otimes \boldsymbol{g}_l\right) = \mathcal{A}^{ijkl}\mathrm{X}_{il}\boldsymbol{g}_j \otimes \boldsymbol{g}_k.
\end{aligned} \tag{5.30}$$

We observe that the basis vectors of the second-order tensor are scalarly multiplied either by the "inner" (right mapping) or "outer" (left mapping) basis vectors of the fourth-order tensor.

5.3 Special Operations with Fourth-Order Tensors

Similarly to second-order tensors one defines also for fourth-order tensors some specific operations which are not generally applicable to conventional vectors in the Euclidean space.

Composition. In analogy with second-order tensors we define the composition of two fourth-order tensors $\mathcal{A}$ and $\mathcal{B}$ denoted by $\mathcal{A} : \mathcal{B}$ as

$$(\mathcal{A} : \mathcal{B}) : \mathbf{X} = \mathcal{A} : (\mathcal{B} : \mathbf{X}), \quad \forall \mathbf{X} \in \mathbf{Lin}^n. \tag{5.31}$$

For the left mapping (5.16) one can thus write

$$\mathbf{Y} : (\mathcal{A} : \mathcal{B}) = (\mathbf{Y} : \mathcal{A}) : \mathcal{B}, \quad \forall \mathbf{Y} \in \mathbf{Lin}^n. \tag{5.32}$$

For the tensor products (5.17) the composition (5.31) further yields

$$(\mathbf{A} \otimes \mathbf{B}) : (\mathbf{C} \otimes \mathbf{D}) = (\mathbf{AC}) \otimes (\mathbf{DB}), \tag{5.33}$$

$$(\mathbf{A} \otimes \mathbf{B}) : (\mathbf{C} \odot \mathbf{D}) = (\mathbf{ACB}) \odot \mathbf{D}, \tag{5.34}$$

$$(\mathbf{A} \odot \mathbf{B}) : (\mathbf{C} \otimes \mathbf{D}) = \mathbf{A} \odot \left(\mathbf{C}^{\mathrm{T}}\mathbf{B}\mathbf{D}^{\mathrm{T}}\right), \tag{5.35}$$

$$(\mathbf{A} \odot \mathbf{B}) : (\mathbf{C} \odot \mathbf{D}) = (\mathbf{B} : \mathbf{C})\, \mathbf{A} \odot \mathbf{D}, \quad \mathbf{A}, \mathbf{B}, \mathbf{C}, \mathbf{D} \in \mathbf{Lin}^n. \tag{5.36}$$

For example, the identity (5.33) can be proved within the following steps

$$\begin{aligned}(\mathbf{A} \otimes \mathbf{B}) : (\mathbf{C} \otimes \mathbf{D}) : \mathbf{X} &= (\mathbf{A} \otimes \mathbf{B}) : (\mathbf{CXD}) \\ &= \mathbf{ACXDB} = (\mathbf{AC}) \otimes (\mathbf{DB}) : \mathbf{X}, \quad \forall \mathbf{X} \in \mathbf{Lin}^n,\end{aligned}$$

where we again take into account the definition of the tensor product (5.17).

For the component representation (5.28) we further obtain

$$\begin{aligned}\mathcal{A} : \mathcal{B} &= \left(\mathcal{A}^{ijkl} \mathbf{g}_i \otimes \mathbf{g}_j \otimes \mathbf{g}_k \otimes \mathbf{g}_l\right) : \left(\mathcal{B}_{pqrt} \mathbf{g}^p \otimes \mathbf{g}^q \otimes \mathbf{g}^r \otimes \mathbf{g}^t\right) \\ &= \mathcal{A}^{ijkl} \mathcal{B}_{jqrk} \mathbf{g}_i \otimes \mathbf{g}^q \otimes \mathbf{g}^r \otimes \mathbf{g}_l.\end{aligned} \tag{5.37}$$

Note that the "inner" basis vectors of the left tensor $\mathcal{A}$ are scalarly multiplied with the "outer" basis vectors of the right tensor $\mathcal{B}$.

The composition of fourth-order tensors also gives rise to the definition of powers as

$$\mathcal{A}^k = \underbrace{\mathcal{A} : \mathcal{A} : \ldots : \mathcal{A}}_{k \text{ times}}, \quad k = 1, 2, \ldots, \quad \mathcal{A}^0 = \mathcal{J}, \tag{5.38}$$

where $\mathcal{J}$ stands for the fourth-order identity tensor to be defined in the next section. By means of (5.33) and (5.36) powers of tensor products (5.17) take the following form

$$(\mathbf{A} \otimes \mathbf{B})^k = \mathbf{A}^k \otimes \mathbf{B}^k, \quad (\mathbf{A} \odot \mathbf{B})^k = (\mathbf{A} : \mathbf{B})^{k-1} \mathbf{A} \odot \mathbf{B}, \quad k = 1, 2, \ldots \tag{5.39}$$

Simple composition with second-order tensors. Let $\mathfrak{D}$ be a fourth-order tensor and $\mathbf{A}$, $\mathbf{B}$ two second-order tensors. One defines a fourth-order tensor $\mathbf{A}\mathfrak{D}\mathbf{B}$ by

$$(\mathbf{A}\mathfrak{D}\mathbf{B}) : \mathbf{X} = \mathbf{A}\,(\mathfrak{D} : \mathbf{X})\,\mathbf{B}, \quad \forall \mathbf{X} \in \mathrm{Lin}^n. \tag{5.40}$$

Thus, we can also write

$$\mathbf{A}\mathfrak{D}\mathbf{B} = (\mathbf{A} \otimes \mathbf{B}) : \mathfrak{D}. \tag{5.41}$$

This operation is very useful for the formulation of tensor differentiation rules to be discussed in the next chapter.

For the tensor products (5.17) we further obtain

$$\mathbf{A}\,(\mathbf{B} \otimes \mathbf{C})\,\mathbf{D} = (\mathbf{A}\mathbf{B}) \otimes (\mathbf{C}\mathbf{D}) = (\mathbf{A} \otimes \mathbf{D}) : (\mathbf{B} \otimes \mathbf{C}), \tag{5.42}$$

$$\mathbf{A}\,(\mathbf{B} \odot \mathbf{C})\,\mathbf{D} = (\mathbf{A}\mathbf{B}\mathbf{D}) \odot \mathbf{C} = (\mathbf{A} \otimes \mathbf{D}) : (\mathbf{B} \odot \mathbf{C}). \tag{5.43}$$

With respect to a basis the simple composition can be given by

$$\begin{aligned} \mathbf{A}\mathfrak{D}\mathbf{B} &= \left(\mathrm{A}_{pq}\boldsymbol{g}^p \otimes \boldsymbol{g}^q\right)\left(\mathfrak{D}^{ijkl}\boldsymbol{g}_i \otimes \boldsymbol{g}_j \otimes \boldsymbol{g}_k \otimes \boldsymbol{g}_l\right)\left(\mathrm{B}_{rs}\boldsymbol{g}^r \otimes \boldsymbol{g}^s\right) \\ &= \mathrm{A}_{pi}\,\mathfrak{D}^{ijkl}\mathrm{B}_{ls}\boldsymbol{g}^p \otimes \boldsymbol{g}_j \otimes \boldsymbol{g}_k \otimes \boldsymbol{g}^s. \end{aligned} \tag{5.44}$$

It is seen that expressed in component form the simple composition of second-order tensors with a fourth-order tensor represents the so-called simple contraction of the classical tensor algebra (see, e.g., [42]).

Transposition. In contrast to second-order tensors allowing for the unique transposition operation one can define for fourth-order tensors various transpositions. We confine our attention here to the following two operations $(\bullet)^{\mathrm{T}}$ and $(\bullet)^{\mathrm{t}}$ defined by

$$\mathcal{A}^{\mathrm{T}} : \mathbf{X} = \mathbf{X} : \mathcal{A}, \quad \mathcal{A}^{\mathrm{t}} : \mathbf{X} = \mathcal{A} : \mathbf{X}^{\mathrm{T}}, \quad \forall \mathbf{X} \in \mathrm{Lin}^n. \tag{5.45}$$

Thus we can also write

$$\mathbf{Y} : \mathcal{A}^{\mathrm{t}} = (\mathbf{Y} : \mathcal{A})^{\mathrm{T}}. \tag{5.46}$$

Indeed, a scalar product with an arbitrary second order tensor $\mathbf{X}$ yields in view of (1.147) and (5.16)

$$\begin{aligned} \left(\mathbf{Y} : \mathcal{A}^{\mathrm{t}}\right) : \mathbf{X} &= \mathbf{Y} : \left(\mathcal{A}^{\mathrm{t}} : \mathbf{X}\right) = \mathbf{Y} : \left(\mathcal{A} : \mathbf{X}^{\mathrm{T}}\right) \\ &= (\mathbf{Y} : \mathcal{A}) : \mathbf{X}^{\mathrm{T}} = (\mathbf{Y} : \mathcal{A})^{\mathrm{T}} : \mathbf{X}, \quad \forall \mathbf{X} \in \mathrm{Lin}^n. \end{aligned}$$

Of special importance is also the following symmetrization operation resulting from the transposition $(\bullet)^{\mathrm{t}}$:

$$\mathcal{F}^{\mathrm{s}} = \frac{1}{2}\left(\mathcal{F} + \mathcal{F}^{\mathrm{t}}\right). \tag{5.47}$$

In view of $(1.153)_1$, $(5.45)_2$ and (5.46) we thus write

$$\mathcal{F}^{\mathrm{s}} : \mathbf{X} = \mathcal{F} : \mathrm{sym}\mathbf{X}, \quad \mathbf{Y} : \mathcal{F}^{\mathrm{s}} = \mathrm{sym}\left(\mathbf{Y} : \mathcal{F}\right). \tag{5.48}$$

Applying the transposition operations to the tensor products (5.17) we have

$$(\mathbf{A} \otimes \mathbf{B})^{\mathrm{T}} = \mathbf{A}^{\mathrm{T}} \otimes \mathbf{B}^{\mathrm{T}}, \quad (\mathbf{A} \odot \mathbf{B})^{\mathrm{T}} = \mathbf{B} \odot \mathbf{A}, \tag{5.49}$$

$$(\mathbf{A} \odot \mathbf{B})^{\mathrm{t}} = \mathbf{A} \odot \mathbf{B}^{\mathrm{T}}, \quad \mathbf{A}, \mathbf{B} \in \mathrm{Lin}^n. \tag{5.50}$$

With the aid of (5.26) and (5.27) we further obtain

$$(\boldsymbol{a} \otimes \boldsymbol{b} \otimes \boldsymbol{c} \otimes \boldsymbol{d})^{\mathrm{T}} = \boldsymbol{b} \otimes \boldsymbol{a} \otimes \boldsymbol{d} \otimes \boldsymbol{c}, \tag{5.51}$$

$$(\boldsymbol{a} \otimes \boldsymbol{b} \otimes \boldsymbol{c} \otimes \boldsymbol{d})^{\mathrm{t}} = \boldsymbol{a} \otimes \boldsymbol{c} \otimes \boldsymbol{b} \otimes \boldsymbol{d}. \tag{5.52}$$

It can also easily be proved that

$$\mathcal{A}^{\mathrm{TT}} = \mathcal{A}, \quad \mathcal{A}^{\mathrm{tt}} = \mathcal{A}, \quad \forall \mathcal{A} \in \mathcal{L}\mathrm{in}^n. \tag{5.53}$$

Note, however, that the transposition operations (5.45) are not commutative with each other so that generally $\mathcal{D}^{\mathrm{Tt}} \neq \mathcal{D}^{\mathrm{tT}}$.

Applied to the composition of fourth-order tensors these transposition operations yield (Exercise 5.6):

$$(\mathcal{A} : \mathcal{B})^{\mathrm{T}} = \mathcal{B}^{\mathrm{T}} : \mathcal{A}^{\mathrm{T}}, \quad (\mathcal{A} : \mathcal{B})^{\mathrm{t}} = \mathcal{A} : \mathcal{B}^{\mathrm{t}}. \tag{5.54}$$

For the tensor products (5.17) we also obtain the following relations (see Exercise 5.7)

$$(\mathbf{A} \otimes \mathbf{B})^{\mathrm{t}} : (\mathbf{C} \otimes \mathbf{D}) = \left[\left(\mathbf{A}\mathbf{D}^{\mathrm{T}}\right) \otimes \left(\mathbf{C}^{\mathrm{T}}\mathbf{B}\right)\right]^{\mathrm{t}}, \tag{5.55}$$

$$(\mathbf{A} \otimes \mathbf{B})^{\mathrm{t}} : (\mathbf{C} \odot \mathbf{D}) = \left(\mathbf{A}\mathbf{C}^{\mathrm{T}}\mathbf{B}\right) \odot \mathbf{D}. \tag{5.56}$$

Scalar product. Similarly to second-order tensors the scalar product of fourth-order tensors can be defined in terms of the basis vectors or tensors. To this end, let us consider two fourth-order tensors $\mathbf{A} \odot \mathbf{B}$ and $\mathbf{C} \odot \mathbf{D}$, where $\mathbf{A}, \mathbf{B}, \mathbf{C}, \mathbf{D} \in \mathrm{Lin}^n$. Then, we set

$$(\mathbf{A} \odot \mathbf{B}) :: (\mathbf{C} \odot \mathbf{D}) = (\mathbf{A} : \mathbf{C})\,(\mathbf{B} : \mathbf{D}). \tag{5.57}$$

As a result of this definition we also obtain in view of (1.141) and (5.23)

$$(\boldsymbol{a} \otimes \boldsymbol{b} \otimes \boldsymbol{c} \otimes \boldsymbol{d}) :: (\boldsymbol{e} \otimes \boldsymbol{f} \otimes \boldsymbol{g} \otimes \boldsymbol{h}) = (\boldsymbol{a} \cdot \boldsymbol{e})\,(\boldsymbol{b} \cdot \boldsymbol{f})\,(\boldsymbol{c} \cdot \boldsymbol{g})\,(\boldsymbol{d} \cdot \boldsymbol{h}). \tag{5.58}$$

For the component representation of fourth-order tensors it finally yields

$$\mathcal{A} :: \mathcal{B} = \left(\mathcal{A}^{ijkl} \boldsymbol{g}_i \otimes \boldsymbol{g}_j \otimes \boldsymbol{g}_k \otimes \boldsymbol{g}_l\right) \\ :: \left(\mathcal{B}_{pqrt} \boldsymbol{g}^p \otimes \boldsymbol{g}^q \otimes \boldsymbol{g}^r \otimes \boldsymbol{g}^t\right) = \mathcal{A}^{ijkl} \mathcal{B}_{ijkl}. \tag{5.59}$$

Using the latter relation one can easily prove that the properties of the scalar product (D.1)–(D.4) hold for fourth-order tensors as well.

5.4 Super-Symmetric Fourth-Order Tensors

On the basis of the transposition operations one defines symmetric and super-symmetric fourth-order tensors. Accordingly, a fourth-order tensor $\mathcal{C}$ is said to be symmetric if (major symmetry)

$$\mathcal{C}^{\mathrm{T}} = \mathcal{C} \tag{5.60}$$

and super-symmetric if additionally (minor symmetry)

$$\mathcal{C}^{\mathrm{t}} = \mathcal{C}. \tag{5.61}$$

In this section we focus on the properties of super-symmetric fourth-order tensors. They constitute a subspace of $\mathcal{L}\mathrm{in}^n$ denoted in the following by $\mathcal{S}\mathrm{sym}^n$. First, we prove that every super-symmetric fourth-order tensor maps an arbitrary (not necessarily symmetric) second-order tensor into a symmetric one so that

$$(\mathcal{C} : \mathbf{X})^{\mathrm{T}} = \mathcal{C} : \mathbf{X}, \quad \forall \mathcal{C} \in \mathcal{S}\mathrm{sym}^n, \quad \forall \mathbf{X} \in \mathbf{Lin}^n. \tag{5.62}$$

Indeed, in view of (5.45), (5.46), (5.60) and (5.61) we have

$$(\mathcal{C} : \mathbf{X})^{\mathrm{T}} = \left(\mathbf{X} : \mathcal{C}^{\mathrm{T}}\right)^{\mathrm{T}} = (\mathbf{X} : \mathcal{C})^{\mathrm{T}} = \mathbf{X} : \mathcal{C}^{\mathrm{t}} = \mathbf{X} : \mathcal{C} = \mathbf{X} : \mathcal{C}^{\mathrm{T}} = \mathcal{C} : \mathbf{X}.$$

Next, we deal with representations of super-symmetric fourth-order tensors and study the properties of their components. Let $\mathcal{F} = \{\mathbf{F}_1, \mathbf{F}_2, \ldots, \mathbf{F}_{n^2}\}$ be an arbitrary basis of $\mathbf{Lin}^n$ and $\mathcal{F}' = \left\{\mathbf{F}^1, \mathbf{F}^2, \ldots, \mathbf{F}^{n^2}\right\}$ the corresponding dual basis such that

$$\mathbf{F}_p : \mathbf{F}^q = \delta_p^q, \quad p, q = 1, 2, \ldots, n^2. \tag{5.63}$$

According to Theorem 5.1 we first write

$$\mathcal{C} = \mathcal{C}^{pq} \mathbf{F}_p \odot \mathbf{F}_q. \tag{5.64}$$

Taking (5.60) into account and in view of $(5.49)_2$ we infer that

$$\mathcal{C}^{pq} = \mathcal{C}^{qp}, \quad p \neq q;\ p, q = 1, 2, \ldots, n^2. \tag{5.65}$$

Let now $\mathbf{F}_p = \mathbf{M}_p$ $(p = 1, 2, \ldots, m)$ and $\mathbf{F}_q = \mathbf{W}_{q-m}$ $\left(q = m+1, \ldots, n^2\right)$ be bases of $\mathbf{Sym}^n$ and $\mathbf{Skew}^n$ (Sect. 1.9), respectively, where $m = \frac{1}{2}n\,(n+1)$. In view of $(5.45)_2$ and (5.61)

$$\mathcal{C} : \mathbf{W}^t = \mathcal{C} : \left(\mathbf{W}^t\right)^{\mathrm{T}} = -\mathcal{C} : \mathbf{W}^t = \mathbf{0}, \quad t = 1, 2, \ldots, \frac{1}{2}n\,(n-1) \tag{5.66}$$

so that

$$\mathcal{C}^{pr} = \mathcal{C}^{rp} = \mathbf{F}^p : \mathcal{C} : \mathbf{F}^r = 0, \quad p = 1, 2, \ldots, n^2;\ r = m+1, \ldots, n^2 \tag{5.67}$$

and consequently

$$\mathcal{C} = \sum_{p,q=1}^{m} \mathcal{C}^{pq} \mathbf{M}_p \odot \mathbf{M}_q, \quad m = \frac{1}{2}n\,(n+1). \tag{5.68}$$

Keeping (5.65) in mind we can also write by analogy with (1.156)

$$\mathcal{C} = \sum_{p=1}^{m} \mathcal{C}^{pp} \mathbf{M}_p \odot \mathbf{M}_p + \sum_{\substack{p,q=1 \\ p>q}}^{m} \mathcal{C}^{pq} \left(\mathbf{M}_p \odot \mathbf{M}_q + \mathbf{M}_q \odot \mathbf{M}_p\right). \tag{5.69}$$

Therefore, every super-symmetric fourth-order tensor can be represented with respect to the basis $\frac{1}{2}\left(\mathbf{M}_p \odot \mathbf{M}_q + \mathbf{M}_q \odot \mathbf{M}_p\right)$, where $\mathbf{M}_q \in \mathbf{Sym}^n$ and $p \geq q = 1, 2, \ldots, \frac{1}{2}n\,(n+1)$. Thus, we infer that the dimension of $\mathcal{S}\mathrm{sym}^n$ is $\frac{1}{2}m\,(m+1) = \frac{1}{8}n^2\,(n+1)^2 + \frac{1}{4}n\,(n+1)$. We also observe that $\mathcal{S}\mathrm{sym}^n$ can be considered as the set of all linear mappings within $\mathbf{Sym}^n$.

Applying Theorem 5.2 we can also represent a super-symmetric tensor by $\mathcal{C} = \mathcal{C}^{ijkl} \boldsymbol{g}_i \otimes \boldsymbol{g}_j \otimes \boldsymbol{g}_k \otimes \boldsymbol{g}_l$. In this case, (5.51) and (5.52) require that (Exercise 5.8)

$$\mathcal{C}^{ijkl} = \mathcal{C}^{jilk} = \mathcal{C}^{ikjl} = \mathcal{C}^{ljki} = \mathcal{C}^{klij}. \tag{5.70}$$

Thus, we can also write

$$\begin{aligned}
\mathcal{C} &= \mathcal{C}^{ijkl} \left(\boldsymbol{g}_i \otimes \boldsymbol{g}_l\right) \odot \left(\boldsymbol{g}_j \otimes \boldsymbol{g}_k\right) \\
&= \frac{1}{4}\mathcal{C}^{ijkl} \left(\boldsymbol{g}_i \otimes \boldsymbol{g}_l + \boldsymbol{g}_l \otimes \boldsymbol{g}_i\right) \odot \left(\boldsymbol{g}_j \otimes \boldsymbol{g}_k + \boldsymbol{g}_k \otimes \boldsymbol{g}_j\right) \\
&= \frac{1}{4}\mathcal{C}^{ijkl} \left(\boldsymbol{g}_j \otimes \boldsymbol{g}_k + \boldsymbol{g}_k \otimes \boldsymbol{g}_j\right) \odot \left(\boldsymbol{g}_i \otimes \boldsymbol{g}_l + \boldsymbol{g}_l \otimes \boldsymbol{g}_i\right).
\end{aligned} \tag{5.71}$$

Finally, we briefly consider the eigenvalue problem for super-symmetric fourth-order tensors. It is defined as

$$\mathcal{C} : \mathbf{M} = \Lambda \mathbf{M}, \quad \mathcal{C} \in \mathcal{S}\mathrm{sym}^n, \quad \mathbf{M} \neq \mathbf{0}, \tag{5.72}$$

where Λ and $\mathbf{M} \in \mathbf{Sym}^n$ denote the eigenvalue and the corresponding eigentensor, respectively. The spectral decomposition of $\mathcal{C}$ can be given similarly to symmetric second-order tensors (4.61) by

$$\mathcal{C} = \sum_{p=1}^{m} \Lambda_p \mathbf{M}_p \odot \mathbf{M}_p, \tag{5.73}$$

where again $m = \frac{1}{2}n(n+1)$ and

$$\mathbf{M}_p : \mathbf{M}_q = \delta_{pq}, \quad p, q = 1, 2, \ldots, m. \tag{5.74}$$

5.5 Special Fourth-Order Tensors

Identity tensor. The fourth-order identity tensor $\mathcal{I}$ is defined by

$$\mathcal{I} : \mathbf{X} = \mathbf{X}, \quad \forall \mathbf{X} \in \mathbf{Lin}^n. \tag{5.75}$$

It is seen that $\mathcal{I}$ is a symmetric (but not super-symmetric) fourth-order tensor such that $\mathcal{I}^{\mathrm{T}} = \mathcal{I}$. Indeed,

$$\mathbf{X} : \mathcal{I} = \mathbf{X}, \quad \forall \mathbf{X} \in \mathbf{Lin}^n. \tag{5.76}$$

With the aid of $(5.17)_1$ the fourth-order identity tensor can be represented by

$$\mathcal{I} = \mathbf{I} \otimes \mathbf{I}. \tag{5.77}$$

Thus, with the aid of (1.91) or alternatively by using (5.29) one obtains

$$\mathcal{I} = \boldsymbol{g}_i \otimes \boldsymbol{g}^i \otimes \boldsymbol{g}_j \otimes \boldsymbol{g}^j. \tag{5.78}$$

An alternative representation for $\mathcal{I}$ in terms of eigenprojections $\mathbf{P}_i$ $(i = 1, 2, \ldots, s)$ of an arbitrary second-order tensor results from (5.77) and (4.46) as

$$\mathcal{I} = \sum_{i,j=1}^{s} \mathbf{P}_i \otimes \mathbf{P}_j. \tag{5.79}$$

For the composition with other fourth-order tensors we can also write

$$\mathcal{I} : \mathcal{A} = \mathcal{A} : \mathcal{I} = \mathcal{A}, \quad \forall \mathcal{A} \in \mathcal{L}\mathrm{in}^n. \tag{5.80}$$

Transposition tensor. The transposition of second-order tensors represents a linear mapping and can therefore be expressed in terms of a fourth-order tensor. This tensor denoted by $\mathcal{T}$ is referred to as the transposition tensor. Thus,

$$\mathcal{T} : \mathbf{X} = \mathbf{X}^{\mathrm{T}}, \quad \forall \mathbf{X} \in \mathbf{Lin}^{n}. \tag{5.81}$$

One can easily show that (Exercise 5.9)

$$\mathbf{Y} : \mathcal{T} = \mathbf{Y}^{\mathrm{T}}, \quad \forall \mathbf{Y} \in \mathbf{Lin}^{n}. \tag{5.82}$$

Hence, the transposition tensor is symmetric such that $\mathcal{T} = \mathcal{T}^{\mathrm{T}}$. By virtue of $(5.45)_2$ and (5.75), $\mathcal{T}$ can further be expressed in terms of the identity tensor by

$$\mathcal{T} = \mathcal{I}^{\mathrm{t}}. \tag{5.83}$$

Indeed,

$$\mathcal{I}^{\mathrm{t}} : \mathbf{X} = \mathcal{I} : \mathbf{X}^{\mathrm{T}} = \mathbf{X}^{\mathrm{T}} = \mathcal{T} : \mathbf{X}, \quad \forall \mathbf{X} \in \mathbf{Lin}^{n}.$$

Considering (5.52) and (5.77)–(5.79) in (5.83) we thus obtain

$$\mathcal{T} = (\mathbf{I} \otimes \mathbf{I})^{\mathrm{t}} = \sum_{i,j=1}^{s} \left(\mathbf{P}_i \otimes \mathbf{P}_j\right)^{\mathrm{t}} = \boldsymbol{g}_i \otimes \boldsymbol{g}_j \otimes \boldsymbol{g}^i \otimes \boldsymbol{g}^j. \tag{5.84}$$

The transposition tensor can further be characterized by the following identities (see Exercise 5.10)

$$\mathcal{A} : \mathcal{T} = \mathcal{A}^{\mathrm{t}}, \quad \mathcal{T} : \mathcal{A} = \mathcal{A}^{\mathrm{TtT}}, \quad \mathcal{T} : \mathcal{T} = \mathcal{I}, \quad \forall \mathcal{A} \in \mathcal{L}\mathrm{in}^{n}. \tag{5.85}$$

Super-symmetric identity tensor. The identity tensor (5.77) is symmetric but not super-symmetric. For this reason, it is useful to define a special identity tensor within $\mathcal{S}\mathrm{sym}^n$. This super-symmetric tensor maps every symmetric second-order tensor into itself like the identity tensor (5.77). It can be expressed by

$$\mathcal{I}^{\mathrm{s}} = \frac{1}{2}(\mathcal{I} + \mathcal{T}) = (\mathbf{I} \otimes \mathbf{I})^{\mathrm{s}}. \tag{5.86}$$

However, in contrast to the identity tensor $\mathcal{I}$ (5.77), the super-symmetric identity tensor $\mathcal{I}^{\mathrm{s}}$ (5.86) maps any arbitrary (not necessarily symmetric) second-order tensor into its symmetric part so that in view of (5.48)

$$\mathcal{I}^{\mathrm{s}} : \mathbf{X} = \mathbf{X} : \mathcal{I}^{\mathrm{s}} = \mathrm{sym}\mathbf{X}, \quad \forall \mathbf{X} \in \mathbf{Lin}^{n}. \tag{5.87}$$

Spherical, deviatoric and trace projection tensors. The spherical and deviatoric part of a second-order tensor are defined as a linear mapping (1.163) and can thus be expressed by

$$\mathrm{sph}\mathbf{A} = \mathcal{P}_{\mathrm{sph}} : \mathbf{A}, \quad \mathrm{dev}\mathbf{A} = \mathcal{P}_{\mathrm{dev}} : \mathbf{A}, \tag{5.88}$$

where the fourth-order tensors $\mathcal{P}_{\mathrm{sph}}$ and $\mathcal{P}_{\mathrm{dev}}$ are called the spherical and deviatoric projection tensors, respectively. In view of (1.163) they are given by

$$\mathcal{P}_{\mathrm{sph}} = \frac{1}{n}\mathbf{I} \odot \mathbf{I}, \quad \mathcal{P}_{\mathrm{dev}} = \mathcal{I} - \frac{1}{n}\mathbf{I} \odot \mathbf{I}, \tag{5.89}$$

where $\mathbf{I} \odot \mathbf{I}$ represents the so-called trace projection tensor. Indeed,

$$\mathbf{I} \odot \mathbf{I} : \mathbf{X} = \mathbf{I}\mathrm{tr}\mathbf{X}, \quad \forall \mathbf{X} \in \mathbf{Lin}^n. \tag{5.90}$$

According to $(5.49)_2$ and (5.50), the spherical and trace projection tensors are super-symmetric. The spherical and deviatoric projection tensors are furthermore characterized by the properties:

$$\begin{aligned} \mathcal{P}_{\mathrm{dev}} : \mathcal{P}_{\mathrm{dev}} &= \mathcal{P}_{\mathrm{dev}}, \quad \mathcal{P}_{\mathrm{sph}} : \mathcal{P}_{\mathrm{sph}} = \mathcal{P}_{\mathrm{sph}}, \\ \mathcal{P}_{\mathrm{dev}} : \mathcal{P}_{\mathrm{sph}} &= \mathcal{P}_{\mathrm{sph}} : \mathcal{P}_{\mathrm{dev}} = \mathcal{O}. \end{aligned} \tag{5.91}$$

Example 5.2. Elasticity tensor for the generalized Hooke's law. The generalized Hooke's law is written as

$$\boldsymbol{\sigma} = 2G\boldsymbol{\epsilon} + \lambda \mathrm{tr}\,(\boldsymbol{\epsilon})\,\mathbf{I} = 2G\mathrm{dev}\boldsymbol{\epsilon} + \left(\lambda + \frac{2}{3}G\right)\mathrm{tr}\,(\boldsymbol{\epsilon})\,\mathbf{I}, \tag{5.92}$$

where G and λ denote the so-called Lamé constants. The corresponding super-symmetric elasticity tensor takes the form

$$\mathcal{C} = 2G\mathcal{I}^{\mathrm{s}} + \lambda\mathbf{I} \odot \mathbf{I} = 2G\mathcal{P}^{\mathrm{s}}_{\mathrm{dev}} + (3\lambda + 2G)\,\mathcal{P}_{\mathrm{sph}}. \tag{5.93}$$

Exercises

5.1. Prove relations (5.20) and (5.21).

5.2. Prove relations (5.22).

5.3. Prove relations (5.42) and (5.43).

5.4. Prove relations (5.49)–(5.52).

5.5. Prove that $\mathcal{A}^{\mathrm{Tt}} \neq \mathcal{A}^{\mathrm{tT}}$ for $\mathcal{A} = \boldsymbol{a} \otimes \boldsymbol{b} \otimes \boldsymbol{c} \otimes \boldsymbol{d}$.

5.6. Prove identities (5.54).

5.7. Verify relations (5.55) and (5.56).

5.8. Prove relations (5.70) for the components of a super-symmetric fourth-order tensor using (5.51) and (5.52).

5.9. Prove relation (5.82) using (5.16) and (5.81).

5.10. Verify the properties of the transposition tensor (5.85).

5.11. Prove that the fourth-order tensor of the form

$$\mathcal{C} = (\mathbf{M}_1 \otimes \mathbf{M}_2 + \mathbf{M}_2 \otimes \mathbf{M}_1)^{\mathrm{s}}$$

is super-symmetric if $\mathbf{M}_1, \mathbf{M}_2 \in \mathbf{Sym}^n$.

5.12. Calculate eigenvalues and eigentensors of the following super-symmetric fourth-order tensors for $n = 3$: (a) $\mathcal{J}^{\mathrm{s}}$ (5.86), (b) $\mathcal{P}_{\mathrm{sph}}$ $(5.89)_1$, (c) $\mathcal{P}^{\mathrm{s}}_{\mathrm{dev}}$ $(5.89)_2$, (d) $\mathcal{C}$ (5.93).

Chapter 6
Analysis of Tensor Functions

6.1 Scalar-Valued Isotropic Tensor Functions

Let us consider a real scalar-valued function $f(\mathbf{A}_1, \mathbf{A}_2, \ldots, \mathbf{A}_l)$ of second-order tensors $\mathbf{A}_k \in \mathbf{Lin}^n$ $(k = 1, 2, \ldots, l)$. The function f is said to be isotropic if

$$f\left(\mathbf{QA}_1\mathbf{Q}^{\mathrm{T}}, \mathbf{QA}_2\mathbf{Q}^{\mathrm{T}}, \ldots, \mathbf{QA}_l\mathbf{Q}^{\mathrm{T}}\right) \\ = f(\mathbf{A}_1, \mathbf{A}_2, \ldots, \mathbf{A}_l), \quad \forall \mathbf{Q} \in \mathbf{Orth}^n. \tag{6.1}$$

Example 6.1. Consider the function $f(\mathbf{A}, \mathbf{B}) = \mathrm{tr}(\mathbf{AB})$. Since in view of (1.135) and (1.151)

$$\begin{aligned} f\left(\mathbf{QAQ}^{\mathrm{T}}, \mathbf{QBQ}^{\mathrm{T}}\right) &= \mathrm{tr}\left(\mathbf{QAQ}^{\mathrm{T}}\mathbf{QBQ}^{\mathrm{T}}\right) \\ &= \mathrm{tr}\left(\mathbf{QABQ}^{\mathrm{T}}\right) = \mathrm{tr}\left(\mathbf{ABQ}^{\mathrm{T}}\mathbf{Q}\right) \\ &= \mathrm{tr}(\mathbf{AB}) = f(\mathbf{A}, \mathbf{B}), \quad \forall \mathbf{Q} \in \mathbf{Orth}^n, \end{aligned}$$

this function is isotropic according to the definition (6.1). In contrast, the function $f(\mathbf{A}) = \mathrm{tr}(\mathbf{AL})$, where $\mathbf{L}$ denotes a second-order tensor, is not isotropic. Indeed,

$$f\left(\mathbf{QAQ}^{\mathrm{T}}\right) = \mathrm{tr}\left(\mathbf{QAQ}^{\mathrm{T}}\mathbf{L}\right) \neq \mathrm{tr}(\mathbf{AL}).$$

Scalar-valued isotropic tensor functions are also called isotropic invariants of the tensors $\mathbf{A}_k$ $(k = 1, 2, \ldots, l)$. For such a tensor system one can construct, in principle, an unlimited number of isotropic invariants. However, for every finite system of tensors one can find a finite number of isotropic invariants in terms of which all other isotropic invariants can be expressed (Hilbert's theorem). This system of invariants is called functional basis of the tensors $\mathbf{A}_k$ $(k = 1, 2, \ldots, l)$. For one and the same system of tensors there exist many functional bases. A functional basis is called irreducible if none of its elements can be expressed in a unique form in terms of the remaining invariants.

M. Itskov, *Tensor Algebra and Tensor Analysis for Engineers*, Mathematical Engineering, DOI 10.1007/978-3-642-30879-6_6,

First, we focus on isotropic functions of one second-order tensor

$$f\left(\mathbf{QAQ}^{\mathrm{T}}\right) = f\left(\mathbf{A}\right), \quad \forall \mathbf{Q} \in \mathbf{Orth}^n, \quad \mathbf{A} \in \mathbf{Lin}^n. \tag{6.2}$$

One can show that the principal traces $\mathrm{tr}\mathbf{A}^k$, principal invariants $\mathrm{I}_{\mathbf{A}}^{(k)}$ and eigenvalues λ_k, $(k = 1, 2, \ldots, n)$ of the tensor $\mathbf{A}$ represent its isotropic tensor functions. Indeed, for the principal traces we can write by virtue of (1.151)

$$\begin{aligned}\mathrm{tr}\left(\mathbf{QAQ}^{\mathrm{T}}\right)^k &= \mathrm{tr}\Bigg(\underbrace{\mathbf{QAQ}^{\mathrm{T}}\mathbf{QAQ}^{\mathrm{T}}\ldots\mathbf{QAQ}^{\mathrm{T}}}_{k \text{ times}}\Bigg) = \mathrm{tr}\left(\mathbf{QA}^k\mathbf{Q}^{\mathrm{T}}\right)\\ &= \mathrm{tr}\left(\mathbf{A}^k\mathbf{Q}^{\mathrm{T}}\mathbf{Q}\right) = \mathrm{tr}\mathbf{A}^k, \quad \forall \mathbf{Q} \in \mathbf{Orth}^n.\end{aligned} \tag{6.3}$$

The principal invariants are uniquely expressed in terms of the principal traces by means of Newton's identities (4.26), while the eigenvalues are, in turn, defined by the principal invariants as solutions of the characteristic equation (4.21) with the characteristic polynomial given by (4.19).

Further, we prove that both the eigenvalues λ_k, principal invariants $\mathrm{I}_{\mathbf{M}}^{(k)}$ and principal traces $\mathrm{tr}\mathbf{M}^k$ $(k = 1, 2, \ldots, n)$ of one symmetric tensor $\mathbf{M} \in \mathbf{Sym}^n$ form its functional bases (see also [46]). To this end, we consider two arbitrary symmetric second-order tensors $\mathbf{M}_1, \mathbf{M}_2 \in \mathbf{Sym}^n$ with the same eigenvalues. Then, the spectral representation (4.61) takes the form

$$\mathbf{M}_1 = \sum_{i=1}^{n} \lambda_i \boldsymbol{n}_i \otimes \boldsymbol{n}_i, \quad \mathbf{M}_2 = \sum_{i=1}^{n} \lambda_i \boldsymbol{m}_i \otimes \boldsymbol{m}_i, \tag{6.4}$$

where according to (4.63) both the eigenvectors $\boldsymbol{n}_i$ and $\boldsymbol{m}_i$ form orthonormal bases such that $\boldsymbol{n}_i \cdot \boldsymbol{n}_j = \delta_{ij}$ and $\boldsymbol{m}_i \cdot \boldsymbol{m}_j = \delta_{ij}$ $(i, j = 1, 2, \ldots, n)$. Now, we consider the orthogonal tensor

$$\mathbf{Q} = \sum_{i=1}^{n} \boldsymbol{m}_i \otimes \boldsymbol{n}_i. \tag{6.5}$$

Indeed,

$$\begin{aligned}\mathbf{QQ}^{\mathrm{T}} &= \left(\sum_{i=1}^{n} \boldsymbol{m}_i \otimes \boldsymbol{n}_i\right)\left(\sum_{j=1}^{n} \boldsymbol{n}_j \otimes \boldsymbol{m}_j\right)\\ &= \sum_{i,j=1}^{n} \delta_{ij} \boldsymbol{m}_i \otimes \boldsymbol{m}_j = \sum_{i=1}^{n} \boldsymbol{m}_i \otimes \boldsymbol{m}_i = \mathbf{I}.\end{aligned}$$

By use of (1.121), (6.4) and (6.5) we further obtain

$$\mathbf{Q}\mathbf{M}_1\mathbf{Q}^{\mathrm{T}} = \left(\sum_{i=1}^{n} \boldsymbol{m}_i \otimes \boldsymbol{n}_i\right) \left(\sum_{j=1}^{n} \lambda_j \boldsymbol{n}_j \otimes \boldsymbol{n}_j\right) \left(\sum_{k=1}^{n} \boldsymbol{n}_k \otimes \boldsymbol{m}_k\right)$$

$$= \sum_{i,j,k=1}^{n} \delta_{ij}\delta_{jk}\lambda_j \boldsymbol{m}_i \otimes \boldsymbol{m}_k = \sum_{i=1}^{n} \lambda_i \boldsymbol{m}_i \otimes \boldsymbol{m}_i = \mathbf{M}_2. \tag{6.6}$$

Hence,

$$f(\mathbf{M}_1) = f\left(\mathbf{Q}\mathbf{M}_1\mathbf{Q}^{\mathrm{T}}\right) = f(\mathbf{M}_2). \tag{6.7}$$

Thus, f takes the same value for all symmetric tensors with pairwise equal eigenvalues. This means that an isotropic tensor function of a symmetric tensor is uniquely defined in terms of its eigenvalues, principal invariants or principal traces because the latter ones are, in turn, uniquely defined by the eigenvalues according to (4.24) and (4.25). This implies the following representations

$$f(\mathbf{M}) = \widehat{f}\left(\mathrm{I}_{\mathbf{M}}^{(1)}, \mathrm{I}_{\mathbf{M}}^{(2)}, \ldots, \mathrm{I}_{\mathbf{M}}^{(n)}\right) = \hat{f}(\lambda_1, \lambda_2, \ldots, \lambda_n)$$

$$= \tilde{f}\left(\mathrm{tr}\mathbf{M}, \mathrm{tr}\mathbf{M}^2, \ldots, \mathrm{tr}\mathbf{M}^n\right), \quad \mathbf{M} \in \mathbf{Sym}^n. \tag{6.8}$$

Example 6.2. Strain energy function of an isotropic hyperelastic material. A material is said to be hyperelastic if it is characterized by the existence of a strain energy ψ defined as a function, for example, of the right Cauchy-Green tensor $\mathbf{C}$. For isotropic materials this strain energy function obeys the condition

$$\psi\left(\mathbf{Q}\mathbf{C}\mathbf{Q}^{\mathrm{T}}\right) = \psi(\mathbf{C}), \quad \forall \mathbf{Q} \in \mathbf{Orth}^3. \tag{6.9}$$

By means of (6.8) this function can be expressed by

$$\psi(\mathbf{C}) = \widehat{\psi}(\mathrm{I_C}, \mathrm{II_C}, \mathrm{III_C}) = \hat{\psi}(\lambda_1, \lambda_2, \lambda_3) = \tilde{\psi}\left(\mathrm{tr}\mathbf{C}, \mathrm{tr}\mathbf{C}^2, \mathrm{tr}\mathbf{C}^3\right), \tag{6.10}$$

where λ_i denote the so-called principal stretches. They are expressed in terms of the eigenvalues Λ_i $(i = 1, 2, 3)$ of the right Cauchy-Green tensor $\mathbf{C} = \sum_{i=1}^{3} \Lambda_i \mathbf{P}_i$ as $\lambda_i = \sqrt{\Lambda_i}$. For example, the strain energy function of the so-called Mooney-Rivlin material is given in terms of the first and second principal invariants by

$$\psi(\mathbf{C}) = c_1(\mathrm{I_C} - 3) + c_2(\mathrm{II_C} - 3), \tag{6.11}$$

where c_1 and c_2 represent material constants. In contrast, the strain energy function of the Ogden material [30] is defined in terms of the principal stretches by

$$\psi(\mathbf{C}) = \sum_{r=1}^{m} \frac{\mu_r}{\alpha_r}\left(\lambda_1^{\alpha_r} + \lambda_2^{\alpha_r} + \lambda_3^{\alpha_r} - 3\right), \tag{6.12}$$

where μ_r, α_r $(r = 1, 2, \ldots, m)$ denote material constants.

For isotropic functions (6.1) of a finite number l of arbitrary second-order tensors the functional basis is obtained only for three-dimensional space. In order to represent this basis, the tensor arguments are split according to (1.152) into a symmetric and a skew-symmetric part respectively as follows:

$$\mathbf{M}_i = \mathrm{sym}\mathbf{A}_i = \frac{1}{2}\left(\mathbf{A}_i + \mathbf{A}_i^{\mathrm{T}}\right), \quad \mathbf{W}_i = \mathrm{skew}\mathbf{A}_i = \frac{1}{2}\left(\mathbf{A}_i - \mathbf{A}_i^{\mathrm{T}}\right). \tag{6.13}$$

In this manner, every isotropic tensor function can be given in terms of a finite number of symmetric tensors $\mathbf{M}_i \in \mathbf{Sym}^3$ $(i = 1, 2, \ldots, m)$ and skew-symmetric tensors $\mathbf{W}_i \in \mathbf{Skew}^3$ $(i = 1, 2, \ldots, w)$ as

$$f = \hat{f}\left(\mathbf{M}_1, \mathbf{M}_2, \ldots, \mathbf{M}_m, \mathbf{W}_1, \mathbf{W}_2, \ldots, \mathbf{W}_w\right). \tag{6.14}$$

An irreducible functional basis of such a system of tensors is proved to be given by (see [2, 33, 41])

$$\begin{gathered}
\mathrm{tr}\mathbf{M}_i,\ \mathrm{tr}\mathbf{M}_i^2,\ \mathrm{tr}\mathbf{M}_i^3, \\
\mathrm{tr}\left(\mathbf{M}_i\mathbf{M}_j\right),\ \mathrm{tr}\left(\mathbf{M}_i^2\mathbf{M}_j\right),\ \mathrm{tr}\left(\mathbf{M}_i\mathbf{M}_j^2\right),\ \mathrm{tr}\left(\mathbf{M}_i^2\mathbf{M}_j^2\right),\ \mathrm{tr}\left(\mathbf{M}_i\mathbf{M}_j\mathbf{M}_k\right), \\
\mathrm{tr}\mathbf{W}_p^2,\ \mathrm{tr}\left(\mathbf{W}_p\mathbf{W}_q\right),\ \mathrm{tr}\left(\mathbf{W}_p\mathbf{W}_q\mathbf{W}_r\right), \\
\mathrm{tr}\left(\mathbf{M}_i\mathbf{W}_p^2\right),\ \mathrm{tr}\left(\mathbf{M}_i^2\mathbf{W}_p^2\right),\ \mathrm{tr}\left(\mathbf{M}_i^2\mathbf{W}_p^2\mathbf{M}_i\mathbf{W}_p\right),\ \mathrm{tr}\left(\mathbf{M}_i\mathbf{W}_p\mathbf{W}_q\right), \\
\mathrm{tr}\left(\mathbf{M}_i\mathbf{W}_p^2\mathbf{W}_q\right),\ \mathrm{tr}\left(\mathbf{M}_i\mathbf{W}_p\mathbf{W}_q^2\right),\ \mathrm{tr}\left(\mathbf{M}_i\mathbf{M}_j\mathbf{W}_p\right), \\
\mathrm{tr}\left(\mathbf{M}_i\mathbf{W}_p^2\mathbf{M}_j\mathbf{W}_p\right),\ \mathrm{tr}\left(\mathbf{M}_i^2\mathbf{M}_j\mathbf{W}_p\right),\ \mathrm{tr}\left(\mathbf{M}_i\mathbf{M}_j^2\mathbf{W}_p\right), \\
i < j < k = 1, 2, \ldots, m, \quad p < q < r = 1, 2, \ldots, w.
\end{gathered} \tag{6.15}$$

For illustration of this result we consider some examples.

Example 6.3. Functional basis of one skew-symmetric second-order tensor $\mathbf{W} \in \mathbf{Skew}^3$. With the aid of (6.15) and (4.91) we obtain the basis consisting of only one invariant

$$\mathrm{tr}\mathbf{W}^2 = -2\mathrm{II}_{\mathbf{W}} = -\left\|\mathbf{W}\right\|^2. \tag{6.16}$$

Example 6.4. Functional basis of an arbitrary second-order tensor $\mathbf{A} \in \mathbf{Lin}^3$. By means of (6.15) one can write the following functional basis of $\mathbf{A}$

$$\begin{gathered}
\mathrm{tr}\mathbf{M},\ \mathrm{tr}\mathbf{M}^2,\ \mathrm{tr}\mathbf{M}^3, \\
\mathrm{tr}\mathbf{W}^2,\ \mathrm{tr}\left(\mathbf{M}\mathbf{W}^2\right),\ \mathrm{tr}\left(\mathbf{M}^2\mathbf{W}^2\right),\ \mathrm{tr}\left(\mathbf{M}^2\mathbf{W}^2\mathbf{M}\mathbf{W}\right),
\end{gathered} \tag{6.17}$$

where $\mathbf{M} = \mathrm{sym}\mathbf{A}$ and $\mathbf{W} = \mathrm{skew}\mathbf{A}$. Inserting representations (6.13) into (6.17) the functional basis of $\mathbf{A}$ can be rewritten as (see Exercise 6.2)

$$\mathrm{tr}\mathbf{A},\ \mathrm{tr}\mathbf{A}^2,\ \mathrm{tr}\mathbf{A}^3,\ \mathrm{tr}\left(\mathbf{A}\mathbf{A}^{\mathrm{T}}\right),\ \mathrm{tr}\left(\mathbf{A}\mathbf{A}^{\mathrm{T}}\right)^2,\ \mathrm{tr}\left(\mathbf{A}^2\mathbf{A}^{\mathrm{T}}\right),$$
$$\mathrm{tr}\left[\left(\mathbf{A}^{\mathrm{T}}\right)^2\mathbf{A}^2\mathbf{A}^{\mathrm{T}}\mathbf{A}-\mathbf{A}^2\left(\mathbf{A}^{\mathrm{T}}\right)^2\mathbf{A}\mathbf{A}^{\mathrm{T}}\right]. \tag{6.18}$$

Example 6.5. Functional basis of two symmetric second-order tensors $\mathbf{M}_1, \mathbf{M}_2 \in \mathbf{Sym}^3$. According to (6.15) the functional basis includes in this case the following ten invariants

$$\mathrm{tr}\mathbf{M}_1,\ \mathrm{tr}\mathbf{M}_1^2,\ \mathrm{tr}\mathbf{M}_1^3,\ \mathrm{tr}\mathbf{M}_2,\ \mathrm{tr}\mathbf{M}_2^2,\ \mathrm{tr}\mathbf{M}_2^3,$$
$$\mathrm{tr}\left(\mathbf{M}_1\mathbf{M}_2\right),\ \mathrm{tr}\left(\mathbf{M}_1^2\mathbf{M}_2\right),\ \mathrm{tr}\left(\mathbf{M}_1\mathbf{M}_2^2\right),\ \mathrm{tr}\left(\mathbf{M}_1^2\mathbf{M}_2^2\right). \tag{6.19}$$

6.2 Scalar-Valued Anisotropic Tensor Functions

A real scalar-valued function $f(\mathbf{A}_1, \mathbf{A}_2, \ldots, \mathbf{A}_l)$ of second-order tensors $\mathbf{A}_k \in \mathbf{Lin}^n$ $(k = 1, 2, \ldots, l)$ is said to be anisotropic if it is invariant only with respect to a subset of all orthogonal transformations:

$$f\left(\mathbf{Q}\mathbf{A}_1\mathbf{Q}^{\mathrm{T}}, \mathbf{Q}\mathbf{A}_2\mathbf{Q}^{\mathrm{T}}, \ldots, \mathbf{Q}\mathbf{A}_l\mathbf{Q}^{\mathrm{T}}\right)$$
$$= f\left(\mathbf{A}_1, \mathbf{A}_2, \ldots, \mathbf{A}_l\right), \quad \forall \mathbf{Q} \in \mathbf{Sorth}^n \subset \mathbf{Orth}^n. \tag{6.20}$$

The subset $\mathbf{Sorth}^n$ represents a group called symmetry group. In continuum mechanics, anisotropic properties of materials are characterized by their symmetry group. The largest symmetry group $\mathbf{Orth}^3$ (in three-dimensional space) includes all orthogonal transformations and is referred to as isotropic. In contrast, the smallest symmetry group consists of only two elements $\mathbf{I}$ and $-\mathbf{I}$ and is called triclinic.

Example 6.6. Transversely isotropic material symmetry. In this case the material is characterized by symmetry with respect to one selected direction referred to as principal material direction. Properties of a transversely isotropic material remain unchanged by rotations about, and reflections from the planes orthogonal or parallel to, this direction. Introducing a unit vector $\boldsymbol{l}$ in the principal direction we can write

$$\mathbf{Q}\boldsymbol{l} = \pm\boldsymbol{l}, \quad \forall \mathbf{Q} \in \mathfrak{g}_t, \tag{6.21}$$

where $\mathfrak{g}_t \subset \mathbf{Orth}^3$ denotes the transversely isotropic symmetry group. With the aid of a special tensor

$$\mathbf{L} = \boldsymbol{l} \otimes \boldsymbol{l}, \tag{6.22}$$

called structural tensor, condition (6.21) can be represented as

$$\mathbf{Q}\mathbf{L}\mathbf{Q}^{\mathrm{T}} = \mathbf{L}, \quad \forall \mathbf{Q} \in \mathfrak{g}_t. \tag{6.23}$$

Hence, the transversely isotropic symmetry group can be defined by

$$\mathfrak{g}_t = \left\{ \mathbf{Q} \in \mathrm{Orth}^3 : \mathbf{Q}\mathbf{L}\mathbf{Q}^{\mathrm{T}} = \mathbf{L} \right\}. \tag{6.24}$$

A strain energy function ψ of a transversely isotropic material is invariant with respect to all orthogonal transformations within $\mathfrak{g}_t$. Using a representation in terms of the right Cauchy-Green tensor $\mathbf{C}$ this leads to the following condition:

$$\psi\left(\mathbf{Q}\mathbf{C}\mathbf{Q}^{\mathrm{T}}\right) = \psi\left(\mathbf{C}\right), \quad \forall \mathbf{Q} \in \mathfrak{g}_t. \tag{6.25}$$

It can be shown that this condition is a priori satisfied if the strain energy function can be represented as an isotropic function of both $\mathbf{C}$ and $\mathbf{L}$ so that

$$\hat{\psi}\left(\mathbf{Q}\mathbf{C}\mathbf{Q}^{\mathrm{T}}, \mathbf{Q}\mathbf{L}\mathbf{Q}^{\mathrm{T}}\right) = \hat{\psi}\left(\mathbf{C}, \mathbf{L}\right), \quad \forall \mathbf{Q} \in \mathrm{Orth}^3. \tag{6.26}$$

Indeed,

$$\hat{\psi}\left(\mathbf{C}, \mathbf{L}\right) = \hat{\psi}\left(\mathbf{Q}\mathbf{C}\mathbf{Q}^{\mathrm{T}}, \mathbf{Q}\mathbf{L}\mathbf{Q}^{\mathrm{T}}\right) = \hat{\psi}\left(\mathbf{Q}\mathbf{C}\mathbf{Q}^{\mathrm{T}}, \mathbf{L}\right), \quad \forall \mathbf{Q} \in \mathfrak{g}_t. \tag{6.27}$$

With the aid of the functional basis (6.19) and taking into account the identities

$$\mathbf{L}^k = \mathbf{L}, \quad \mathrm{tr}\mathbf{L}^k = 1, \quad k = 1, 2, \ldots \tag{6.28}$$

resulting from (6.22) we can thus represent the transversely isotropic function in terms of the five invariants by (see also [43])

$$\psi = \hat{\psi}\left(\mathbf{C}, \mathbf{L}\right) = \tilde{\psi}\left[\mathrm{tr}\mathbf{C}, \mathrm{tr}\mathbf{C}^2, \mathrm{tr}\mathbf{C}^3, \mathrm{tr}\left(\mathbf{C}\mathbf{L}\right), \mathrm{tr}\left(\mathbf{C}^2\mathbf{L}\right)\right]. \tag{6.29}$$

The above procedure can be generalized for an arbitrary anisotropic symmetry group $\mathfrak{g}$. Let $\mathbf{L}_i$ $(i = 1, 2, \ldots, m)$ be a set of second-order tensors which uniquely define $\mathfrak{g}$ by

$$\mathfrak{g} = \left\{ \mathbf{Q} \in \mathrm{Orth}^n : \mathbf{Q}\mathbf{L}_i\mathbf{Q}^{\mathrm{T}} = \mathbf{L}_i, \ i = 1, 2, \ldots, m \right\}. \tag{6.30}$$

In continuum mechanics the tensors $\mathbf{L}_i$ are called structural tensors since they lay down the material or structural symmetry.

It is seen that the isotropic tensor function

$$f\left(\mathbf{Q}\mathbf{A}_i\mathbf{Q}^{\mathrm{T}}, \mathbf{Q}\mathbf{L}_j\mathbf{Q}^{\mathrm{T}}\right) = f\left(\mathbf{A}_i, \mathbf{L}_j\right), \quad \forall \mathbf{Q} \in \mathrm{Orth}^n, \tag{6.31}$$

where we use the abbreviated notation

$$f\left(\mathbf{A}_i, \mathbf{L}_j\right) = f\left(\mathbf{A}_1, \mathbf{A}_2, \ldots, \mathbf{A}_l, \mathbf{L}_1, \mathbf{L}_2, \ldots, \mathbf{L}_m\right), \tag{6.32}$$

is anisotropic with respect to the arguments $\mathbf{A}_i$ $(i = 1, 2, \ldots, l)$ so that

$$f\left(\mathbf{Q}\mathbf{A}_i\mathbf{Q}^{\mathrm{T}}\right) = f\left(\mathbf{A}_i\right), \quad \forall \mathbf{Q} \in \mathfrak{g}. \tag{6.33}$$

Indeed, by virtue of (6.30) and (6.31) we have

$$f\left(\mathbf{A}_i, \mathbf{L}_j\right) = f\left(\mathbf{Q}\mathbf{A}_i\mathbf{Q}^{\mathrm{T}}, \mathbf{Q}\mathbf{L}_j\mathbf{Q}^{\mathrm{T}}\right) = f\left(\mathbf{Q}\mathbf{A}_i\mathbf{Q}^{\mathrm{T}}, \mathbf{L}_j\right), \quad \forall \mathbf{Q} \in \mathfrak{g}. \tag{6.34}$$

Thus, every isotropic invariant of the tensor system $\mathbf{A}_i$ $(i = 1, 2, \ldots, l)$, $\mathbf{L}_j$ $(j = 1, 2, \ldots, m)$ represents an anisotropic invariant of the tensors $\mathbf{A}_i$ $(i = 1, 2, \ldots, l)$ in the sense of definition (6.20). Conversely, one can show that for every anisotropic function (6.33) there exists an equivalent isotropic function of the tensor system $\mathbf{A}_i$ $(i = 1, 2, \ldots, l)$, $\mathbf{L}_j$ $(j = 1, 2, \ldots, m)$. In order to prove this statement we consider a new tensor function defined by

$$\hat{f}\left(\mathbf{A}_i, \mathbf{X}_j\right) = f\left(\mathbf{Q}'\mathbf{A}_i\mathbf{Q}'^{\mathrm{T}}\right), \tag{6.35}$$

where the tensor $\mathbf{Q}' \in \mathbf{Orth}^n$ results from the condition:

$$\mathbf{Q}'\mathbf{X}_j\mathbf{Q}'^{\mathrm{T}} = \mathbf{L}_j, \quad j = 1, 2, \ldots, m. \tag{6.36}$$

Thus, the function $\hat{f}$ is defined only over such tensors $\mathbf{X}_j$ that can be obtained from the structural tensors $\mathbf{L}_j$ $(j = 1, 2, \ldots, m)$ by the transformation

$$\mathbf{X}_j = \mathbf{Q}'^{\mathrm{T}}\mathbf{L}_j\mathbf{Q}', \quad j = 1, 2, \ldots, m, \tag{6.37}$$

where $\mathbf{Q}'$ is an arbitrary orthogonal tensor.

Further, one can show that the so-defined function (6.35) is isotropic. Indeed,

$$\hat{f}\left(\mathbf{Q}\mathbf{A}_i\mathbf{Q}^{\mathrm{T}}, \mathbf{Q}\mathbf{X}_j\mathbf{Q}^{\mathrm{T}}\right) = f\left(\mathbf{Q}''\mathbf{Q}\mathbf{A}_i\mathbf{Q}^{\mathrm{T}}\mathbf{Q}''^{\mathrm{T}}\right), \quad \forall \mathbf{Q} \in \mathbf{Orth}^n, \tag{6.38}$$

where according to (6.36)

$$\mathbf{Q}''\mathbf{Q}\mathbf{X}_j\mathbf{Q}^{\mathrm{T}}\mathbf{Q}''^{\mathrm{T}} = \mathbf{L}_j, \quad \mathbf{Q}'' \in \mathbf{Orth}^n. \tag{6.39}$$

Inserting (6.37) into (6.39) yields

$$\mathbf{Q}''\mathbf{Q}\mathbf{Q}'^{\mathrm{T}}\mathbf{L}_j\mathbf{Q}'\mathbf{Q}^{\mathrm{T}}\mathbf{Q}''^{\mathrm{T}} = \mathbf{L}_j, \tag{6.40}$$

so that

$$\mathbf{Q}^* = \mathbf{Q}''\mathbf{Q}\mathbf{Q}'^{\mathrm{T}} \in \mathfrak{g}. \tag{6.41}$$

Hence, we can write

$$f\left(\mathbf{Q}''\mathbf{Q}\mathbf{A}_i\mathbf{Q}^{\mathrm{T}}\mathbf{Q}''^{\mathrm{T}}\right) = f\left(\mathbf{Q}^*\mathbf{Q}'\mathbf{A}_i\mathbf{Q}'^{\mathrm{T}}\mathbf{Q}^{*\mathrm{T}}\right)$$
$$= f\left(\mathbf{Q}'\mathbf{A}_i\mathbf{Q}'^{\mathrm{T}}\right) = \hat{f}\left(\mathbf{A}_i, \mathbf{X}_j\right)$$

and consequently in view of (6.38)

$$\hat{f}\left(\mathbf{Q}\mathbf{A}_i\mathbf{Q}^{\mathrm{T}}, \mathbf{Q}\mathbf{X}_j\mathbf{Q}^{\mathrm{T}}\right) = \hat{f}\left(\mathbf{A}_i, \mathbf{X}_j\right), \quad \forall \mathbf{Q} \in \mathbf{Orth}^n. \tag{6.42}$$

Thus, we have proved the following theorem [50].

Theorem 6.1. *A scalar-valued function $f(\mathbf{A}_i)$ is invariant within the symmetry group $\mathfrak{g}$ defined by (6.30) if and only if there exists an isotropic function $\hat{f}\left(\mathbf{A}_i, \mathbf{L}_j\right)$ such that*

$$f(\mathbf{A}_i) = \hat{f}\left(\mathbf{A}_i, \mathbf{L}_j\right). \tag{6.43}$$

6.3 Derivatives of Scalar-Valued Tensor Functions

Let us again consider a scalar-valued tensor function $f(\mathbf{A}) : \mathbf{Lin}^n \mapsto \mathbb{R}$. This function is said to be differentiable in a neighborhood of $\mathbf{A}$ if there exists a tensor $f(\mathbf{A})_{,\mathbf{A}} \in \mathbf{Lin}^n$, such that

$$\left.\frac{\mathrm{d}}{\mathrm{d}t} f(\mathbf{A} + t\mathbf{X})\right|_{t=0} = f(\mathbf{A})_{,\mathbf{A}} : \mathbf{X}, \quad \forall \mathbf{X} \in \mathbf{Lin}^n. \tag{6.44}$$

This definition implies that the directional derivative (also called Gateaux derivative) $\left.\frac{\mathrm{d}}{\mathrm{d}t} f(\mathbf{A} + t\mathbf{X})\right|_{t=0}$ exists and is continuous at $\mathbf{A}$. The tensor $f(\mathbf{A})_{,\mathbf{A}}$ is referred to as the derivative or the gradient of the tensor function $f(\mathbf{A})$.

In order to obtain a direct expression for $f(\mathbf{A})_{,\mathbf{A}}$ we represent the tensors $\mathbf{A}$ and $\mathbf{X}$ in (6.44) with respect to an arbitrary basis, say $\boldsymbol{g}_i \otimes \boldsymbol{g}^j$ $(i, j = 1, 2, \ldots, n)$. Then, using the chain rule one can write

$$\left.\frac{\mathrm{d}}{\mathrm{d}t} f(\mathbf{A} + t\mathbf{X})\right|_{t=0} = \left.\frac{\mathrm{d}}{\mathrm{d}t} f\left[\left(\mathrm{A}^i_{\cdot j} + t\mathrm{X}^i_{\cdot j}\right)\boldsymbol{g}_i \otimes \boldsymbol{g}^j\right]\right|_{t=0} = \frac{\partial f}{\partial \mathrm{A}^i_{\cdot j}}\mathrm{X}^i_{\cdot j}.$$

Comparing this result with (6.44) yields

$$f(\mathbf{A})_{,\mathbf{A}} = \frac{\partial f}{\partial \mathrm{A}^i_{\cdot j}}\boldsymbol{g}^i \otimes \boldsymbol{g}_j = \frac{\partial f}{\partial \mathrm{A}_{ij}}\boldsymbol{g}_i \otimes \boldsymbol{g}_j = \frac{\partial f}{\partial \mathrm{A}^{ij}}\boldsymbol{g}^i \otimes \boldsymbol{g}^j = \frac{\partial f}{\partial \mathrm{A}_{i\cdot}^{\cdot j}}\boldsymbol{g}_i \otimes \boldsymbol{g}^j. \tag{6.45}$$

If the function $f(\mathbf{A})$ is defined not on all linear transformations but only on a subset $\mathbf{Slin}^n \subset \mathbf{Lin}^n$, the directional derivative (6.44) does not, however, yield a unique result for $f(\mathbf{A})_{,\mathbf{A}}$. In this context, let us consider for example scalar-valued functions of symmetric tensors: $f(\mathbf{M}) : \mathbf{Sym}^n \mapsto \mathbb{R}$. In this case, the directional derivative (6.44) defines $f(\mathbf{M})_{,\mathbf{M}}$ only up to an arbitrary skew-symmetric component $\mathbf{W}$. Indeed,

$$f(\mathbf{M})_{,\mathbf{M}} : \mathbf{X} = [f(\mathbf{M})_{,\mathbf{M}} + \mathbf{W}] : \mathbf{X}, \quad \forall \mathbf{W} \in \mathbf{Skew}^n, \ \forall \mathbf{X} \in \mathbf{Sym}^n. \tag{6.46}$$

In this relation, $\mathbf{X}$ is restricted to symmetric tensors because the tensor $\mathbf{M} + t\mathbf{X}$ appearing in the directional derivative (6.44) must belong to the definition domain of the function f for all real values of t.

To avoid this non-uniqueness we will assume that the derivative $f(\mathbf{A})_{,\mathbf{A}}$ belongs to the same subset $\mathbf{Slin}^n \subset \mathbf{Lin}^n$ as its argument $\mathbf{A} \in \mathbf{Slin}^n$. In particular, for symmetric tensor functions it implies that

$$f(\mathbf{M})_{,\mathbf{M}} \in \mathbf{Sym}^n \quad \text{for} \quad \mathbf{M} \in \mathbf{Sym}^n. \tag{6.47}$$

In order to calculate the derivative of a symmetric tensor function satisfying the condition (6.47) one can apply the following procedure. First, the definition domain of the function f is notionally extended to all linear transformations $\mathbf{Lin}^n$. Applying then the directional derivative (6.44) one obtains a unique result for the tensor $f_{,\mathbf{M}}$ which is finally symmetrized. For the derivative with respect to a symmetric part (1.153) of a tensor argument this procedure can be written by

$$f(\mathrm{sym}\mathbf{A})_{,\mathrm{sym}\mathbf{A}} = \mathrm{sym}\left[f(\mathbf{A})_{,\mathbf{A}}\right], \quad \mathbf{A} \in \mathbf{Lin}^n. \tag{6.48}$$

The problem with the non-uniqueness appears likewise by using the component representation (6.45) for the gradient of symmetric tensor functions. Indeed, in this case $\mathrm{M}^{ij} = \mathrm{M}^{ji}$ $(i \neq j = 1, 2, \ldots, n)$, so that only $n(n+1)/2$ among all n^2 components of the tensor argument $\mathbf{M} \in \mathbf{Sym}^n$ are independent. Thus, according to (1.156)

$$\mathbf{M} = \sum_{i=1}^{n} \mathrm{M}^{ii} \boldsymbol{g}_i \otimes \boldsymbol{g}_i + \sum_{\substack{i,j=1 \\ j<i}}^{n} \mathrm{M}^{ij} \left(\boldsymbol{g}_i \otimes \boldsymbol{g}_j + \boldsymbol{g}_j \otimes \boldsymbol{g}_i\right), \quad \mathbf{M} \in \mathbf{Sym}^n. \tag{6.49}$$

Hence, instead of (6.45) we obtain

$$\begin{aligned} f(\mathbf{M})_{,\mathbf{M}} &= \frac{1}{2} \sum_{\substack{i,j=1 \\ j \leq i}}^{n} \frac{\partial f}{\partial \mathrm{M}^{ij}} \left(\boldsymbol{g}^i \otimes \boldsymbol{g}^j + \boldsymbol{g}^j \otimes \boldsymbol{g}^i\right) \\ &= \frac{1}{2} \sum_{\substack{i,j=1 \\ j \leq i}}^{n} \frac{\partial f}{\partial \mathrm{M}_{ij}} \left(\boldsymbol{g}_i \otimes \boldsymbol{g}_j + \boldsymbol{g}_j \otimes \boldsymbol{g}_i\right), \quad \mathbf{M} \in \mathbf{Sym}^n. \end{aligned} \tag{6.50}$$

It is seen that the derivative is taken here only with respect to the independent components of the symmetric tensor argument; the resulting tensor is then symmetrized.

Example 6.7. Derivative of the quadratic norm $\|\mathbf{A}\| = \sqrt{\mathbf{A}:\mathbf{A}}$:

$$\frac{\mathrm{d}}{\mathrm{d}t}\left[(\mathbf{A}+t\mathbf{X}):(\mathbf{A}+t\mathbf{X})\right]^{1/2}\Big|_{t=0}$$

$$= \frac{\mathrm{d}}{\mathrm{d}t}\left[\mathbf{A}:\mathbf{A}+2t\mathbf{A}:\mathbf{X}+t^2\mathbf{X}:\mathbf{X}\right]^{1/2}\Big|_{t=0}$$

$$= \frac{2\mathbf{A}:\mathbf{X}+2t\mathbf{X}:\mathbf{X}}{2\left[\mathbf{A}:\mathbf{A}+2t\mathbf{A}:\mathbf{X}+t^2\mathbf{X}:\mathbf{X}\right]^{1/2}}\Bigg|_{t=0} = \frac{\mathbf{A}}{\|\mathbf{A}\|}:\mathbf{X}.$$

Thus,

$$\|\mathbf{A}\|_{,\mathbf{A}} = \frac{\mathbf{A}}{\|\mathbf{A}\|}. \tag{6.51}$$

The same result can also be obtained using (6.45). Indeed, let $\mathbf{A} = \mathrm{A}_{ij}\boldsymbol{g}^i \otimes \boldsymbol{g}^j$. Then,

$$\|\mathbf{A}\| = \sqrt{\mathbf{A}:\mathbf{A}} = \sqrt{\left(\mathrm{A}_{ij}\boldsymbol{g}^i \otimes \boldsymbol{g}^j\right):\left(\mathrm{A}_{kl}\boldsymbol{g}^k \otimes \boldsymbol{g}^l\right)} = \sqrt{\mathrm{A}_{ij}\mathrm{A}_{kl}g^{ik}g^{jl}}.$$

Utilizing the identity

$$\frac{\partial \mathrm{A}_{ij}}{\partial \mathrm{A}_{pq}} = \delta_i^p \delta_j^q, \quad i,j,p,q = 1,2,\ldots,n$$

we further write

$$\|\mathbf{A}\|_{,\mathbf{A}} = \frac{\partial\sqrt{\mathrm{A}_{ij}\mathrm{A}_{kl}g^{ik}g^{jl}}}{\partial \mathrm{A}_{pq}}\boldsymbol{g}_p \otimes \boldsymbol{g}_q$$

$$= \frac{1}{2\|\mathbf{A}\|}\left(\mathrm{A}_{kl}g^{ik}g^{jl}\boldsymbol{g}_i \otimes \boldsymbol{g}_j + \mathrm{A}_{ij}g^{ik}g^{jl}\boldsymbol{g}_k \otimes \boldsymbol{g}_l\right)$$

$$= \frac{1}{2\|\mathbf{A}\|}2\mathrm{A}_{kl}g^{ik}g^{jl}\boldsymbol{g}_i \otimes \boldsymbol{g}_j = \frac{1}{\|\mathbf{A}\|}\mathrm{A}_{kl}\boldsymbol{g}^k \otimes \boldsymbol{g}^l = \frac{\mathbf{A}}{\|\mathbf{A}\|}.$$

Example 6.8. Derivatives of the principal traces $\mathrm{tr}\mathbf{A}^k$ $(k = 1,2,\ldots)$:

$$\frac{\mathrm{d}}{\mathrm{d}t}\left[\mathrm{tr}\,(\mathbf{A}+t\mathbf{X})^k\right]\Big|_{t=0} = \frac{\mathrm{d}}{\mathrm{d}t}\left[(\mathbf{A}+t\mathbf{X})^k:\mathbf{I}\right]\Big|_{t=0} = \frac{\mathrm{d}}{\mathrm{d}t}\left[(\mathbf{A}+t\mathbf{X})^k\right]\Big|_{t=0}:\mathbf{I}$$

$$= \frac{\mathrm{d}}{\mathrm{d}t}\left[\underbrace{(\mathbf{A}+t\mathbf{X})(\mathbf{A}+t\mathbf{X})\ldots(\mathbf{A}+t\mathbf{X})}_{k\text{ times}}\right]\Bigg|_{t=0}:\mathbf{I}$$

$$= \frac{\mathrm{d}}{\mathrm{d}t}\left[\mathbf{A}^k + t\sum_{i=0}^{k-1}\mathbf{A}^i\mathbf{X}\mathbf{A}^{k-1-i} + t^2\ldots\right]\Bigg|_{t=0} : \mathbf{I}$$

$$= \sum_{i=0}^{k-1}\mathbf{A}^i\mathbf{X}\mathbf{A}^{k-1-i} : \mathbf{I} = k\left(\mathbf{A}^{k-1}\right)^{\mathrm{T}} : \mathbf{X}.$$

Thus,

$$\left(\mathrm{tr}\mathbf{A}^k\right)_{,\mathbf{A}} = k\left(\mathbf{A}^{k-1}\right)^{\mathrm{T}}. \tag{6.52}$$

In the special case $k = 1$ we obtain

$$(\mathrm{tr}\mathbf{A})_{,\mathbf{A}} = \mathbf{I}. \tag{6.53}$$

Example 6.9. Derivatives of $\mathrm{tr}\left(\mathbf{A}^k\mathbf{L}\right)$ $(k = 1, 2, \ldots)$ with respect to $\mathbf{A}$, where $\mathbf{L}$ is independent of $\mathbf{A}$:

$$\frac{\mathrm{d}}{\mathrm{d}t}\left[(\mathbf{A}+t\mathbf{X})^k : \mathbf{L}^{\mathrm{T}}\right]\Bigg|_{t=0} = \frac{\mathrm{d}}{\mathrm{d}t}\left[(\mathbf{A}+t\mathbf{X})^k\right]\Bigg|_{t=0} : \mathbf{L}^{\mathrm{T}}$$

$$= \sum_{i=0}^{k-1}\mathbf{A}^i\mathbf{X}\mathbf{A}^{k-1-i} : \mathbf{L}^{\mathrm{T}} = \sum_{i=0}^{k-1}\left(\mathbf{A}^{\mathrm{T}}\right)^i\mathbf{L}^{\mathrm{T}}\left(\mathbf{A}^{\mathrm{T}}\right)^{k-1-i} : \mathbf{X}.$$

Hence,

$$\mathrm{tr}\left(\mathbf{A}^k\mathbf{L}\right)_{,\mathbf{A}} = \sum_{i=0}^{k-1}\left(\mathbf{A}^i\mathbf{L}\mathbf{A}^{k-1-i}\right)^{\mathrm{T}}. \tag{6.54}$$

In the special case $k = 1$ we have

$$\mathrm{tr}(\mathbf{A}\mathbf{L})_{,\mathbf{A}} = \mathbf{L}^{\mathrm{T}}. \tag{6.55}$$

It is seen that the derivative of $\mathrm{tr}\left(\mathbf{A}^k\mathbf{L}\right)$ is not in general symmetric even if the tensor argument $\mathbf{A}$ is. Applying (6.48) we can write in this case

$$\mathrm{tr}\left(\mathbf{M}^k\mathbf{L}\right)_{,\mathbf{M}} = \mathrm{sym}\left[\sum_{i=0}^{k-1}\left(\mathbf{M}^i\mathbf{L}\mathbf{M}^{k-1-i}\right)^{\mathrm{T}}\right] = \sum_{i=0}^{k-1}\mathbf{M}^i\left(\mathrm{sym}\mathbf{L}\right)\mathbf{M}^{k-1-i}, \tag{6.56}$$

where $\mathbf{M} \in \mathbf{Sym}^n$.

Of special importance are the following, respectively, chain and product rule of differentiation

$$u\left[v(\mathbf{A})\right]_{,\mathbf{A}} = \frac{\mathrm{d}u}{\mathrm{d}v}v_{,\mathbf{A}}, \tag{6.57}$$

$$\left[f(\mathbf{A})\,g(\mathbf{A})\right]_{,\mathbf{A}} = g(\mathbf{A})\,f(\mathbf{A})_{,\mathbf{A}} + f(\mathbf{A})\,g(\mathbf{A})_{,\mathbf{A}}, \tag{6.58}$$

which can easily be proved by using the formalism of the directional derivative (6.44). Indeed, we can write for example for (6.57)

$$\frac{\mathrm{d}}{\mathrm{d}t}u\left[v\left(\mathbf{A}+t\mathbf{X}\right)\right]\Big|_{t=0}=\frac{\mathrm{d}u}{\mathrm{d}v}\frac{\mathrm{d}}{\mathrm{d}t}v\left(\mathbf{A}+t\mathbf{X}\right)\Big|_{t=0}=\frac{\mathrm{d}u}{\mathrm{d}v}v,_{\mathbf{A}}:\mathbf{X}.$$

Example 6.10. Derivatives of the principal invariants $\mathrm{I}_{\mathbf{A}}^{(k)}$ $(k=1,2,\ldots,n)$ *of* a second-order tensor $\mathbf{A}\in\mathbf{Lin}^n$. By virtue of the representations (4.26) and using (6.52), (6.58) we obtain

$$\begin{aligned}
\mathrm{I}_{\mathbf{A}}^{(1)},_{\mathbf{A}} &= \left(\mathrm{tr}\mathbf{A}\right),_{\mathbf{A}}=\mathbf{I},\\
\mathrm{I}_{\mathbf{A}}^{(2)},_{\mathbf{A}} &= \frac{1}{2}\left(\mathrm{I}_{\mathbf{A}}^{(1)}\mathrm{tr}\mathbf{A}-\mathrm{tr}\mathbf{A}^2\right),_{\mathbf{A}}=\mathrm{I}_{\mathbf{A}}^{(1)}\mathbf{I}-\mathbf{A}^{\mathrm{T}},\\
\mathrm{I}_{\mathbf{A}}^{(3)},_{\mathbf{A}} &= \frac{1}{3}\left(\mathrm{I}_{\mathbf{A}}^{(2)}\mathrm{tr}\mathbf{A}-\mathrm{I}_{\mathbf{A}}^{(1)}\mathrm{tr}\mathbf{A}^2+\mathrm{tr}\mathbf{A}^3\right),_{\mathbf{A}}\\
&= \frac{1}{3}\left[\mathrm{tr}\mathbf{A}\left(\mathrm{I}_{\mathbf{A}}^{(1)}\mathbf{I}-\mathbf{A}^{\mathrm{T}}\right)+\mathrm{I}_{\mathbf{A}}^{(2)}\mathbf{I}-\left(\mathrm{tr}\mathbf{A}^2\right)\mathbf{I}-2\mathrm{I}_{\mathbf{A}}^{(1)}\mathbf{A}^{\mathrm{T}}+3\left(\mathbf{A}^{\mathrm{T}}\right)^2\right]\\
&= \left[\mathbf{A}^2-\mathrm{I}_{\mathbf{A}}^{(1)}\mathbf{A}+\mathrm{I}_{\mathbf{A}}^{(2)}\mathbf{I}\right]^{\mathrm{T}},\ldots
\end{aligned}\tag{6.59}$$

Herein, one can observe the following regularity [46]

$$\mathrm{I}_{\mathbf{A}}^{(k)},_{\mathbf{A}}=\sum_{i=0}^{k-1}(-1)^i\,\mathrm{I}_{\mathbf{A}}^{(k-1-i)}\left(\mathbf{A}^{\mathrm{T}}\right)^i=-\mathrm{I}_{\mathbf{A}}^{(k-1)},_{\mathbf{A}}\mathbf{A}^{\mathrm{T}}+\mathrm{I}_{\mathbf{A}}^{(k-1)}\mathbf{I},\quad k=1,2,\ldots,\tag{6.60}$$

where we again set $\mathrm{I}_{\mathbf{A}}^{(0)}=1$. The above identity can be proved by mathematical induction (see also [7]). To this end, we first assume that it holds for all natural numbers smaller than some $k+1$ as

$$\mathrm{I}_{\mathbf{A}}^{(l)},_{\mathbf{A}}=\mathbf{Y}_l,\quad l=0,1,\ldots,k,\tag{6.61}$$

where the abbreviation

$$\mathbf{Y}_k=\sum_{i=0}^{k-1}(-1)^i\,\mathrm{I}_{\mathbf{A}}^{(k-1-i)}\left(\mathbf{A}^{\mathrm{T}}\right)^i\tag{6.62}$$

is used. Then,

$$\mathbf{Y}_{k+1}=\sum_{i=0}^{k}(-1)^i\,\mathrm{I}_{\mathbf{A}}^{(k-i)}\left(\mathbf{A}^{\mathrm{T}}\right)^i=-\mathbf{Y}_k\mathbf{A}^{\mathrm{T}}+\mathrm{I}_{\mathbf{A}}^{(k)}\mathbf{I}=-\mathrm{I}_{\mathbf{A}}^{(k)},_{\mathbf{A}}\mathbf{A}^{\mathrm{T}}+\mathrm{I}_{\mathbf{A}}^{(k)}\mathbf{I}.\tag{6.63}$$

Further, according to (4.26), (6.52) and (6.58)

$$
\begin{aligned}
\mathrm{I}_{\mathbf{A}}^{(k)},_{\mathbf{A}} &= \frac{1}{k}\left[\sum_{i=1}^{k}(-1)^{i-1}\,\mathrm{I}_{\mathbf{A}}^{(k-i)}\mathrm{tr}\mathbf{A}^{i}\right],_{\mathbf{A}} \\
&= \frac{1}{k}\sum_{i=1}^{k}(-1)^{i-1}\,i\,\mathrm{I}_{\mathbf{A}}^{(k-i)}\left(\mathbf{A}^{\mathrm{T}}\right)^{i-1} + \frac{1}{k}\sum_{i=1}^{k-1}(-1)^{i-1}\,\mathrm{I}_{\mathbf{A}}^{(k-i)},_{\mathbf{A}}\,\mathrm{tr}\mathbf{A}^{i}. \quad (6.64)
\end{aligned}
$$

Inserting this result as well as (4.26) into the last expression (6.63) we obtain

$$
\begin{aligned}
\mathbf{Y}_{k+1} &= -\frac{1}{k}\sum_{i=1}^{k}(-1)^{i-1}\,i\,\mathrm{I}_{\mathbf{A}}^{(k-i)}\left(\mathbf{A}^{\mathrm{T}}\right)^{i} - \left[\sum_{i=1}^{k-1}(-1)^{i-1}\,\mathrm{I}_{\mathbf{A}}^{(k-i)},_{\mathbf{A}}\,\mathrm{tr}\mathbf{A}^{i}\right]\frac{\mathbf{A}^{\mathrm{T}}}{k} \\
&\quad + \frac{\mathbf{I}}{k}\left[\sum_{i=1}^{k}(-1)^{i-1}\,\mathrm{I}_{\mathbf{A}}^{(k-i)}\mathrm{tr}\mathbf{A}^{i}\right].
\end{aligned}
$$

Adding $\mathbf{Y}_{k+1}/k$ to both sides of this equality and using for $\mathbf{Y}_{k+1}$ the first expression in (6.63) we further obtain keeping in mind that $\mathrm{I}_{\mathbf{A}}^{0},_{\mathbf{A}} = \mathbf{0}$

$$
\begin{aligned}
\frac{k+1}{k}\mathbf{Y}_{k+1} &= \frac{1}{k}\sum_{i=0}^{k}(-1)^{i}\,i\,\mathrm{I}_{\mathbf{A}}^{(k-i)}\left(\mathbf{A}^{\mathrm{T}}\right)^{i} + \frac{1}{k}\sum_{i=0}^{k}(-1)^{i}\,\mathrm{I}_{\mathbf{A}}^{(k-i)}\left(\mathbf{A}^{\mathrm{T}}\right)^{i} \\
&\quad + \frac{1}{k}\left[\sum_{i=1}^{k}(-1)^{i-1}\left(-\mathrm{I}_{\mathbf{A}}^{(k-i)},_{\mathbf{A}}\,\mathbf{A}^{\mathrm{T}} + \mathrm{I}_{\mathbf{A}}^{(k-i)}\mathbf{I}\right)\mathrm{tr}\mathbf{A}^{i}\right].
\end{aligned}
$$

Taking again (6.60) and (6.64) into account we can write

$$
\begin{aligned}
\frac{k+1}{k}\mathbf{Y}_{k+1} &= \frac{1}{k}\sum_{i=0}^{k}(-1)^{i}\,(i+1)\,\mathrm{I}_{\mathbf{A}}^{(k-i)}\left(\mathbf{A}^{\mathrm{T}}\right)^{i} \\
&\quad + \frac{1}{k}\left[\sum_{i=1}^{k}(-1)^{i-1}\,\mathrm{I}_{\mathbf{A}}^{(k+1-i)},_{\mathbf{A}}\,\mathrm{tr}\mathbf{A}^{i}\right] \\
&= \frac{1}{k}\sum_{i=1}^{k+1}(-1)^{i-1}\,i\,\mathrm{I}_{\mathbf{A}}^{(k-i)}\left(\mathbf{A}^{\mathrm{T}}\right)^{i-1} \\
&\quad + \frac{1}{k}\left[\sum_{i=1}^{k}(-1)^{i-1}\,\mathrm{I}_{\mathbf{A}}^{(k+1-i)},_{\mathbf{A}}\,\mathrm{tr}\mathbf{A}^{i}\right] = \frac{k+1}{k}\mathrm{I}_{\mathbf{A}}^{(k+1)},_{\mathbf{A}}.
\end{aligned}
$$

Hence,

$$
\mathrm{I}_{\mathbf{A}}^{(k+1)},_{\mathbf{A}} = \mathbf{Y}_{k+1},
$$

which immediately implies that (6.60) holds for $k+1$ as well. Thereby, representation (6.60) is proved.

For invertible tensors one can get a simpler representation for the derivative of the last invariant $\mathrm{I}_{\mathbf{A}}^{(n)}$. This representation results from the Cayley-Hamilton theorem (4.95) as follows

$$\mathrm{I}_{\mathbf{A}}^{(n)},_{\mathbf{A}} \mathbf{A}^{\mathrm{T}} = \left[\sum_{i=0}^{n-1} (-1)^i \, \mathrm{I}_{\mathbf{A}}^{(n-1-i)} \left(\mathbf{A}^{\mathrm{T}}\right)^i \right] \mathbf{A}^{\mathrm{T}} = \sum_{i=1}^{n} (-1)^{i-1} \, \mathrm{I}_{\mathbf{A}}^{(n-i)} \left(\mathbf{A}^{\mathrm{T}}\right)^i$$

$$= \sum_{i=0}^{n} (-1)^{i-1} \, \mathrm{I}_{\mathbf{A}}^{(n-i)} \left(\mathbf{A}^{\mathrm{T}}\right)^i + \mathrm{I}_{\mathbf{A}}^{(n)} \mathbf{I} = \mathrm{I}_{\mathbf{A}}^{(n)} \mathbf{I}.$$

Thus,

$$\mathrm{I}_{\mathbf{A}}^{(n)},_{\mathbf{A}} = \mathrm{I}_{\mathbf{A}}^{(n)} \mathbf{A}^{-\mathrm{T}}, \quad \mathbf{A} \in \mathbf{Inv}^n. \tag{6.65}$$

Example 6.11. Derivatives of the eigenvalues λ_i. First, we show that simple eigenvalues of a second-order tensor $\mathbf{A}$ are differentiable. To this end, we consider the directional derivative (6.44) of an eigenvalue λ:

$$\left. \frac{\mathrm{d}}{\mathrm{d}t} \lambda \left(\mathbf{A} + t\mathbf{X}\right) \right|_{t=0}. \tag{6.66}$$

Herein, $\lambda(t)$ represents an implicit function defined through the characteristic equation

$$\det\left(\mathbf{A} + t\mathbf{X} - \lambda \mathbf{I}\right) = p\left(\lambda, t\right) = 0. \tag{6.67}$$

This equation can be written out in the polynomial form (4.19) with respect to powers of λ. The coefficients of this polynomial are principal invariants of the tensor $\mathbf{A} + t\mathbf{X}$. According to the results of the previous example these invariants are differentiable with respect to $\mathbf{A} + t\mathbf{X}$ and therefore also with respect to t. For this reason, the function $p(\lambda, t)$ is differentiable both with respect to λ and t. For a simple eigenvalue $\lambda_0 = \lambda(0)$ we can further write (see also [27])

$$p\left(\lambda_0, 0\right) = 0, \qquad \left. \frac{\partial p\left(\lambda, 0\right)}{\partial \lambda} \right|_{\lambda=\lambda_0} \neq 0. \tag{6.68}$$

According to the implicit function theorem (see, e.g., [5]), the above condition ensures differentiability of the function $\lambda(t)$ at $t = 0$. Thus, the directional derivative (6.66) exists and is continuous at $\mathbf{A}$. It can be expressed by

$$\left. \frac{\mathrm{d}}{\mathrm{d}t} \lambda \left(\mathbf{A} + t\mathbf{X}\right) \right|_{t=0} = - \left. \frac{\partial p / \partial t}{\partial p / \partial \lambda} \right|_{t=0, \lambda=\lambda_0}. \tag{6.69}$$

In order to represent the derivative $\lambda_i,_{\mathbf{A}}$ we first consider the spectral representation (4.43) of the tensor $\mathbf{A}$ with pairwise distinct eigenvalues

$$\mathbf{A} = \sum_{i=1}^{n} \lambda_i \mathbf{P}_i, \tag{6.70}$$

where $\mathbf{P}_i$ $(i = 1, 2, \ldots, n)$ denote the eigenprojections. They can uniquely be determined from the equation system

$$\mathbf{A}^k = \sum_{i=1}^{n} \lambda_i^k \mathbf{P}_i, \quad k = 0, 1, \ldots, n-1 \tag{6.71}$$

resulting from (4.47). Applying the Vieta theorem to the tensor $\mathbf{A}^l$ $(l = 1, 2, \ldots, n)$ we further obtain relation (4.25) written as

$$\mathrm{tr}\mathbf{A}^l = \sum_{i=1}^{n} \lambda_i^l, \quad l = 1, 2, \ldots, n. \tag{6.72}$$

Differentiation of (6.72) with respect to $\mathbf{A}$ further yields by virtue of (6.52) and (6.57)

$$l \left(\mathbf{A}^{\mathrm{T}}\right)^{l-1} = l \sum_{i=1}^{n} \lambda_i^{l-1} \lambda_i,_{\mathbf{A}}, \quad l = 1, 2, \ldots, n$$

and consequently

$$\mathbf{A}^k = \sum_{i=1}^{n} \lambda_i^k \left(\lambda_i,_{\mathbf{A}}\right)^{\mathrm{T}}, \quad k = 0, 1, \ldots, n-1. \tag{6.73}$$

Comparing the linear equation systems (6.71) and (6.73) we notice that

$$\lambda_i,_{\mathbf{A}} = \mathbf{P}_i^{\mathrm{T}}. \tag{6.74}$$

Finally, the Sylvester formula (4.55) results in the expression

$$\lambda_i,_{\mathbf{A}} = \delta_{1n} \mathbf{I} + \prod_{\substack{j=1 \\ j \neq i}}^{n} \frac{\mathbf{A}^{\mathrm{T}} - \lambda_j \mathbf{I}}{\lambda_i - \lambda_j}. \tag{6.75}$$

It is seen that the solution (6.75) holds even if the remainder eigenvalues λ_j $(j = 1, 2, \ldots, i-1, i+1, \ldots, n)$ of the tensor $\mathbf{A}$ are not simple. In this case (6.75) transforms to

$$\lambda_{i,\mathbf{A}} = \delta_{1n}\mathbf{I} + \prod_{\substack{j=1\\ j\neq i}}^{s} \frac{\mathbf{A}^{\mathrm{T}} - \lambda_j \mathbf{I}}{\lambda_i - \lambda_j}, \tag{6.76}$$

where s denotes the number of pairwise distinct eigenvalues λ_i $(i = 1, 2, \ldots, s)$.

Let us further consider a scalar-valued tensor function $f(\mathbf{A}) : \mathbf{Lin}^n \mapsto \mathbb{R}$, where $\mathbf{A} = \mathbf{A}(t)$ itself is a differentiable tensor-valued function of a real variable t (see Sect. 2.1). Of special interest is the derivative of the composite function $f(\mathbf{A}(t))$. Using $(2.15)_1$ we write

$$\frac{\mathrm{d}f}{\mathrm{d}t} = \left.\frac{\mathrm{d}f(\mathbf{A}(t+s))}{\mathrm{d}s}\right|_{s=0} = \left.\frac{\mathrm{d}f\left(\mathbf{A}(t) + s\frac{\mathrm{d}\mathbf{A}}{\mathrm{d}t} + s\mathbf{O}(s)\right)}{\mathrm{d}s}\right|_{s=0}.$$

Introducing auxiliary functions $s_1(s) = s$ and $s_2(s) = s$ and applying the formalism of the directional derivative (6.44) we further obtain

$$\begin{aligned}
\frac{\mathrm{d}f}{\mathrm{d}t} &= \left.\frac{\mathrm{d}f\left(\mathbf{A}(t) + s_1\frac{\mathrm{d}\mathbf{A}}{\mathrm{d}t} + s_1\mathbf{O}(s_2)\right)}{\mathrm{d}s}\right|_{s=0} \\
&= \left.\frac{\partial f\left(\mathbf{A}(t) + s_1\frac{\mathrm{d}\mathbf{A}}{\mathrm{d}t} + s_1\mathbf{O}(s_2)\right)}{\partial s_1}\frac{\mathrm{d}s_1}{\mathrm{d}s}\right|_{s=0} \\
&\quad + \left.\frac{\partial f\left(\mathbf{A}(t) + s_1\frac{\mathrm{d}\mathbf{A}}{\mathrm{d}t} + s_1\mathbf{O}(s_2)\right)}{\partial s_2}\frac{\mathrm{d}s_2}{\mathrm{d}s}\right|_{s=0} \\
&= f_{,\mathbf{A}} : \left.\left[\frac{\mathrm{d}\mathbf{A}}{\mathrm{d}t} + \mathbf{O}(s_2)\right]\right|_{s_2=0} + \left.\frac{\partial f\left(\mathbf{A}(t) + s_1\frac{\mathrm{d}\mathbf{A}}{\mathrm{d}t} + s_1\mathbf{O}(s_2)\right)}{\partial s_2}\right|_{s_1=s_2=0} \\
&= f_{,\mathbf{A}} : \frac{\mathrm{d}\mathbf{A}}{\mathrm{d}t} + \left.\frac{\partial f(\mathbf{A}(t))}{\partial s_2}\right|_{s_2=0}.
\end{aligned}$$

This finally leads to the result

$$\frac{\mathrm{d}f(\mathbf{A}(t))}{\mathrm{d}t} = f_{,\mathbf{A}} : \frac{\mathrm{d}\mathbf{A}}{\mathrm{d}t}. \tag{6.77}$$

Example 6.12. Constitutive relations for hyperelastic materials with isochoric-volumetric split of the strain energy function. For such materials the strain energy function is represented in terms of the right Cauchy-Green tensor $\mathbf{C}$ by

$$\psi(\mathbf{C}) = \bar{\psi}\left(\bar{\mathbf{C}}\right) + U(J), \tag{6.78}$$

where

$$J = \sqrt{\mathrm{III}_{\mathbf{C}}}, \quad \bar{\mathbf{C}} = J^{-2/3}\mathbf{C} \tag{6.79}$$

describe the volumetric and isochoric parts of deformation, respectively. Constitutive relations for hyperelastic materials can be expressed in terms of the strain energy function by (see e.g. [46])

$$\mathbf{S} = 2\frac{\partial \psi}{\partial \mathbf{C}}, \tag{6.80}$$

where **S** denotes the second Piola-Kirchhoff stress tensor. Insertion of (6.78) yields

$$\mathbf{S} = \mathbf{S}_{\mathrm{iso}} + \mathbf{S}_{\mathrm{vol}}, \tag{6.81}$$

where

$$\mathbf{S}_{\mathrm{iso}} = 2\bar{\psi}\left(\bar{\mathbf{C}}\right)_{,\mathbf{C}}, \quad \mathbf{S}_{\mathrm{vol}} = 2U\left(J\right)_{,\mathbf{C}}. \tag{6.82}$$

In order to express these derivatives we first obtain

$$J^{\alpha}{}_{,\mathbf{C}} = \mathrm{III}_{\mathbf{C}}^{\alpha/2}{}_{,\mathbf{C}} = \frac{\alpha}{2}\mathrm{III}_{\mathbf{C}}^{\alpha/2-1}\mathrm{III}_{\mathbf{C}}\mathbf{C}^{-1} = \frac{\alpha}{2}J^{\alpha}\mathbf{C}^{-1} \tag{6.83}$$

using (6.65), (6.57) and $(6.79)_1$ and taking symmetry of **C** into account. As the next step, we calculate the directional derivative of $\bar{\mathbf{C}}$ by virtue of (2.4)

$$\begin{aligned}
\frac{\mathrm{d}}{\mathrm{d}t}\bar{\mathbf{C}}\left(\mathbf{C}+t\mathbf{X}\right)\Big|_{t=0} &= \frac{\mathrm{d}}{\mathrm{d}t}\left(\mathbf{C}+t\mathbf{X}\right)\left[\det\left(\mathbf{C}+t\mathbf{X}\right)\right]^{-1/3}\Big|_{t=0} \\
&= \frac{\mathrm{d}}{\mathrm{d}t}\left(\mathbf{C}+t\mathbf{X}\right)\Big|_{t=0} J^{-2/3} + \mathbf{C}\frac{\mathrm{d}}{\mathrm{d}t}\left[J\left(\mathbf{C}+t\mathbf{X}\right)\right]^{-2/3}\Big|_{t=0} \\
&= J^{-2/3}\mathbf{X} - \frac{1}{3}J^{-2/3}\mathbf{C}\left(\mathbf{C}^{-1}:\mathbf{X}\right) = \mathcal{P}_{\mathrm{iso}}:\mathbf{X},
\end{aligned} \tag{6.84}$$

where

$$\mathcal{P}_{\mathrm{iso}} = J^{-2/3}\left[\mathcal{I}^{\mathrm{s}} - \frac{1}{3}\mathbf{C}\odot\mathbf{C}^{-1}\right] \tag{6.85}$$

denotes the isochoric projection tensor. To the directional derivative of $\bar{\psi}\left(\bar{\mathbf{C}}\right)$ we further apply (6.77) as follows

$$\frac{\mathrm{d}}{\mathrm{d}t}\bar{\psi}\left[\bar{\mathbf{C}}\left(\mathbf{C}+t\mathbf{X}\right)\right]\Big|_{t=0} = \bar{\psi}_{,\bar{\mathbf{C}}}:\frac{\mathrm{d}}{\mathrm{d}t}\bar{\mathbf{C}}\left(\mathbf{C}+t\mathbf{X}\right)\Big|_{t=0} = \bar{\psi}_{,\bar{\mathbf{C}}}:\mathcal{P}_{\mathrm{iso}}:\mathbf{X}. \tag{6.86}$$

Inserting these results in (6.82) we finally obtain

$$\mathbf{S}_{\mathrm{iso}} = \bar{\mathbf{S}}:\mathcal{P}_{\mathrm{iso}} = J^{-2/3}\left[\bar{\mathbf{S}} - \frac{1}{3}\left(\bar{\mathbf{S}}:\mathbf{C}\right)\mathbf{C}^{-1}\right], \tag{6.87}$$

$$\mathbf{S}_{\mathrm{vol}} = U' J \mathbf{C}^{-1}, \tag{6.88}$$

where

$$\bar{\mathbf{S}} = 2\bar{\psi},_{\bar{\mathbf{C}}} . \tag{6.89}$$

6.4 Tensor-Valued Isotropic and Anisotropic Tensor Functions

A tensor-valued function $g(\mathbf{A}_1, \mathbf{A}_2, \ldots, \mathbf{A}_l) \in \mathbf{Lin}^n$ of a tensor system $\mathbf{A}_k \in \mathbf{Lin}^n$ $(k = 1, 2, \ldots, l)$ is called anisotropic if

$$g\left(\mathbf{Q}\mathbf{A}_1\mathbf{Q}^{\mathrm{T}}, \mathbf{Q}\mathbf{A}_2\mathbf{Q}^{\mathrm{T}}, \ldots, \mathbf{Q}\mathbf{A}_l\mathbf{Q}^{\mathrm{T}}\right) \\ = \mathbf{Q}g(\mathbf{A}_1, \mathbf{A}_2, \ldots, \mathbf{A}_l)\mathbf{Q}^{\mathrm{T}}, \quad \forall \mathbf{Q} \in \mathrm{Sorth}^n \subset \mathbf{Orth}^n. \tag{6.90}$$

For isotropic tensor-valued tensor functions the above identity holds for all orthogonal transformations so that $\mathrm{Sorth}^n = \mathbf{Orth}^n$.

As a starting point for the discussion of tensor-valued tensor functions we again consider isotropic functions of one argument. In this case,

$$g\left(\mathbf{Q}\mathbf{A}\mathbf{Q}^{\mathrm{T}}\right) = \mathbf{Q}g(\mathbf{A})\mathbf{Q}^{\mathrm{T}}, \quad \forall \mathbf{Q} \in \mathbf{Orth}^n. \tag{6.91}$$

For example, one can easily show that the polynomial function (1.113) and the exponential function (1.114) introduced in Chap. 1 are isotropic. Indeed, for a tensor polynomial $g(\mathbf{A}) = \sum_{k=0}^{m} a_k \mathbf{A}^k$ we have (see also Exercise 1.34)

$$\begin{aligned} g\left(\mathbf{Q}\mathbf{A}\mathbf{Q}^{\mathrm{T}}\right) &= \sum_{k=0}^{m} a_k \left(\mathbf{Q}\mathbf{A}\mathbf{Q}^{\mathrm{T}}\right)^k = \sum_{k=0}^{m} a_k \left(\underbrace{\mathbf{Q}\mathbf{A}\mathbf{Q}^{\mathrm{T}}\mathbf{Q}\mathbf{A}\mathbf{Q}^{\mathrm{T}} \ldots \mathbf{Q}\mathbf{A}\mathbf{Q}^{\mathrm{T}}}_{k \text{ times}}\right) \\ &= \sum_{k=0}^{m} a_k \left(\mathbf{Q}\mathbf{A}^k\mathbf{Q}^{\mathrm{T}}\right) = \mathbf{Q}\left(\sum_{k=0}^{m} a_k \mathbf{A}^k\right)\mathbf{Q}^{\mathrm{T}} \\ &= \mathbf{Q}g(\mathbf{A})\mathbf{Q}^{\mathrm{T}}, \quad \forall \mathbf{Q} \in \mathbf{Orth}^n. \end{aligned} \tag{6.92}$$

Of special interest are isotropic functions of a symmetric tensor. First, we prove that the tensors $g(\mathbf{M})$ and $\mathbf{M} \in \mathbf{Sym}^n$ are coaxial i.e. have the eigenvectors in common. To this end, we represent $\mathbf{M}$ in the spectral form (4.61) by

$$\mathbf{M} = \sum_{i=1}^{n} \lambda_i \boldsymbol{b}_i \otimes \boldsymbol{b}_i, \tag{6.93}$$

where $\boldsymbol{b}_i \cdot \boldsymbol{b}_j = \delta_{ij}$ $(i, j = 1, 2, \ldots, n)$. Further, we choose an arbitrary eigenvector, say $\boldsymbol{b}_k$, and show that it simultaneously represents an eigenvector of $g(\mathbf{M})$.

Indeed, let

$$\mathbf{Q} = 2\boldsymbol{b}_k \otimes \boldsymbol{b}_k - \mathbf{I} = \boldsymbol{b}_k \otimes \boldsymbol{b}_k + \sum_{\substack{i=1 \\ i \neq k}}^{n} (-1)\, \boldsymbol{b}_i \otimes \boldsymbol{b}_i \tag{6.94}$$

bearing in mind that $\mathbf{I} = \sum_{i=1}^{n} \boldsymbol{b}_i \otimes \boldsymbol{b}_i$ in accordance with (1.92). The tensor $\mathbf{Q}$ (6.94) is orthogonal since

$$\mathbf{Q}\mathbf{Q}^{\mathrm{T}} = (2\boldsymbol{b}_k \otimes \boldsymbol{b}_k - \mathbf{I})\,(2\boldsymbol{b}_k \otimes \boldsymbol{b}_k - \mathbf{I}) = 4\boldsymbol{b}_k \otimes \boldsymbol{b}_k - 2\boldsymbol{b}_k \otimes \boldsymbol{b}_k - 2\boldsymbol{b}_k \otimes \boldsymbol{b}_k + \mathbf{I} = \mathbf{I}$$

and symmetric as well. One of its eigenvalues is equal to 1 while all the other ones are -1. Thus, we can write

$$\mathbf{Q}\mathbf{M} = (2\boldsymbol{b}_k \otimes \boldsymbol{b}_k - \mathbf{I})\,\mathbf{M} = 2\lambda_k \boldsymbol{b}_k \otimes \boldsymbol{b}_k - \mathbf{M} = \mathbf{M}\,(2\boldsymbol{b}_k \otimes \boldsymbol{b}_k - \mathbf{I}) = \mathbf{M}\mathbf{Q}$$

and consequently

$$\mathbf{Q}\mathbf{M}\mathbf{Q}^{\mathrm{T}} = \mathbf{M}. \tag{6.95}$$

Since the function $g(\mathbf{M})$ is isotropic

$$g(\mathbf{M}) = g\left(\mathbf{Q}\mathbf{M}\mathbf{Q}^{\mathrm{T}}\right) = \mathbf{Q}g(\mathbf{M})\,\mathbf{Q}^{\mathrm{T}}$$

and therefore

$$\mathbf{Q}g(\mathbf{M}) = g(\mathbf{M})\,\mathbf{Q}. \tag{6.96}$$

Mapping the vector $\boldsymbol{b}_k$ by both sides of this identity yields in view of (6.94)

$$\mathbf{Q}g(\mathbf{M})\,\boldsymbol{b}_k = g(\mathbf{M})\,\boldsymbol{b}_k. \tag{6.97}$$

It is seen that the vector $g(\mathbf{M})\,\boldsymbol{b}_k$ is an eigenvector of $\mathbf{Q}$ (6.94) associated with the eigenvalue 1. Since it is the simple eigenvalue

$$g(\mathbf{M})\,\boldsymbol{b}_k = \gamma_k \boldsymbol{b}_k, \tag{6.98}$$

where γ_k is some real number. Hence, $\boldsymbol{b}_k$ represents the right eigenvector of $g(\mathbf{M})$. Forming the left mapping of $\boldsymbol{b}_k$ by (6.96) one can similarly show that $\boldsymbol{b}_k$ is also the left eigenvector of $g(\mathbf{M})$, which implies the symmetry of the tensor $g(\mathbf{M})$.

Now, we are in a position to prove the following representation theorem [36,46].

Theorem 6.2. *A tensor-valued tensor function* $g(\mathbf{M})$, $\mathbf{M} \in \mathbf{Sym}^n$ *is isotropic if and only if it allows the following representation*

$$g(\mathbf{M}) = \varphi_0 \mathbf{I} + \varphi_1 \mathbf{M} + \varphi_2 \mathbf{M}^2 + \ldots + \varphi_{n-1} \mathbf{M}^{n-1} = \sum_{i=0}^{n-1} \varphi_i \mathbf{M}^i, \tag{6.99}$$

where φ_i are isotropic invariants (isotropic scalar functions) of $\mathbf{M}$ *and can therefore be expressed as functions of its principal invariants by*

$$\varphi_i = \widehat{\varphi}_i \left(\mathrm{I}_{\mathbf{M}}^{(1)}, \mathrm{I}_{\mathbf{M}}^{(2)}, \ldots, \mathrm{I}_{\mathbf{M}}^{(n)} \right), \quad i = 0, 1, \ldots, n-1. \tag{6.100}$$

Proof. We have already proved that the tensors $g(\mathbf{M})$ and $\mathbf{M}$ have eigenvectors in common. Thus, according to (6.93)

$$g(\mathbf{M}) = \sum_{i=1}^{n} \gamma_i \boldsymbol{b}_i \otimes \boldsymbol{b}_i, \tag{6.101}$$

where $\gamma_i = \gamma_i(\mathbf{M})$. Hence (see Exercise 6.1(e)),

$$g\left(\mathbf{QMQ}^{\mathrm{T}}\right) = \sum_{i=1}^{n} \gamma_i \left(\mathbf{QMQ}^{\mathrm{T}}\right) \mathbf{Q} \left(\boldsymbol{b}_i \otimes \boldsymbol{b}_i\right) \mathbf{Q}^{\mathrm{T}}. \tag{6.102}$$

Since the function $g(\mathbf{M})$ is isotropic we have

$$\begin{aligned} g\left(\mathbf{QMQ}^{\mathrm{T}}\right) &= \mathbf{Q} g(\mathbf{M}) \mathbf{Q}^{\mathrm{T}} \\ &= \sum_{i=1}^{n} \gamma_i (\mathbf{M}) \mathbf{Q} \left(\boldsymbol{b}_i \otimes \boldsymbol{b}_i\right) \mathbf{Q}^{\mathrm{T}}, \quad \forall \mathbf{Q} \in \mathbf{Orth}^n. \end{aligned} \tag{6.103}$$

Comparing (6.102) with (6.103) we conclude that

$$\gamma_i \left(\mathbf{QMQ}^{\mathrm{T}}\right) = \gamma_i (\mathbf{M}), \quad i = 1, \ldots, n, \quad \forall \mathbf{Q} \in \mathbf{Orth}^n. \tag{6.104}$$

Thus, the eigenvalues of the tensor $g(\mathbf{M})$ represent isotropic (scalar-valued) functions of $\mathbf{M}$. Collecting repeated eigenvalues of $g(\mathbf{M})$ we can further rewrite (6.101) in terms of the eigenprojections $\mathbf{P}_i$ $(i = 1, 2, \ldots, s)$ by

$$g(\mathbf{M}) = \sum_{i=1}^{s} \gamma_i \mathbf{P}_i, \tag{6.105}$$

where s $(1 \le s \le n)$ denotes the number of pairwise distinct eigenvalues of $g(\mathbf{M})$. Using the representation of the eigenprojections (4.56) based on the Sylvester formula (4.55) we can write

$$\mathbf{P}_i = \sum_{r=0}^{s-1} \alpha_i^{(r)} (\lambda_1, \lambda_2, \ldots, \lambda_s) \mathbf{M}^r, \quad i = 1, 2, \ldots, s. \tag{6.106}$$

Inserting this result into (6.105) yields the representation (sufficiency):

$$g(\mathbf{M}) = \sum_{i=0}^{s-1} \varphi_i \mathbf{M}^i, \tag{6.107}$$

where the functions φ_i $(i = 0, 1, 2, \ldots, s-1)$ are given according to (6.8) and (6.104) by (6.100). The necessity is evident. Indeed, the function (6.99) is isotropic since in view of (6.92)

$$\begin{aligned} g\left(\mathbf{QMQ}^{\mathrm{T}}\right) &= \sum_{i=0}^{n-1} \varphi_i \left(\mathbf{QMQ}^{\mathrm{T}}\right) \mathbf{QM}^i\mathbf{Q}^{\mathrm{T}} \\ &= \mathbf{Q}\left[\sum_{i=0}^{n-1} \varphi_i (\mathbf{M}) \mathbf{M}^i\right] \mathbf{Q}^{\mathrm{T}} = \mathbf{Q}g(\mathbf{M})\mathbf{Q}^{\mathrm{T}}, \ \forall \mathbf{Q} \in \mathbf{Orth}^n. \end{aligned} \tag{6.108}$$

Example 6.13. Constitutive relations for isotropic materials. For isotropic materials the second Piola-Kirchhoff stress tensor **S** represents an isotropic function of the right Cauchy-Green tensor **C** so that

$$\mathbf{S}\left(\mathbf{QCQ}^{\mathrm{T}}\right) = \mathbf{QS}(\mathbf{C})\mathbf{Q}^{\mathrm{T}}, \quad \forall \mathbf{Q} \in \mathbf{Orth}^3. \tag{6.109}$$

Thus, according to the representation theorem

$$\mathbf{S}(\mathbf{C}) = \alpha_0 \mathbf{I} + \alpha_1 \mathbf{C} + \alpha_2 \mathbf{C}^2, \tag{6.110}$$

where $\alpha_i = \alpha_i(\mathbf{C})$ $(i = 0, 1, 2)$ are some scalar-valued isotropic functions of **C**. The same result can be obtained for isotropic hyperelastic materials by considering the representation of the strain energy function (6.10) in the relation (6.80). Indeed, using the chain rule of differentiation and keeping in mind that the tensor **C** is symmetric we obtain by means of (6.52)

$$\mathbf{S} = 2\sum_{k=1}^{3} \frac{\partial \tilde{\psi}}{\partial \mathrm{tr}\mathbf{C}^k} \frac{\partial \mathrm{tr}\mathbf{C}^k}{\partial \mathbf{C}} = 2\sum_{k=1}^{3} k \frac{\partial \tilde{\psi}}{\partial \mathrm{tr}\mathbf{C}^k} \mathbf{C}^{k-1}, \tag{6.111}$$

so that $\alpha_i(\mathbf{C}) = 2(i+1)\,\partial\tilde{\psi}/\partial\mathrm{tr}\mathbf{C}^{i+1}$ $(i = 0, 1, 2)$.

Let us further consider a linearly elastic material characterized by a linear stress-strain response. In this case, the relation (6.110) reduces to

$$\mathbf{S}(\mathbf{C}) = \varphi(\mathbf{C})\mathbf{I} + c\mathbf{C}, \tag{6.112}$$

where c is a material constant and $\varphi(\mathbf{C})$ represents an isotropic scalar-valued function linear in **C**. In view of (6.15) this function can be expressed by

$$\varphi(\mathbf{C}) = a + b\mathrm{tr}\mathbf{C}, \tag{6.113}$$

where a and b are again material constants. Assuming that the reference configuration, in which $\mathbf{C} = \mathbf{I}$, is stress free, yields $a + 3b + c = 0$ and consequently

$$\mathbf{S}(\mathbf{C}) = (-c - 3b + b\mathrm{tr}\mathbf{C})\,\mathbf{I} + c\mathbf{C} = b\,(\mathrm{tr}\mathbf{C} - 3)\,\mathbf{I} + c\,(\mathbf{C} - \mathbf{I})\,.$$

Introducing further the so-called Green-Lagrange strain tensor defined by

$$\tilde{\mathbf{E}} = \frac{1}{2}(\mathbf{C} - \mathbf{I}) \tag{6.114}$$

we finally obtain

$$\mathbf{S}\left(\tilde{\mathbf{E}}\right) = 2b\left(\mathrm{tr}\tilde{\mathbf{E}}\right)\mathbf{I} + 2c\tilde{\mathbf{E}}. \tag{6.115}$$

The material described by the linear constitutive relation (6.115) is referred to as St.Venant-Kirchhoff material. The corresponding material constants $2b$ and $2c$ are called Lamé constants. The strain energy function resulting in the constitutive law (6.115) by (6.80) or equivalently by $\mathbf{S} = \partial\psi/\partial\tilde{\mathbf{E}}$ is of the form

$$\psi\left(\tilde{\mathbf{E}}\right) = b\mathrm{tr}^2\tilde{\mathbf{E}} + c\mathrm{tr}\tilde{\mathbf{E}}^2. \tag{6.116}$$

For isotropic functions of an arbitrary tensor system $\mathbf{A}_k \in \mathbf{Lin}^n$ $(k = 1, 2, \ldots, l)$ the representations are obtained only for the three-dimensional space. One again splits tensor arguments into symmetric $\mathbf{M}_i \in \mathbf{Sym}^3$ $(i = 1, 2, \ldots, m)$ and skew-symmetric tensors $\mathbf{W}_j \in \mathbf{Skew}^3$ $(j = 1, 2, \ldots, w)$ according to (6.13). Then, all isotropic tensor-valued functions of these tensors can be represented as linear combinations of the following terms (see [33, 41]), where the coefficients represent scalar-valued isotropic functions of the same tensor arguments.

Symmetric generators:

$$\begin{gathered}
\mathbf{I},\\
\mathbf{M}_i,\quad \mathbf{M}_i^2,\quad \mathbf{M}_i\mathbf{M}_j + \mathbf{M}_j\mathbf{M}_i,\quad \mathbf{M}_i^2\mathbf{M}_j + \mathbf{M}_j\mathbf{M}_i^2,\quad \mathbf{M}_i\mathbf{M}_j^2 + \mathbf{M}_j^2\mathbf{M}_i,\\
\mathbf{W}_p^2,\quad \mathbf{W}_p\mathbf{W}_q + \mathbf{W}_q\mathbf{W}_p,\quad \mathbf{W}_p^2\mathbf{W}_q - \mathbf{W}_q\mathbf{W}_p^2,\quad \mathbf{W}_p\mathbf{W}_q^2 - \mathbf{W}_q^2\mathbf{W}_p,\\
\mathbf{M}_i\mathbf{W}_p - \mathbf{W}_p\mathbf{M}_i,\quad \mathbf{W}_p\mathbf{M}_i\mathbf{W}_p,\quad \mathbf{M}_i^2\mathbf{W}_p - \mathbf{W}_p\mathbf{M}_i^2,\\
\mathbf{W}_p\mathbf{M}_i\mathbf{W}_p^2 - \mathbf{W}_p^2\mathbf{M}_i\mathbf{W}_p.
\end{gathered} \tag{6.117}$$

Skew-symmetric generators:

$$\begin{gathered}
\mathbf{W}_p,\quad \mathbf{W}_p\mathbf{W}_q - \mathbf{W}_q\mathbf{W}_p,\\
\mathbf{M}_i\mathbf{M}_j - \mathbf{M}_j\mathbf{M}_i,\quad \mathbf{M}_i^2\mathbf{M}_j - \mathbf{M}_j\mathbf{M}_i^2,\quad \mathbf{M}_i\mathbf{M}_j^2 - \mathbf{M}_j^2\mathbf{M}_i,\\
\mathbf{M}_i\mathbf{M}_j\mathbf{M}_i^2 - \mathbf{M}_i^2\mathbf{M}_j\mathbf{M}_i,\quad \mathbf{M}_j\mathbf{M}_i\mathbf{M}_j^2 - \mathbf{M}_j^2\mathbf{M}_i\mathbf{M}_j,
\end{gathered}$$

$$\mathbf{M}_i\mathbf{M}_j\mathbf{M}_k + \mathbf{M}_j\mathbf{M}_k\mathbf{M}_i + \mathbf{M}_k\mathbf{M}_i\mathbf{M}_j - \mathbf{M}_j\mathbf{M}_i\mathbf{M}_k - \mathbf{M}_k\mathbf{M}_j\mathbf{M}_i - \mathbf{M}_i\mathbf{M}_k\mathbf{M}_j,$$
$$\mathbf{M}_i\mathbf{W}_p + \mathbf{W}_p\mathbf{M}_i, \quad \mathbf{M}_i\mathbf{W}_p^2 - \mathbf{W}_p^2\mathbf{M}_i,$$
$$i < j < k = 1, 2, \ldots, m, \quad p < q = 1, 2, \ldots, w. \tag{6.118}$$

For anisotropic tensor-valued tensor functions one utilizes the procedure applied for scalar-valued functions. It is based on the following theorem [50] (cf. Theorem 6.1).

Theorem 6.3 (Rychlewski's theorem). *A tensor-valued function* $g(\mathbf{A}_i)$ *is anisotropic with the symmetry group* $\mathrm{Sorth}^n = \mathfrak{g}$ *defined by (6.30) if and only if there exists an isotropic tensor-valued function* $\hat{g}(\mathbf{A}_i, \mathbf{L}_j)$ *such that*

$$g(\mathbf{A}_i) = \hat{g}(\mathbf{A}_i, \mathbf{L}_j). \tag{6.119}$$

Proof. Let us define a new tensor-valued function by

$$\hat{g}(\mathbf{A}_i, \mathbf{X}_j) = \mathbf{Q}'^{\mathrm{T}} g\left(\mathbf{Q}'\mathbf{A}_i\mathbf{Q}'^{\mathrm{T}}\right)\mathbf{Q}', \tag{6.120}$$

where the tensor $\mathbf{Q}' \in \mathbf{Orth}^n$ results from the condition (6.36). The further proof is similar to Theorem 6.1 (Exercise 6.13).

Example 6.14. Constitutive relations for a transversely isotropic elastic material. For illustration of the above results we construct a general constitutive equation for an elastic transversely isotropic material. The transversely isotropic material symmetry is defined by one structural tensor $\mathbf{L}$ (6.22) according to (6.24). The second Piola-Kirchhoff stress tensor $\mathbf{S}$ is a transversely isotropic function of the right Cauchy-Green tensor $\mathbf{C}$. According to Rychlewski's theorem $\mathbf{S}$ can be represented as an isotropic tensor function of $\mathbf{C}$ and $\mathbf{L}$ by

$$\mathbf{S} = \mathbf{S}(\mathbf{C}, \mathbf{L}), \tag{6.121}$$

such that

$$\mathbf{S}\left(\mathbf{QCQ}^{\mathrm{T}}, \mathbf{QLQ}^{\mathrm{T}}\right) = \mathbf{QS}(\mathbf{C}, \mathbf{L})\mathbf{Q}^{\mathrm{T}}, \quad \forall \mathbf{Q} \in \mathbf{Orth}^3. \tag{6.122}$$

This ensures that the condition of the material symmetry is fulfilled a priori since

$$\mathbf{S}\left(\mathbf{QCQ}^{\mathrm{T}}, \mathbf{L}\right) = \mathbf{S}\left(\mathbf{QCQ}^{\mathrm{T}}, \mathbf{QLQ}^{\mathrm{T}}\right) = \mathbf{QS}(\mathbf{C}, \mathbf{L})\mathbf{Q}^{\mathrm{T}}, \ \forall \mathbf{Q} \in \mathfrak{g}_t. \tag{6.123}$$

Keeping in mind that $\mathbf{S}$, $\mathbf{C}$ and $\mathbf{L}$ are symmetric tensors we can write by virtue of $(6.28)_1$ and (6.117)

$$\mathbf{S}(\mathbf{C}, \mathbf{L}) = \alpha_0\mathbf{I} + \alpha_1\mathbf{L} + \alpha_2\mathbf{C}$$
$$+ \alpha_3\mathbf{C}^2 + \alpha_4(\mathbf{CL} + \mathbf{LC}) + \alpha_5\left(\mathbf{C}^2\mathbf{L} + \mathbf{LC}^2\right). \tag{6.124}$$

The coefficients α_i $(i = 0, 1, \ldots, 5)$ represent scalar-valued isotropic tensor functions of $\mathbf{C}$ and $\mathbf{L}$ so that similar to (6.29)

$$\alpha_i\left(\mathbf{C}, \mathbf{L}\right) = \hat{\alpha}_i\left[\mathrm{tr}\mathbf{C}, \mathrm{tr}\mathbf{C}^2, \mathrm{tr}\mathbf{C}^3, \mathrm{tr}\left(\mathbf{C}\mathbf{L}\right), \mathrm{tr}\left(\mathbf{C}^2\mathbf{L}\right)\right]. \tag{6.125}$$

For comparison we derive the constitutive equations for a hyperelastic transversely isotropic material. To this end, we utilize the general representation for the transversely isotropic strain energy function (6.29). By the chain rule of differentiation and with the aid of (6.52) and (6.54) we obtain

$$\begin{aligned} \mathbf{S} = {} & 2\frac{\partial\tilde{\psi}}{\partial\mathrm{tr}\mathbf{C}}\mathbf{I} + 4\frac{\partial\tilde{\psi}}{\partial\mathrm{tr}\mathbf{C}^2}\mathbf{C} + 6\frac{\partial\tilde{\psi}}{\partial\mathrm{tr}\mathbf{C}^3}\mathbf{C}^2 \\ & + 2\frac{\partial\tilde{\psi}}{\partial\mathrm{tr}\left(\mathbf{C}\mathbf{L}\right)}\mathbf{L} + 2\frac{\partial\tilde{\psi}}{\partial\mathrm{tr}\left(\mathbf{C}^2\mathbf{L}\right)}\left(\mathbf{C}\mathbf{L} + \mathbf{L}\mathbf{C}\right) \end{aligned} \tag{6.126}$$

and finally

$$\mathbf{S} = \alpha_0\mathbf{I} + \alpha_1\mathbf{L} + \alpha_2\mathbf{C} + \alpha_3\mathbf{C}^2 + \alpha_4\left(\mathbf{C}\mathbf{L} + \mathbf{L}\mathbf{C}\right). \tag{6.127}$$

Comparing (6.124) and (6.127) we observe that the representation for the hyperelastic transversely isotropic material does not include the last term in (6.124) with $\mathbf{C}^2\mathbf{L} + \mathbf{L}\mathbf{C}^2$. Thus, the constitutive equations containing this term correspond to an elastic but not hyperelastic transversely isotropic material. The latter material cannot be described by a strain energy function.

6.5 Derivatives of Tensor-Valued Tensor Functions

The derivative of a tensor-valued tensor function can be defined in a similar fashion to (6.44). A function $g(\mathbf{A}) : \mathbf{Lin}^n \mapsto \mathbf{Lin}^n$ is said to be differentiable in a neighborhood of $\mathbf{A}$ if there exists a fourth-order tensor $g(\mathbf{A})_{,\mathbf{A}} \in \mathcal{L}\mathrm{in}^n$ (called the derivative), such that

$$\left.\frac{\mathrm{d}}{\mathrm{d}t}g\left(\mathbf{A} + t\mathbf{X}\right)\right|_{t=0} = g\left(\mathbf{A}\right)_{,\mathbf{A}} : \mathbf{X}, \quad \forall\mathbf{X} \in \mathbf{Lin}^n. \tag{6.128}$$

The above definition implies that the directional derivative $\left.\frac{\mathrm{d}}{\mathrm{d}t}g\left(\mathbf{A} + t\mathbf{X}\right)\right|_{t=0}$ exists and is continuous at $\mathbf{A}$.

Similarly to (6.45) we can obtain a direct relation for the fourth-order tensor $g(\mathbf{A})_{,\mathbf{A}}$. To this end, we represent the tensors $\mathbf{A}$, $\mathbf{X}$ and $\mathbf{G} = g(\mathbf{A})$ with respect to an arbitrary basis in $\mathbf{Lin}^n$, say $\boldsymbol{g}_i \otimes \boldsymbol{g}^j$ $(i, j = 1, 2, \ldots, n)$. Applying the chain rule of differentiation we can write

$$\frac{\mathrm{d}}{\mathrm{d}t} g(\mathbf{A} + t\mathbf{X})\bigg|_{t=0} = \frac{\mathrm{d}}{\mathrm{d}t} \left\{ \mathrm{G}^{i}_{\cdot j} \left[\left(\mathrm{A}^{k}_{\cdot l} + t\mathrm{X}^{k}_{\cdot l} \right) \boldsymbol{g}_k \otimes \boldsymbol{g}^l \right] \boldsymbol{g}_i \otimes \boldsymbol{g}^j \right\} \bigg|_{t=0}$$
$$= \frac{\partial \mathrm{G}^{i}_{\cdot j}}{\partial \mathrm{A}^{k}_{\cdot l}} \mathrm{X}^{k}_{\cdot l} \boldsymbol{g}_i \otimes \boldsymbol{g}^j . \quad (6.129)$$

In view of $(5.30)_1$ and (6.128) this results in the following representations

$$g,_{\mathbf{A}} = \frac{\partial \mathrm{G}^{i}_{\cdot j}}{\partial \mathrm{A}^{k}_{\cdot l}} \boldsymbol{g}_i \otimes \boldsymbol{g}^k \otimes \boldsymbol{g}_l \otimes \boldsymbol{g}^j = \frac{\partial \mathrm{G}^{i}_{\cdot j}}{\partial \mathrm{A}_{k\cdot}^{\ \ l}} \boldsymbol{g}_i \otimes \boldsymbol{g}_k \otimes \boldsymbol{g}^l \otimes \boldsymbol{g}^j$$
$$= \frac{\partial \mathrm{G}^{i}_{\cdot j}}{\partial \mathrm{A}^{kl}} \boldsymbol{g}_i \otimes \boldsymbol{g}^k \otimes \boldsymbol{g}^l \otimes \boldsymbol{g}^j = \frac{\partial \mathrm{G}^{i}_{\cdot j}}{\partial \mathrm{A}_{kl}} \boldsymbol{g}_i \otimes \boldsymbol{g}_k \otimes \boldsymbol{g}_l \otimes \boldsymbol{g}^j . \quad (6.130)$$

For functions defined only on a subset $\mathbf{Slin}^n \subset \mathbf{Lin}^n$ the directional derivative (6.128) again does not deliver a unique result. Similarly to scalar-valued functions this problem can be avoided defining the fourth-order tensor $g(\mathbf{A}),_{\mathbf{A}}$ as a linear mapping on $\mathbf{Slin}^n$. Of special interest in this context are symmetric tensor functions. In this case, using (5.47) and applying the procedure described in Sect. 6.3 we can write

$$g(\mathrm{sym}\mathbf{A}),_{\mathrm{sym}\mathbf{A}} = \left[g(\mathbf{A}),_{\mathbf{A}} \right]^{\mathrm{s}}, \quad \mathbf{A} \in \mathbf{Lin}^n . \quad (6.131)$$

The component representation (6.130) can be given for symmetric tensor functions by

$$g(\mathbf{M}),_{\mathbf{M}} = \frac{1}{2} \sum_{\substack{k,l=1 \\ l \le k}}^{n} \frac{\partial \mathrm{G}^{i}_{\cdot j}}{\partial \mathrm{M}^{kl}} \boldsymbol{g}_i \otimes \left(\boldsymbol{g}^k \otimes \boldsymbol{g}^l + \boldsymbol{g}^l \otimes \boldsymbol{g}^k \right) \otimes \boldsymbol{g}^j$$
$$= \frac{1}{2} \sum_{\substack{k,l=1 \\ l \le k}}^{n} \frac{\partial \mathrm{G}^{i}_{\cdot j}}{\partial \mathrm{M}_{kl}} \boldsymbol{g}_i \otimes \left(\boldsymbol{g}_k \otimes \boldsymbol{g}_l + \boldsymbol{g}_l \otimes \boldsymbol{g}_k \right) \otimes \boldsymbol{g}^j , \quad (6.132)$$

where $\mathbf{M} \in \mathbf{Sym}^n$.

Example 6.15. Derivative of the power function $\mathbf{A}^k$ $(k = 1, 2, \ldots)$. The directional derivative (6.128) of the power function yields

$$\frac{\mathrm{d}}{\mathrm{d}t} (\mathbf{A} + t\mathbf{X})^k \bigg|_{t=0} = \frac{\mathrm{d}}{\mathrm{d}t} \left(\mathbf{A}^k + t \sum_{i=0}^{k-1} \mathbf{A}^i \mathbf{X} \mathbf{A}^{k-1-i} + t^2 \ldots \right) \bigg|_{t=0}$$
$$= \sum_{i=0}^{k-1} \mathbf{A}^i \mathbf{X} \mathbf{A}^{k-1-i} . \quad (6.133)$$

Bearing $(5.17)_1$ and (6.128) in mind we finally obtain

$$\mathbf{A}^k{}_{,\mathbf{A}} = \sum_{i=0}^{k-1} \mathbf{A}^i \otimes \mathbf{A}^{k-1-i}, \quad \mathbf{A} \in \mathbf{Lin}^n. \tag{6.134}$$

In the special case $k = 1$ it leads to the identity

$$\mathbf{A}_{,\mathbf{A}} = \mathfrak{I}, \quad \mathbf{A} \in \mathbf{Lin}^n. \tag{6.135}$$

For power functions of symmetric tensors application of (6.131) yields

$$\mathbf{M}^k{}_{,\mathbf{M}} = \sum_{i=0}^{k-1} \left(\mathbf{M}^i \otimes \mathbf{M}^{k-1-i}\right)^{\mathrm{s}}, \quad \mathbf{M} \in \mathbf{Sym}^n \tag{6.136}$$

and consequently

$$\mathbf{M}_{,\mathbf{M}} = \mathfrak{I}^{\mathrm{s}}, \quad \mathbf{M} \in \mathbf{Sym}^n. \tag{6.137}$$

Example 6.16. Derivative of the transposed tensor $\mathbf{A}^{\mathrm{T}}$. In this case, we can write

$$\frac{\mathrm{d}}{\mathrm{d}t} (\mathbf{A} + t\mathbf{X})^{\mathrm{T}}\Big|_{t=0} = \frac{\mathrm{d}}{\mathrm{d}t} \left(\mathbf{A}^{\mathrm{T}} + t\mathbf{X}^{\mathrm{T}}\right)\Big|_{t=0} = \mathbf{X}^{\mathrm{T}}.$$

On use of (5.81) this yields

$$\mathbf{A}^{\mathrm{T}}{}_{,\mathbf{A}} = \mathfrak{T}. \tag{6.138}$$

Example 6.17. Derivative of the inverse tensor $\mathbf{A}^{-1}$, where $\mathbf{A} \in \mathbf{Inv}^n$. Consider the directional derivative of the identity $\mathbf{A}^{-1}\mathbf{A} = \mathbf{I}$. It delivers:

$$\frac{\mathrm{d}}{\mathrm{d}t} (\mathbf{A} + t\mathbf{X})^{-1} (\mathbf{A} + t\mathbf{X})\Big|_{t=0} = \mathbf{0}.$$

Applying the product rule of differentiation (2.9) and using (6.133) we further write

$$\frac{\mathrm{d}}{\mathrm{d}t} (\mathbf{A} + t\mathbf{X})^{-1}\Big|_{t=0} \mathbf{A} + \mathbf{A}^{-1}\mathbf{X} = \mathbf{0}$$

and finally

$$\frac{\mathrm{d}}{\mathrm{d}t} (\mathbf{A} + t\mathbf{X})^{-1}\Big|_{t=0} = -\mathbf{A}^{-1}\mathbf{X}\mathbf{A}^{-1}.$$

Hence, in view of $(5.17)_1$

$$\mathbf{A}^{-1}{}_{,\mathbf{A}} = -\mathbf{A}^{-1} \otimes \mathbf{A}^{-1}. \tag{6.139}$$

The calculation of the derivative of tensor functions can be simplified by means of differentiation rules. One of them is the following composition rule. Let

$\mathbf{G} = g(\mathbf{A})$ and $\mathbf{H} = h(\mathbf{A})$ be two arbitrary differentiable tensor-valued functions of $\mathbf{A}$. Then,

$$(\mathbf{GH})_{,\mathbf{A}} = \mathbf{G}_{,\mathbf{A}}\mathbf{H} + \mathbf{G}\mathbf{H}_{,\mathbf{A}}. \tag{6.140}$$

For the proof we again apply the directional derivative (6.128) taking (2.9) and (5.40) into account

$$\begin{aligned}
(\mathbf{GH})_{,\mathbf{A}} : \mathbf{X} &= \frac{\mathrm{d}}{\mathrm{d}t}\left[g(\mathbf{A}+t\mathbf{X})\,h(\mathbf{A}+t\mathbf{X})\right]\Big|_{t=0} \\
&= \frac{\mathrm{d}}{\mathrm{d}t}g(\mathbf{A}+t\mathbf{X})\Big|_{t=0}\mathbf{H} + \mathbf{G}\frac{\mathrm{d}}{\mathrm{d}t}h(\mathbf{A}+t\mathbf{X})\Big|_{t=0} \\
&= (\mathbf{G}_{,\mathbf{A}} : \mathbf{X})\,\mathbf{H} + \mathbf{G}\,(\mathbf{H}_{,\mathbf{A}} : \mathbf{X}) \\
&= (\mathbf{G}_{,\mathbf{A}}\mathbf{H} + \mathbf{G}\mathbf{H}_{,\mathbf{A}}) : \mathbf{X}, \quad \forall \mathbf{X} \in \mathbf{Lin}^n.
\end{aligned}$$

Example 6.18. The right and left Cauchy-Green tensors are given in terms of the deformation gradient $\mathbf{F}$ respectively by

$$\mathbf{C} = \mathbf{F}^{\mathrm{T}}\mathbf{F}, \quad \mathbf{b} = \mathbf{F}\mathbf{F}^{\mathrm{T}}. \tag{6.141}$$

Of special interest in continuum mechanics is the derivative of these tensors with respect to $\mathbf{F}$. With the aid of the product rule (6.140) and using (5.42), (5.77), (5.84), $(5.85)_1$, (6.135) and (6.138) we obtain

$$\mathbf{C}_{,\mathbf{F}} = \mathbf{F}^{\mathrm{T}}{}_{,\mathbf{F}}\,\mathbf{F} + \mathbf{F}^{\mathrm{T}}\mathbf{F}_{,\mathbf{F}} = \mathcal{T}\mathbf{F} + \mathbf{F}^{\mathrm{T}}\mathcal{I} = (\mathbf{I} \otimes \mathbf{F})^{\mathrm{t}} + \mathbf{F}^{\mathrm{T}} \otimes \mathbf{I}, \tag{6.142}$$

$$\mathbf{b}_{,\mathbf{F}} = \mathbf{F}_{,\mathbf{F}}\,\mathbf{F}^{\mathrm{T}} + \mathbf{F}\mathbf{F}^{\mathrm{T}}{}_{,\mathbf{F}} = \mathcal{I}\mathbf{F}^{\mathrm{T}} + \mathbf{F}\mathcal{T} = \mathbf{I} \otimes \mathbf{F}^{\mathrm{T}} + (\mathbf{F} \otimes \mathbf{I})^{\mathrm{t}}. \tag{6.143}$$

Further product rules of differentiation of tensor functions can be written as

$$(f\mathbf{G})_{,\mathbf{A}} = \mathbf{G} \odot f_{,\mathbf{A}} + f\,\mathbf{G}_{,\mathbf{A}}, \tag{6.144}$$

$$(\mathbf{G} : \mathbf{H})_{,\mathbf{A}} = \mathbf{H} : \mathbf{G}_{,\mathbf{A}} + \mathbf{G} : \mathbf{H}_{,\mathbf{A}}, \tag{6.145}$$

where $f = \hat{f}(\mathbf{A})$, $\mathbf{G} = g(\mathbf{A})$ and $\mathbf{H} = h(\mathbf{A})$ are again a scalar-valued and two tensor-valued differentiable tensor functions, respectively. The proof is similar to (6.140) (see Exercise 6.15).

Example 6.19. With the aid of the above differentiation rules we can easily express the derivatives of the spherical and deviatoric parts (1.163) of a second-order tensor by (compare with (5.89))

$$\mathrm{sph}\mathbf{A}_{,\mathbf{A}} = \left[\frac{1}{n}\mathrm{tr}(\mathbf{A})\,\mathbf{I}\right]_{,\mathbf{A}} = \frac{1}{n}\mathbf{I} \odot \mathbf{I} = \mathcal{P}_{\mathrm{sph}}, \tag{6.146}$$

$$\mathrm{dev}\mathbf{A},_{\mathbf{A}} = \left[\mathbf{A} - \frac{1}{n}\mathrm{tr}\,(\mathbf{A})\,\mathbf{I}\right],_{\mathbf{A}} = \mathfrak{J} - \frac{1}{n}\mathbf{I}\odot\mathbf{I} = \mathcal{P}_{\mathrm{dev}}. \tag{6.147}$$

In a similar way we can also express the derivative of the isochoric part of the deformation $(6.79)_2$ as

$$\begin{aligned}\bar{\mathbf{C}},_{\mathbf{C}} &= \left(J^{-2/3}\mathbf{C}\right),_{\mathbf{C}} = J^{-2/3}\mathbf{C},_{\mathbf{C}} + \mathbf{C}\odot\left(J^{-2/3}\right),_{\mathbf{C}} \\ &= J^{-2/3}\left(\mathfrak{J}^{\mathrm{s}} - \frac{1}{3}\mathbf{C}\odot\mathbf{C}^{-1}\right) = \mathcal{P}_{\mathrm{iso}},\end{aligned} \tag{6.148}$$

where relation (6.83) is utilized.

Example 6.20. Tangent moduli of hyperelastic isotropic and transversely isotropic materials. The tangent moduli are defined by (see, e.g., [30])

$$\mathcal{C} = \frac{\partial\mathbf{S}}{\partial\tilde{\mathbf{E}}} = 2\frac{\partial\mathbf{S}}{\partial\mathbf{C}}, \tag{6.149}$$

where $\tilde{\mathbf{E}}$ denotes the Green-Lagrange strain tensor defined in (6.114). For hyperelastic materials this definition implies in view of (6.80) the representation

$$\mathcal{C} = \frac{\partial^2\psi}{\partial\tilde{\mathbf{E}}\partial\tilde{\mathbf{E}}} = 4\frac{\partial^2\psi}{\partial\mathbf{C}\partial\mathbf{C}}. \tag{6.150}$$

For a hyperelastic isotropic material we thus obtain by virtue of (6.136), (6.144), (6.10) or (6.111)

$$\begin{aligned}\mathcal{C} &= 4\sum_{k,l=1}^{3} kl\frac{\partial^2\tilde{\psi}}{\partial\mathrm{tr}\mathbf{C}^k\,\partial\mathrm{tr}\mathbf{C}^l}\mathbf{C}^{k-1}\odot\mathbf{C}^{l-1} \\ &\quad + 8\frac{\partial\tilde{\psi}}{\partial\mathrm{tr}\mathbf{C}^2}\mathfrak{J}^{\mathrm{s}} + 12\frac{\partial\tilde{\psi}}{\partial\mathrm{tr}\mathbf{C}^3}\left(\mathbf{C}\otimes\mathbf{I} + \mathbf{I}\otimes\mathbf{C}\right)^{\mathrm{s}}.\end{aligned} \tag{6.151}$$

For a hyperelastic transversely isotropic material the above procedure yields with the aid of (6.126)

$$\begin{aligned}\mathcal{C} &= 4\sum_{k,l=1}^{3} kl\frac{\partial^2\tilde{\psi}}{\partial\mathrm{tr}\mathbf{C}^k\,\partial\mathrm{tr}\mathbf{C}^l}\mathbf{C}^{k-1}\odot\mathbf{C}^{l-1} + 4\frac{\partial^2\tilde{\psi}}{\partial\mathrm{tr}\,(\mathbf{CL})\,\partial\mathrm{tr}\,(\mathbf{CL})}\mathbf{L}\odot\mathbf{L} \\ &\quad + 4\frac{\partial^2\tilde{\psi}}{\partial\mathrm{tr}\,(\mathbf{C}^2\mathbf{L})\,\partial\mathrm{tr}\,(\mathbf{C}^2\mathbf{L})}\left(\mathbf{CL} + \mathbf{LC}\right)\odot\left(\mathbf{CL} + \mathbf{LC}\right) \\ &\quad + 4\sum_{k}^{3} k\frac{\partial^2\tilde{\psi}}{\partial\mathrm{tr}\mathbf{C}^k\,\partial\mathrm{tr}\,(\mathbf{CL})}\left(\mathbf{C}^{k-1}\odot\mathbf{L} + \mathbf{L}\odot\mathbf{C}^{k-1}\right)\end{aligned}$$

$$+4\sum_{k}^{3} k \frac{\partial^2 \tilde{\psi}}{\partial \mathrm{tr}\mathbf{C}^k \partial \mathrm{tr}\left(\mathbf{C}^2\mathbf{L}\right)} \left[\mathbf{C}^{k-1} \odot (\mathbf{CL}+\mathbf{LC}) + (\mathbf{CL}+\mathbf{LC}) \odot \mathbf{C}^{k-1}\right]$$

$$+4\frac{\partial^2 \tilde{\psi}}{\partial \mathrm{tr}\left(\mathbf{CL}\right) \partial \mathrm{tr}\left(\mathbf{C}^2\mathbf{L}\right)} \left[\mathbf{L} \odot (\mathbf{CL}+\mathbf{LC}) + (\mathbf{CL}+\mathbf{LC}) \odot \mathbf{L}\right] + 8\frac{\partial \tilde{\psi}}{\partial \mathrm{tr}\mathbf{C}^2}\mathfrak{I}^{\mathrm{s}}$$

$$+12\frac{\partial \tilde{\psi}}{\partial \mathrm{tr}\mathbf{C}^3}\left(\mathbf{C}\otimes\mathbf{I}+\mathbf{I}\otimes\mathbf{C}\right)^{\mathrm{s}} + 4\frac{\partial \tilde{\psi}}{\partial \mathrm{tr}\left(\mathbf{C}^2\mathbf{L}\right)}\left(\mathbf{L}\otimes\mathbf{I}+\mathbf{I}\otimes\mathbf{L}\right)^{\mathrm{s}}. \tag{6.152}$$

Example 6.21. Tangent moduli of hyperelastic materials with isochoric-volumetric split of the strain energy function can be obtained by inserting (6.87) and (6.88) into (6.81) and (6.149). Thus, applying product rules of differentiation (6.140), (6.144) and (6.145) and using (6.83) we can write

$$\mathcal{C} = 2\mathbf{S},_{\mathbf{C}} = \mathcal{C}_{\mathrm{iso}} + \mathcal{C}_{\mathrm{vol}}, \tag{6.153}$$

where

$$\mathcal{C}_{\mathrm{vol}} = 2\mathbf{S}_{\mathrm{vol}},_{\mathbf{C}} = \left(U''J^2 + U'J\right)\mathbf{C}^{-1}\odot\mathbf{C}^{-1} - 2U'J\left(\mathbf{C}^{-1}\otimes\mathbf{C}^{-1}\right)^{\mathrm{s}}, \tag{6.154}$$

$$\begin{aligned}\mathcal{C}_{\mathrm{iso}} = 2\mathbf{S}_{\mathrm{iso}},_{\mathbf{C}} = &-\frac{2}{3}J^{-2/3}\left[\bar{\mathbf{S}} - \frac{1}{3}\left(\bar{\mathbf{S}}:\mathbf{C}\right)\mathbf{C}^{-1}\right]\odot\mathbf{C}^{-1} \\ &+2J^{-2/3}\left[\bar{\mathbf{S}},_{\bar{\mathbf{C}}}:\mathcal{P}_{\mathrm{iso}} + \frac{1}{3}\left(\bar{\mathbf{S}}:\mathbf{C}\right)\left(\mathbf{C}^{-1}\otimes\mathbf{C}^{-1}\right)^{\mathrm{s}}\right. \\ &\left.-\frac{1}{3}\mathbf{C}^{-1}\odot\left(\mathbf{C}:\bar{\mathbf{S}},_{\bar{\mathbf{C}}}:\mathcal{P}_{\mathrm{iso}}+\bar{\mathbf{S}}\right)\right].\end{aligned}$$

Using the abbreviation

$$\bar{\mathcal{C}} = 2\bar{\mathbf{S}},_{\bar{\mathbf{C}}} = 4\bar{\psi},_{\bar{\mathbf{C}}\bar{\mathbf{C}}} \tag{6.155}$$

the last expression can finally be simplified as follows

$$\begin{aligned}\mathcal{C}_{\mathrm{iso}} = &\mathcal{P}_{\mathrm{iso}}^{\mathrm{T}}:\bar{\mathcal{C}}:\mathcal{P}_{\mathrm{iso}} - \frac{2}{3}J^{-2/3}\left(\bar{\mathbf{S}}\odot\mathbf{C}^{-1}+\mathbf{C}^{-1}\odot\bar{\mathbf{S}}\right) \\ &+\frac{2}{3}\left(\bar{\mathbf{S}}:\bar{\mathbf{C}}\right)\left[\left(\mathbf{C}^{-1}\otimes\mathbf{C}^{-1}\right)^{\mathrm{s}}+\frac{1}{3}\mathbf{C}^{-1}\odot\mathbf{C}^{-1}\right].\end{aligned} \tag{6.156}$$

6.6 Generalized Rivlin's Identities

The Cayley-Hamilton equation (4.95)

$$\mathbf{A}^n - \mathrm{I}_{\mathbf{A}}^{(1)}\mathbf{A}^{n-1} + \mathrm{I}_{\mathbf{A}}^{(2)}\mathbf{A}^{n-2} + \ldots + (-1)^n\,\mathrm{I}_{\mathbf{A}}^{(n)}\mathbf{I} = \mathbf{0} \tag{6.157}$$

represents a universal relation connecting powers of a second-order tensor **A** with its principal invariants. Similar universal relations connecting several second-order tensors might also be useful for example for the representation of isotropic tensor functions or for the solution of tensor equations. Such relations are generally called Rivlin's identities.

In order to formulate the Rivlin identities we first differentiate the Cayley-Hamilton equation (6.157) with respect to **A**. With the aid of (6.60), (6.134) and (6.144) we can write

$$\begin{aligned}\mathcal{O} &= \left[\sum_{k=0}^{n} (-1)^k \, \mathrm{I}_{\mathbf{A}}^{(k)} \mathbf{A}^{n-k}\right]_{,\mathbf{A}} \\ &= \sum_{k=1}^{n} (-1)^k \, \mathbf{A}^{n-k} \odot \left[\sum_{i=1}^{k} (-1)^{i-1} \, \mathrm{I}_{\mathbf{A}}^{(k-i)} \left(\mathbf{A}^{\mathrm{T}}\right)^{i-1}\right] \\ &\quad + \sum_{k=0}^{n-1} (-1)^k \, \mathrm{I}_{\mathbf{A}}^{(k)} \left[\sum_{i=1}^{n-k} \mathbf{A}^{n-k-i} \otimes \mathbf{A}^{i-1}\right].\end{aligned}$$

Substituting in the last row the summation index $k + i$ by k and using (5.42) and (5.43) we further obtain

$$\sum_{k=1}^{n} \mathbf{A}^{n-k} \sum_{i=1}^{k} (-1)^{k-i} \, \mathrm{I}_{\mathbf{A}}^{(k-i)} \left[\mathbf{I} \odot \left(\mathbf{A}^{\mathrm{T}}\right)^{i-1} - \mathbf{I} \otimes \mathbf{A}^{i-1}\right] = \mathcal{O}. \tag{6.158}$$

Mapping an arbitrary second-order tensor **B** by both sides of this equation yields an identity written in terms of second-order tensors [11]

$$\sum_{k=1}^{n} \mathbf{A}^{n-k} \sum_{i=1}^{k} (-1)^{k-i} \, \mathrm{I}_{\mathbf{A}}^{(k-i)} \left[\mathrm{tr}\left(\mathbf{A}^{i-1}\mathbf{B}\right) \mathbf{I} - \mathbf{B}\mathbf{A}^{i-1}\right] = \mathbf{0}. \tag{6.159}$$

This relation is referred to as the generalized Rivlin's identity. Indeed, in the special case of three-dimensional space ($n = 3$) it takes the form

$$\begin{aligned}&\mathbf{ABA} + \mathbf{A}^2\mathbf{B} + \mathbf{BA}^2 - \mathrm{tr}\,(\mathbf{A})\,(\mathbf{AB} + \mathbf{BA}) - \mathrm{tr}\,(\mathbf{B})\,\mathbf{A}^2 \\ &\quad - \left[\mathrm{tr}\,(\mathbf{AB}) - \mathrm{tr}\mathbf{A}\mathrm{tr}\mathbf{B}\right]\mathbf{A} + \frac{1}{2}\left[\mathrm{tr}^2\mathbf{A} - \mathrm{tr}\mathbf{A}^2\right]\mathbf{B} \\ &\quad - \left\{\mathrm{tr}\left(\mathbf{A}^2\mathbf{B}\right) - \mathrm{tr}\mathbf{A}\,\mathrm{tr}\,(\mathbf{AB}) + \frac{1}{2}\mathrm{tr}\mathbf{B}\left[\mathrm{tr}^2\mathbf{A} - \mathrm{tr}\mathbf{A}^2\right]\right\}\mathbf{I} = \mathbf{0},\end{aligned} \tag{6.160}$$

originally obtained by Rivlin [35] by means of matrix calculations.

Differentiating (6.159) again with respect to $\mathbf{A}$ delivers

$$\mathbb{O} = \sum_{k=1}^{n-1}\sum_{j=1}^{n-k}\left(\mathbf{A}^{n-k-j} \otimes \mathbf{A}^{j-1}\right)\sum_{i=1}^{k}(-1)^{k-i}\,\mathrm{I}_{\mathbf{A}}^{(k-i)}\left[\mathrm{tr}\left(\mathbf{A}^{i-1}\mathbf{B}\right)\mathbf{I} - \mathbf{B}\mathbf{A}^{i-1}\right]$$

$$+ \sum_{k=2}^{n}\sum_{i=1}^{k-1}(-1)^{k-i}\,\mathbf{A}^{n-k}\left[\mathrm{tr}\left(\mathbf{A}^{i-1}\mathbf{B}\right)\mathbf{I} - \mathbf{B}\mathbf{A}^{i-1}\right]$$

$$\odot\left[\sum_{j=1}^{k-i}(-1)^{j-1}\,\mathrm{I}_{\mathbf{A}}^{(k-i-j)}\left(\mathbf{A}^{\mathrm{T}}\right)^{j-1}\right]$$

$$+ \sum_{k=2}^{n}\sum_{i=2}^{k}(-1)^{k-i}\,\mathrm{I}_{\mathbf{A}}^{(k-i)}\mathbf{A}^{n-k}\odot\left[\sum_{j=1}^{i-1}\left(\mathbf{A}^{j-1}\mathbf{B}\mathbf{A}^{i-1-j}\right)^{\mathrm{T}}\right]$$

$$- \sum_{k=2}^{n}\sum_{i=2}^{k}(-1)^{k-i}\,\mathrm{I}_{\mathbf{A}}^{(k-i)}\mathbf{A}^{n-k}\mathbf{B}\left[\sum_{j=1}^{i-1}\left(\mathbf{A}^{i-j-1} \otimes \mathbf{A}^{j-1}\right)\right].$$

Changing the summation indices and summation order we obtain

$$\sum_{i=1}^{n-1}\sum_{k=i+1}^{n}\sum_{j=1}^{k-i}(-1)^{k-i-j}\,\mathrm{I}_{\mathbf{A}}^{(k-i-j)}\mathbf{A}^{n-k}\Big\{\mathbf{I}\otimes\left[\mathrm{tr}\left(\mathbf{A}^{j-1}\mathbf{B}\right)\mathbf{A}^{i-1}\right.$$

$$\left. - \mathbf{A}^{i-1}\mathbf{B}\mathbf{A}^{j-1}\right] - \left[\mathrm{tr}\left(\mathbf{A}^{i-1}\mathbf{B}\right)\mathbf{I} - \mathbf{B}\mathbf{A}^{i-1}\right]\odot\left(\mathbf{A}^{\mathrm{T}}\right)^{j-1}$$

$$+ \mathbf{I}\odot\left(\mathbf{A}^{i-1}\mathbf{B}\mathbf{A}^{j-1}\right)^{\mathrm{T}} - \mathbf{B}\mathbf{A}^{j-1}\otimes\mathbf{A}^{i-1}\Big\} = \mathbb{O}. \tag{6.161}$$

The second-order counterpart of this relation can be obtained by mapping another arbitrary second-order tensor $\mathbf{C} \in \mathbf{Lin}^n$ as [11]

$$\sum_{i=1}^{n-1}\sum_{k=i+1}^{n}\sum_{j=1}^{k-i}(-1)^{k-i-j}\,\mathrm{I}_{\mathbf{A}}^{(k-i-j)}\mathbf{A}^{n-k}\Big\{\mathrm{tr}\left(\mathbf{A}^{j-1}\mathbf{B}\right)\mathbf{C}\mathbf{A}^{i-1}$$

$$- \mathbf{C}\mathbf{A}^{i-1}\mathbf{B}\mathbf{A}^{j-1} - \left[\mathrm{tr}\left(\mathbf{A}^{i-1}\mathbf{B}\right)\mathbf{I} - \mathbf{B}\mathbf{A}^{i-1}\right]\mathrm{tr}\left(\mathbf{A}^{j-1}\mathbf{C}\right)$$

$$+ \mathrm{tr}\left(\mathbf{A}^{i-1}\mathbf{B}\mathbf{A}^{j-1}\mathbf{C}\right)\mathbf{I} - \mathbf{B}\mathbf{A}^{j-1}\mathbf{C}\mathbf{A}^{i-1}\Big\} = \mathbf{0}. \tag{6.162}$$

In the special case of three-dimensional space ($n = 3$) Eq. (6.162) leads to the well-known identity (see [28, 35, 37])

$$\begin{aligned}&\mathbf{ABC}+\mathbf{ACB}+\mathbf{BCA}+\mathbf{BAC}+\mathbf{CAB}+\mathbf{CBA}-\mathrm{tr}\,(\mathbf{A})\,(\mathbf{BC}+\mathbf{CB})\\&-\mathrm{tr}\,(\mathbf{B})\,(\mathbf{CA}+\mathbf{AC})-\mathrm{tr}\,(\mathbf{C})\,(\mathbf{AB}+\mathbf{BA})+[\mathrm{tr}\,(\mathbf{B})\,\mathrm{tr}\,(\mathbf{C})-\mathrm{tr}\,(\mathbf{BC})]\,\mathbf{A}\\&+[\mathrm{tr}\,(\mathbf{C})\,\mathrm{tr}\,(\mathbf{A})-\mathrm{tr}\,(\mathbf{CA})]\,\mathbf{B}+[\mathrm{tr}\,(\mathbf{A})\,\mathrm{tr}\,(\mathbf{B})-\mathrm{tr}\,(\mathbf{AB})]\,\mathbf{C}\\&-[\mathrm{tr}\,(\mathbf{A})\,\mathrm{tr}\,(\mathbf{B})\,\mathrm{tr}\,(\mathbf{C})-\mathrm{tr}\,(\mathbf{A})\,\mathrm{tr}\,(\mathbf{BC})-\mathrm{tr}\,(\mathbf{B})\,\mathrm{tr}\,(\mathbf{CA})\\&-\mathrm{tr}\,(\mathbf{C})\,\mathrm{tr}\,(\mathbf{AB})+\mathrm{tr}\,(\mathbf{ABC})+\mathrm{tr}\,(\mathbf{ACB})]\,\mathbf{I}=\mathbf{0}.\end{aligned}\tag{6.163}$$

Exercises

6.1. Check isotropy of the following tensor functions:

(a) $f(\mathbf{A}) = \boldsymbol{a}\mathbf{A}\boldsymbol{b}$, where $\boldsymbol{a}, \boldsymbol{b} \in \mathbb{E}^n$,
(b) $f(\mathbf{A}) = \mathrm{A}^{11} + \mathrm{A}^{22} + \mathrm{A}^{33}$,
(c) $f(\mathbf{A}) = \mathrm{A}^{11} + \mathrm{A}^{12} + \mathrm{A}^{13}$, where A^{ij} represent the components of $\mathbf{A} \in \mathbf{Lin}^3$ with respect to an orthonormal basis $\boldsymbol{e}_i$ $(i = 1, 2, 3)$, so that $\mathbf{A} = \mathrm{A}^{ij}\boldsymbol{e}_i \otimes \boldsymbol{e}_j$,
(d) $f(\mathbf{A}) = \det\mathbf{A}$,
(e) $f(\mathbf{A}) = \lambda_{\max}$, where $\lambda_{\max}$ denotes the maximal (in the sense of the norm $\sqrt{\lambda\bar{\lambda}}$) eigenvalue of $\mathbf{A} \in \mathbf{Lin}^n$.

6.2. Prove the alternative representation (6.18) for the functional basis of an arbitrary second-order tensor **A**.

6.3. Prove the product rule of differentiation (6.58) by applying the formalism of the directional derivative (6.44).

6.4. An orthotropic symmetry group $\mathfrak{g}_o$ is described in terms of three structural tensors defined by $\mathbf{L}_i = \boldsymbol{l}_i \otimes \boldsymbol{l}_i$, where $\boldsymbol{l}_i \cdot \boldsymbol{l}_j = \delta_{ij}$ $(i, j = 1, 2, 3)$ are unit vectors along mutually orthogonal principal material directions. Represent the general orthotropic strain energy function

$$\psi\left(\mathbf{QCQ}^{\mathrm{T}}\right) = \psi(\mathbf{C}), \quad \forall\mathbf{Q} \in \mathfrak{g}_o \tag{6.164}$$

in terms of the orthotropic invariants.

6.5. Using the results of Exercise 6.4, derive the constitutive relation for the second Piola-Kirchhoff stress tensor **S** (6.80) and the tangent moduli $\mathfrak{C}$ (6.149) for the general hyperelastic orthotropic material.

6.6. Represent the general constitutive relation for an orthotropic elastic material as a function **S** (**C**).

6.7. A symmetry group $\mathfrak{g}_f$ of a fiber reinforced material with an isotropic matrix is described in terms of structural tensors defined by $\mathbf{L}_i = \boldsymbol{l}_i \otimes \boldsymbol{l}_i$, where the unit vectors $\boldsymbol{l}_i$ $(i = 1, 2, \ldots, k)$ define the directions of fiber families and are not necessarily orthogonal to each other. Represent the strain energy function

$$\psi\left(\mathbf{QCQ}^{\mathrm{T}}\right) = \psi\left(\mathbf{C}\right), \quad \forall \mathbf{Q} \in \mathfrak{g}_f \tag{6.165}$$

of a fiber reinforced material with two families of fibers ($k = 2$).

6.8. Derive the constitutive relation $\mathbf{S} = 2\partial\psi/\partial\mathbf{C} + p\mathbf{C}^{-1}$ and the tangent moduli $\mathcal{C} = 2\partial\mathbf{S}/\partial\mathbf{C}$ for the Mooney-Rivlin material represented by the strain energy function (6.11).

6.9. Derive the constitutive relation for the Ogden material (6.12) in terms of the second Piola-Kirchhoff stress tensor using expression (6.80).

6.10. Show that $\mathrm{tr}\left(\mathbf{CL}_i\mathbf{CL}_j\right)$, where $\mathbf{L}_i$ $(i = 1, 2, 3)$ are structural tensors defined in Exercise 6.4, represents an orthotropic tensor function (orthotropic invariant) of $\mathbf{C}$. Express this function in terms of the orthotropic functional basis obtained in Exercise 6.4.

6.11. The strain energy function of the orthotropic St.Venant-Kirchhoff material is given by

$$\psi\left(\tilde{\mathbf{E}}\right) = \frac{1}{2}\sum_{i,j=1}^{3} a_{ij}\,\mathrm{tr}\left(\tilde{\mathbf{E}}\mathbf{L}_i\right)\mathrm{tr}\left(\tilde{\mathbf{E}}\mathbf{L}_j\right) + \sum_{\substack{i,j=1\\ i\neq j}}^{3} G_{ij}\,\mathrm{tr}\left(\tilde{\mathbf{E}}\mathbf{L}_i\tilde{\mathbf{E}}\mathbf{L}_j\right), \tag{6.166}$$

where $\tilde{\mathbf{E}}$ denotes the Green-Lagrange strain tensor (6.114) and $\mathbf{L}_i$ $(i = 1, 2, 3)$ are the structural tensors defined in Exercise 6.4. $a_{ij} = a_{ji}$ $(i, j = 1, 2, 3)$ and $G_{ij} = G_{ji}$ $(i \neq j = 1, 2, 3)$ represent material constants. Derive the constitutive relation for the second Piola-Kirchhoff stress tensor $\mathbf{S}$ (6.80) and the tangent moduli $\mathcal{C}$ (6.149).

6.12. Show that the function $\psi(\tilde{\mathbf{E}})$ (6.166) becomes transversely isotropic if

$$a_{22} = a_{33}, \quad a_{12} = a_{13}, \quad G_{12} = G_{13}, \quad G_{23} = \frac{1}{2}\left(a_{22} - a_{23}\right) \tag{6.167}$$

and isotropic of the form (6.116) if

$$a_{12} = a_{13} = a_{23} = \lambda, \quad G_{12} = G_{13} = G_{23} = G,$$
$$a_{11} = a_{22} = a_{33} = \lambda + 2G. \tag{6.168}$$

6.13. Complete the proof of Theorem 6.3.

6.14. Express $\mathbf{A}^{-k}{}_{,\mathbf{A}}$, where $k = 1, 2, \ldots$.

6.15. Prove the product rules of differentiation (6.144) and (6.145).

6.16. Write out Rivlin's identity (6.159) for $n = 2$.

Chapter 7
Analytic Tensor Functions

7.1 Introduction

In the previous chapter we discussed isotropic and anisotropic tensor functions and their general representations. Of particular interest in continuum mechanics are isotropic tensor-valued functions of one arbitrary (not necessarily symmetric) tensor. For example, the exponential function of the velocity gradient or other non-symmetric strain rates is very suitable for the formulation of evolution equations in large strain anisotropic plasticity. In this section we focus on a special class of isotropic tensor-valued functions referred here to as analytic tensor functions. In order to specify this class of functions we first deal with the general question how an isotropic tensor-valued function can be defined.

For isotropic functions of diagonalizable tensors the most natural way is the spectral decomposition (4.43)

$$\mathbf{A} = \sum_{i=1}^{s} \lambda_i \mathbf{P}_i, \tag{7.1}$$

so that we may write similarly to (4.48)

$$g(\mathbf{A}) = \sum_{i=1}^{s} g(\lambda_i) \mathbf{P}_i, \tag{7.2}$$

where $g(\lambda_i)$ is an arbitrary (not necessarily polynomial) scalar function defined on the spectrum $\lambda_i \; (i = 1, 2, \ldots, s)$ of the tensor $\mathbf{A}$. Obviously, the so-defined function $g(\mathbf{A})$ is isotropic in the sense of the condition (6.91). Indeed,

$$g\left(\mathbf{Q}\mathbf{A}\mathbf{Q}^{\mathrm{T}}\right) = \sum_{i=1}^{s} g(\lambda_i) \mathbf{Q}\mathbf{P}_i\mathbf{Q}^{\mathrm{T}} = \mathbf{Q} g(\mathbf{A}) \mathbf{Q}^{\mathrm{T}}, \quad \forall \mathbf{Q} \in \mathbf{Orth}^n, \tag{7.3}$$

M. Itskov, *Tensor Algebra and Tensor Analysis for Engineers*, Mathematical Engineering, DOI 10.1007/978-3-642-30879-6_7, © Springer-Verlag Berlin Heidelberg 2013

where we take into account that the spectral decomposition of the tensor $\mathbf{QAQ}^{\mathrm{T}}$ is given by

$$\mathbf{QAQ}^{\mathrm{T}} = \sum_{i=1}^{s} \lambda_i \mathbf{QP}_i \mathbf{Q}^{\mathrm{T}}. \tag{7.4}$$

Example 7.1. Generalized strain measures. The so-called generalized strain measures $\mathbf{E}$ and $\mathbf{e}$ (also known as Hill's strains, [16, 17]) play an important role in kinematics of continuum. They are defined by (7.2) as isotropic tensor-valued functions of the symmetric right and left stretch tensor $\mathbf{U}$ and $\mathbf{v}$ and are referred to as Lagrangian (material) and Eulerian (spatial) strains, respectively. The definition of the generalized strains is based on the spectral representations by

$$\mathbf{U} = \sum_{i=1}^{s} \lambda_i \mathbf{P}_i, \quad \mathbf{v} = \sum_{i=1}^{s} \lambda_i \mathbf{p}_i, \tag{7.5}$$

where $\lambda_i > 0$ are the eigenvalues (referred to as principal stretches) while $\mathbf{P}_i$ and $\mathbf{p}_i$ $(i = 1, 2, \ldots, s)$ denote the corresponding eigenprojections. Accordingly,

$$\mathbf{E} = \sum_{i=1}^{s} f(\lambda_i) \mathbf{P}_i, \quad \mathbf{e} = \sum_{i=1}^{s} f(\lambda_i) \mathbf{p}_i, \tag{7.6}$$

where f is a strictly-increasing scalar function satisfying the conditions $f(1) = 0$ and $f'(1) = 1$. A special class of generalized strain measures specified by

$$\mathbf{E}^{(a)} = \begin{cases} \sum\limits_{i=1}^{s} \dfrac{1}{a}\left(\lambda_i^a - 1\right) \mathbf{P}_i & \text{for } a \neq 0, \\ \sum\limits_{i=1}^{s} \ln(\lambda_i) \mathbf{P}_i & \text{for } a = 0, \end{cases} \tag{7.7}$$

$$\mathbf{e}^{(a)} = \begin{cases} \sum\limits_{i=1}^{s} \dfrac{1}{a}\left(\lambda_i^a - 1\right) \mathbf{p}_i & \text{for } a \neq 0, \\ \sum\limits_{i=1}^{s} \ln(\lambda_i) \mathbf{p}_i & \text{for } a = 0 \end{cases} \tag{7.8}$$

are referred to as Seth's strains [40], where a is a real number. For example, the Green-Lagrange strain tensor (6.114) introduced in Chap. 6 belongs to Seth's strains as $\mathbf{E}^{(2)}$.

Since non-symmetric tensors do not generally admit the spectral decomposition in the diagonal form (7.1), it is necessary to search for other approaches for the definition of the isotropic tensor function $g(\mathbf{A}) : \mathbf{Lin}^n \mapsto \mathbf{Lin}^n$. One of these approaches is the tensor power series of the form

$$g(\mathbf{A}) = a_0 \mathbf{I} + a_1 \mathbf{A} + a_2 \mathbf{A}^2 + \ldots = \sum_{r=0}^{\infty} a_r \mathbf{A}^r. \tag{7.9}$$

Indeed, in view of (6.92)

$$g\left(\mathbf{QAQ}^{\mathrm{T}}\right) = \sum_{r=0}^{\infty} a_r \left(\mathbf{QAQ}^{\mathrm{T}}\right)^r$$
$$= \sum_{r=0}^{\infty} a_r \mathbf{QA}^r\mathbf{Q}^{\mathrm{T}} = \mathbf{Q}g\left(\mathbf{A}\right)\mathbf{Q}^{\mathrm{T}}, \quad \forall \mathbf{Q} \in \mathbf{Orth}^n. \tag{7.10}$$

For example, the exponential tensor function can be defined in terms of the infinite power series (7.9) by (1.114).

One can show that the power series (7.9), provided it converges, represents a generalization of (7.2) to arbitrary (and not necessarily diagonalizable) second-order tensors. Conversely, the isotropic tensor function (7.2) with $g(\lambda)$ analytic on the spectrum of $\mathbf{A}$ can be considered as an extension of infinite power series (7.9) to its non-convergent domain if the latter exists. Indeed, for diagonalizable tensor arguments within the convergence domain of the tensor power series (7.9) both definitions coincide. For example, inserting (7.1) into (7.9) and taking (4.47) into account we have

$$g\left(\mathbf{A}\right) = \sum_{r=0}^{\infty} a_r \left(\sum_{i=1}^{s} \lambda_i \mathbf{P}_i\right)^r = \sum_{r=0}^{\infty} a_r \sum_{i=1}^{s} \lambda_i^r \mathbf{P}_i = \sum_{i=1}^{s} g\left(\lambda_i\right) \mathbf{P}_i \tag{7.11}$$

with the abbreviation

$$g\left(\lambda\right) = \sum_{r=0}^{\infty} a_r \lambda^r, \tag{7.12}$$

so that

$$a_r = \frac{1}{r!} \left.\frac{\partial^r g\left(\lambda\right)}{\partial \lambda^r}\right|_{\lambda=0}. \tag{7.13}$$

The above mentioned convergence requirement vastly restricts the definition domain of many isotropic tensor functions defined in terms of infinite series (7.9). For example, one can show that the power series for the logarithmic tensor function

$$\ln\left(\mathbf{A} + \mathbf{I}\right) = \sum_{r=1}^{\infty} (-1)^{r+1} \frac{\mathbf{A}^r}{r} \tag{7.14}$$

converges for $|\lambda_i| < 1$ $(i = 1, 2, \ldots, s)$ and diverges if $|\lambda_k| > 1$ at least for some k $(1 \le k \le s)$ (see, e.g., [13]).

In order to avoid this convergence problem we consider a tensor function defined by the so-called Dunford-Taylor integral as (see, for example, [25])

$$g\left(\mathbf{A}\right) = \frac{1}{2\pi \mathrm{i}} \oint_{\Gamma} g\left(\zeta\right) \left(\zeta \mathbf{I} - \mathbf{A}\right)^{-1} \mathrm{d}\zeta \tag{7.15}$$

taken on the complex plane over Γ, where Γ represents a closed curve or consists of simple closed curves, the union interior of which includes all the eigenvalues $\lambda_i \in \mathbb{C}$ $(i = 1, 2, \ldots, s)$ of the tensor argument $\mathbf{A}$. $g(\zeta) : \mathbb{C} \mapsto \mathbb{C}$ is an arbitrary scalar function analytic within and on Γ.

One can easily prove that the tensor function (7.15) is isotropic in the sense of the definition (6.91). Indeed, with the aid of (1.133) and (1.134) we obtain (cf. [34])

$$
\begin{aligned}
g\left(\mathbf{QAQ}^{\mathrm{T}}\right) &= \frac{1}{2\pi \mathrm{i}} \oint_{\Gamma} g(\zeta) \left(\zeta \mathbf{I} - \mathbf{QAQ}^{\mathrm{T}}\right)^{-1} \mathrm{d}\zeta \\
&= \frac{1}{2\pi \mathrm{i}} \oint_{\Gamma} g(\zeta) \left[\mathbf{Q}(\zeta \mathbf{I} - \mathbf{A})\mathbf{Q}^{\mathrm{T}}\right]^{-1} \mathrm{d}\zeta \\
&= \frac{1}{2\pi \mathrm{i}} \oint_{\Gamma} g(\zeta) \mathbf{Q} (\zeta \mathbf{I} - \mathbf{A})^{-1} \mathbf{Q}^{\mathrm{T}} \mathrm{d}\zeta \\
&= \mathbf{Q} g(\mathbf{A}) \mathbf{Q}^{\mathrm{T}}, \quad \forall \mathbf{Q} \in \mathbf{Orth}^n.
\end{aligned}
\tag{7.16}
$$

It can be verified that for diagonalizable tensors the Dunford-Taylor integral (7.15) reduces to the spectral decomposition (7.2) and represents therefore its generalization. Indeed, inserting (7.1) into (7.15) delivers

$$
\begin{aligned}
g(\mathbf{A}) &= \frac{1}{2\pi \mathrm{i}} \oint_{\Gamma} g(\zeta) \left(\zeta \mathbf{I} - \sum_{i=1}^{s} \lambda_i \mathbf{P}_i\right)^{-1} \mathrm{d}\zeta \\
&= \frac{1}{2\pi \mathrm{i}} \oint_{\Gamma} g(\zeta) \left[\sum_{i=1}^{s} (\zeta - \lambda_i) \mathbf{P}_i\right]^{-1} \mathrm{d}\zeta \\
&= \frac{1}{2\pi \mathrm{i}} \oint_{\Gamma} g(\zeta) \sum_{i=1}^{s} (\zeta - \lambda_i)^{-1} \mathbf{P}_i \mathrm{d}\zeta \\
&= \sum_{i=1}^{s} \left[\frac{1}{2\pi \mathrm{i}} \oint_{\Gamma} g(\zeta) (\zeta - \lambda_i)^{-1} \mathrm{d}\zeta\right] \mathbf{P}_i = \sum_{i=1}^{s} g(\lambda_i) \mathbf{P}_i,
\end{aligned}
\tag{7.17}
$$

where we keep (4.46) in mind and apply the Cauchy integral formula (see, e.g. [5]). Using this result we can represent, for example, the generalized strain measures (7.6) by

$$
\mathbf{E} = f(\mathbf{U}), \quad \mathbf{e} = f(\mathbf{v}), \tag{7.18}
$$

where the tensor functions $f(\mathbf{U})$ and $f(\mathbf{v})$ are defined by (7.15).

Further, one can show that the Dunford-Taylor integral (7.15) also represents a generalization of tensor power series (7.9). For this purpose, it suffices to verify that (7.15) based on a scalar function $g(\zeta) = \zeta^k$ $(k = 0, 1, 2, \ldots)$ results into the monomial $g(\mathbf{A}) = \mathbf{A}^k$. To this end, we consider in (7.15) the following identity [25]

$$g(\zeta)\mathbf{I} = (\zeta\mathbf{I})^k = (\zeta\mathbf{I} - \mathbf{A} + \mathbf{A})^k = (\zeta\mathbf{I} - \mathbf{A})^k + \ldots + \mathbf{A}^k. \qquad (7.19)$$

Thereby, all terms except of the last one have no pole within Γ and vanish according to the Cauchy theorem (see, e.g., [5]), so that

$$g(\mathbf{A}) = \frac{1}{2\pi\mathrm{i}} \oint_{\Gamma} \left[(\zeta\mathbf{I} - \mathbf{A})^{k-1} + \ldots + \mathbf{A}^k (\zeta\mathbf{I} - \mathbf{A})^{-1} \right] \mathrm{d}\zeta = \mathbf{A}^k. \qquad (7.20)$$

Isotropic tensor functions defined by (7.15) will henceforth be referred to as analytic tensor functions. The above discussed properties of analytic tensor functions can be completed by the following relations (Exercise 7.3)

$$\begin{aligned} g(\mathbf{A}) &= \alpha f(\mathbf{A}) + \beta h(\mathbf{A}), \text{ if } && g(\lambda) = \alpha f(\lambda) + \beta h(\lambda), \\ g(\mathbf{A}) &= f(\mathbf{A}) h(\mathbf{A}), && \text{if } \quad g(\lambda) = f(\lambda) h(\lambda), \\ g(\mathbf{A}) &= f(h(\mathbf{A})), && \text{if } \quad g(\lambda) = f(h(\lambda)). \end{aligned} \qquad (7.21)$$

In the following we will deal with representations for analytic tensor functions and their derivatives.

7.2 Closed-Form Representation for Analytic Tensor Functions and Their Derivatives

Our aim is to obtain the so-called closed form representation for analytic tensor functions and their derivatives. This representation should be given only in terms of finite powers of the tensor argument and its eigenvalues and avoid any reference to the integral over the complex plane or to power series.

We start with the Cayley-Hamilton theorem (4.95) for the tensor $\zeta\mathbf{I} - \mathbf{A}$

$$\sum_{k=0}^{n} (-1)^k \mathrm{I}_{\zeta\mathbf{I}-\mathbf{A}}^{(k)} (\zeta\mathbf{I} - \mathbf{A})^{n-k} = \mathbf{0}. \qquad (7.22)$$

With the aid of the Vieta theorem (4.24) we can write

$$\mathrm{I}_{\zeta\mathbf{I}-\mathbf{A}}^{(0)} = 1, \quad \mathrm{I}_{\zeta\mathbf{I}-\mathbf{A}}^{(k)} = \sum_{i_1<i_2<\ldots<i_k} (\zeta - \lambda_{i_1})(\zeta - \lambda_{i_2}) \ldots (\zeta - \lambda_{i_k}), \qquad (7.23)$$

where $k = 1, 2, \ldots, n$ and the eigenvalues λ_i $(i = 1, 2, \ldots, n)$ of the tensor $\mathbf{A}$ are counted with their multiplicity.

Composing (7.22) with the so-called resolvent of $\mathbf{A}$

$$\mathbf{R}(\zeta) = (\zeta\mathbf{I} - \mathbf{A})^{-1} \tag{7.24}$$

yields

$$\begin{aligned}\mathbf{R}(\zeta) &= \frac{1}{\mathrm{I}^{(n)}_{\zeta\mathbf{I}-\mathbf{A}}} \sum_{k=0}^{n-1} (-1)^{n-k-1} \, \mathrm{I}^{(k)}_{\zeta\mathbf{I}-\mathbf{A}} \, (\zeta\mathbf{I} - \mathbf{A})^{n-k-1} \\ &= \frac{1}{\mathrm{I}^{(n)}_{\zeta\mathbf{I}-\mathbf{A}}} \sum_{k=0}^{n-1} \mathrm{I}^{(k)}_{\zeta\mathbf{I}-\mathbf{A}} \, (\mathbf{A} - \zeta\mathbf{I})^{n-k-1} .\end{aligned} \tag{7.25}$$

Applying the binomial theorem (see, e.g., [5])

$$(\mathbf{A} - \zeta\mathbf{I})^{l} = \sum_{p=0}^{l} (-1)^{l-p} \binom{l}{p} \zeta^{l-p} \mathbf{A}^{p}, \quad l = 1, 2, \ldots, \tag{7.26}$$

where

$$\binom{l}{p} = \frac{l!}{p!\,(l-p)!}, \tag{7.27}$$

we obtain

$$\mathbf{R}(\zeta) = \frac{1}{\mathrm{I}^{(n)}_{\zeta\mathbf{I}-\mathbf{A}}} \sum_{k=0}^{n-1} \mathrm{I}^{(k)}_{\zeta\mathbf{I}-\mathbf{A}} \sum_{p=0}^{n-k-1} (-1)^{n-k-1-p} \binom{n-k-1}{p} \zeta^{n-k-1-p} \mathbf{A}^{p}. \tag{7.28}$$

Rearranging this expression with respect to the powers of the tensor $\mathbf{A}$ delivers

$$\mathbf{R}(\zeta) = \sum_{p=0}^{n-1} \alpha_p \mathbf{A}^{p} \tag{7.29}$$

with

$$\alpha_p = \frac{1}{\mathrm{I}^{(n)}_{\zeta\mathbf{I}-\mathbf{A}}} \sum_{k=0}^{n-p-1} (-1)^{n-k-p-1} \binom{n-k-1}{p} \mathrm{I}^{(k)}_{\zeta\mathbf{I}-\mathbf{A}} \zeta^{n-k-p-1}, \tag{7.30}$$

where $p = 0, 1, \ldots, n-1$. Inserting this result into (7.15) we obtain the following closed-form representation for the tensor function $g(\mathbf{A})$ [21]

$$g(\mathbf{A}) = \sum_{p=0}^{n-1} \varphi_p \mathbf{A}^{p}, \tag{7.31}$$

where

$$\varphi_p = \frac{1}{2\pi \mathrm{i}} \oint_{\Gamma} g(\zeta)\, \alpha_p \mathrm{d}\zeta, \quad p = 0, 1, \ldots, n-1. \tag{7.32}$$

The Cauchy integrals in (7.32) can be calculated with the aid of the residue theorem (see, e.g., [5]). To this end, we first represent the determinant of the tensor $\zeta \mathbf{I} - \mathbf{A}$ in the form

$$\mathrm{I}^{(n)}_{\zeta \mathbf{I} - \mathbf{A}} = \det(\zeta \mathbf{I} - \mathbf{A}) = \prod_{i=1}^{s} (\zeta - \lambda_i)^{r_i}, \tag{7.33}$$

where λ_i denote pairwise distinct eigenvalues with the algebraic multiplicities r_i $(i = 1, 2, \ldots, s)$ such that

$$\sum_{i=1}^{s} r_i = n. \tag{7.34}$$

Thus, inserting (7.30) and (7.33) into (7.32) we obtain

$$\varphi_p = \sum_{i=1}^{s} \frac{1}{(r_i - 1)!} \lim_{\zeta \to \lambda_i} \left\{ \frac{\mathrm{d}^{r_i - 1}}{\mathrm{d}\zeta^{r_i - 1}} \left[g(\zeta)\, \alpha_p(\zeta)\, (\zeta - \lambda_i)^{r_i} \right] \right\}, \tag{7.35}$$

where $p = 1, 2, \ldots, n-1$.

The derivative of the tensor function $g(\mathbf{A})$ can be obtained by direct differentiation of the Dunfod-Taylor integral (7.15). Thus, by use of (6.139) we can write

$$g(\mathbf{A})_{,\mathbf{A}} = \frac{1}{2\pi \mathrm{i}} \oint_{\Gamma} g(\zeta)\, (\zeta \mathbf{I} - \mathbf{A})^{-1} \otimes (\zeta \mathbf{I} - \mathbf{A})^{-1} \mathrm{d}\zeta \tag{7.36}$$

and consequently

$$g(\mathbf{A})_{,\mathbf{A}} = \frac{1}{2\pi \mathrm{i}} \oint_{\Gamma} g(\zeta)\, \mathbf{R}(\zeta) \otimes \mathbf{R}(\zeta)\, \mathrm{d}\zeta. \tag{7.37}$$

Taking (7.29) into account further yields

$$g(\mathbf{A})_{,\mathbf{A}} = \sum_{p,q=0}^{n-1} \eta_{pq} \mathbf{A}^p \otimes \mathbf{A}^q, \tag{7.38}$$

where

$$\eta_{pq} = \eta_{qp} = \frac{1}{2\pi \mathrm{i}} \oint_{\Gamma} g(\zeta)\, \alpha_p(\zeta)\, \alpha_q(\zeta)\, \mathrm{d}\zeta, \quad p, q = 0, 1, \ldots, n-1. \tag{7.39}$$

The residue theorem finally delivers

$$\eta_{pq} = \sum_{i=1}^{s} \frac{1}{(2r_i - 1)!} \lim_{\zeta \to \lambda_i} \left\{ \frac{\mathrm{d}^{2r_i - 1}}{\mathrm{d}\zeta^{2r_i - 1}} \left[g(\zeta)\, \alpha_p(\zeta)\, \alpha_q(\zeta)\, (\zeta - \lambda_i)^{2r_i} \right] \right\}, \tag{7.40}$$

where $p, q = 0, 1, \ldots, n-1$.

7.3 Special Case: Diagonalizable Tensor Functions

For analytic functions of diagonalizable tensors the definitions in terms of the Dunford-Taylor integral (7.15) on the one side and eigenprojections (7.2) on the other side become equivalent. In this special case, one can obtain alternative closed-form representations for analytic tensor functions and their derivatives. To this end, we first derive an alternative representation of the Sylvester formula (4.55). In Sect. 4.4 we have shown that the eigenprojections can be given by (4.52)

$$\mathbf{P}_i = p_i(\mathbf{A}), \quad i = 1, 2, \ldots, s, \tag{7.41}$$

where p_i $(i = 1, 2, \ldots, s)$ are polynomials satisfying the requirements (4.51). Thus, the eigenprojections of a second-order tensor can be considered as its analytic (isotropic) tensor functions. Applying the Dunford-Taylor integral (7.15) we can thus write

$$\mathbf{P}_i = \frac{1}{2\pi \mathrm{i}} \oint_{\Gamma} p_i(\zeta) (\zeta \mathbf{I} - \mathbf{A})^{-1} \mathrm{d}\zeta, \quad i = 1, 2, \ldots, s. \tag{7.42}$$

Similarly to (7.31) and (7.35) we further obtain

$$\mathbf{P}_i = \sum_{p=0}^{n-1} \rho_{ip} \mathbf{A}^p, \quad i = 1, 2, \ldots, s, \tag{7.43}$$

where

$$\rho_{ip} = \sum_{k=1}^{s} \frac{1}{(r_k - 1)!} \lim_{\zeta \to \lambda_k} \left\{ \frac{\mathrm{d}^{r_k - 1}}{\mathrm{d}\zeta^{r_k - 1}} \left[p_i(\zeta) \alpha_p(\zeta) (\zeta - \lambda_k)^{r_k} \right] \right\} \tag{7.44}$$

and α_p $(p = 0, 1, \ldots, n-1)$ are given by (7.30). With the aid of polynomial functions $p_i(\lambda)$ satisfying in addition to (4.51) the following conditions

$$\left. \frac{\mathrm{d}^r}{\mathrm{d}\lambda^r} p_i(\lambda) \right|_{\lambda = \lambda_j} = 0 \quad i, j = 1, 2, \ldots, s; \; r = 1, 2, \ldots, r_i - 1 \tag{7.45}$$

we can simplify (7.44) by

$$\rho_{ip} = \frac{1}{(r_i - 1)!} \lim_{\zeta \to \lambda_i} \left\{ \frac{\mathrm{d}^{r_i - 1}}{\mathrm{d}\zeta^{r_i - 1}} \left[\alpha_p(\zeta) (\zeta - \lambda_i)^{r_i} \right] \right\}. \tag{7.46}$$

Now, inserting (7.43) into (7.2) delivers

$$g(\mathbf{A}) = \sum_{i=1}^{s} g(\lambda_i) \sum_{p=0}^{n-1} \rho_{ip} \mathbf{A}^p. \tag{7.47}$$

In order to obtain an alternative representation for $g(\mathbf{A})_{,\mathbf{A}}$ we again consider the tensor power series (7.9). Direct differentiation of (7.9) with respect to $\mathbf{A}$ delivers with the aid of (6.134)

$$g(\mathbf{A})_{,\mathbf{A}} = \sum_{r=1}^{\infty} a_r \sum_{k=0}^{r-1} \mathbf{A}^{r-1-k} \otimes \mathbf{A}^k. \tag{7.48}$$

Applying the spectral representation (7.1) and taking (4.47) and (7.12) into account we further obtain (see also [19,49])

$$\begin{aligned} g(\mathbf{A})_{,\mathbf{A}} &= \sum_{r=1}^{\infty} a_r \sum_{k=0}^{r-1} \sum_{i,j=1}^{s} \lambda_i^{r-1-k} \lambda_j^k \mathbf{P}_i \otimes \mathbf{P}_j \\ &= \sum_{i=1}^{s} \sum_{r=1}^{\infty} r a_r \lambda_i^{r-1} \mathbf{P}_i \otimes \mathbf{P}_i + \sum_{\substack{i,j=1 \\ j\neq i}}^{s} \sum_{r=1}^{\infty} a_r \frac{\lambda_i^r - \lambda_j^r}{\lambda_i - \lambda_j} \mathbf{P}_i \otimes \mathbf{P}_j \\ &= \sum_{i=1}^{s} g'(\lambda_i) \mathbf{P}_i \otimes \mathbf{P}_i + \sum_{\substack{i,j=1 \\ j\neq i}}^{s} \frac{g(\lambda_i) - g(\lambda_j)}{\lambda_i - \lambda_j} \mathbf{P}_i \otimes \mathbf{P}_j \\ &= \sum_{i,j=1}^{s} G_{ij} \mathbf{P}_i \otimes \mathbf{P}_j, \end{aligned} \tag{7.49}$$

where

$$G_{ij} = \begin{cases} g'(\lambda_i) & \text{if } i = j, \\ \dfrac{g(\lambda_i) - g(\lambda_j)}{\lambda_i - \lambda_j} & \text{if } i \neq j. \end{cases} \tag{7.50}$$

Inserting into (7.49) the alternative representation for the eigenprojections (7.43) yields

$$g(\mathbf{A})_{,\mathbf{A}} = \sum_{i,j=1}^{s} G_{ij} \sum_{p,q=0}^{n-1} \rho_{ip} \rho_{jq} \mathbf{A}^p \otimes \mathbf{A}^q. \tag{7.51}$$

Thus, we again end up with the representation (7.38)

$$g(\mathbf{A})_{,\mathbf{A}} = \sum_{p,q=0}^{n-1} \eta_{pq} \mathbf{A}^p \otimes \mathbf{A}^q, \tag{7.52}$$

where

$$\eta_{pq} = \eta_{qp} = \sum_{i,j=1}^{s} G_{ij} \rho_{ip} \rho_{jq}, \quad p, q = 0, 1, \ldots, n-1. \tag{7.53}$$

Finally, let us focus on the differentiability of eigenprojections. To this end, we represent them by [25] (Exercise 7.5)

$$\mathbf{P}_i = \frac{1}{2\pi \mathrm{i}} \oint_{\Gamma_i} (\zeta \mathbf{I} - \mathbf{A})^{-1} \, \mathrm{d}\zeta, \quad i = 1, 2, \ldots s, \tag{7.54}$$

where the integral is taken on the complex plane over a closed curve Γ_i the interior of which includes only the eigenvalue λ_i. All other eigenvalues of $\mathbf{A}$ lie outside Γ_i. Γ_i does not depend on λ_i as far as this eigenvalue is simple and does not lie directly on Γ_i. Indeed, if λ_i is multiple, a perturbation of $\mathbf{A}$ by $\mathbf{A} + t\mathbf{X}$ can lead to a split of eigenvalues within Γ_i. In this case, (7.54) yields a sum of eigenprojections corresponding to these split eigenvalues which coalesce in λ_i for $t = 0$. Thus, the eigenprojection $\mathbf{P}_i$ corresponding to a simple eigenvalue λ_i is differentiable according to (7.54). Direct differentiation of (7.54) delivers in this case

$$\mathbf{P}_i,_{\mathbf{A}} = \frac{1}{2\pi \mathrm{i}} \oint_{\Gamma_i} (\zeta \mathbf{I} - \mathbf{A})^{-1} \otimes (\zeta \mathbf{I} - \mathbf{A})^{-1} \, \mathrm{d}\zeta, \quad r_i = 1. \tag{7.55}$$

By analogy with (7.38) we thus obtain

$$\mathbf{P}_i,_{\mathbf{A}} = \sum_{p,q=0}^{n-1} \upsilon_{ipq} \mathbf{A}^p \otimes \mathbf{A}^q, \tag{7.56}$$

where

$$\upsilon_{ipq} = \upsilon_{iqp} = \frac{1}{2\pi \mathrm{i}} \oint_{\Gamma_i} \alpha_p(\zeta) \, \alpha_q(\zeta) \, \mathrm{d}\zeta, \quad p, q = 0, 1, \ldots, n-1. \tag{7.57}$$

By the residue theorem we further write

$$\upsilon_{ipq} = \lim_{\zeta \to \lambda_i} \left\{ \frac{\mathrm{d}}{\mathrm{d}\zeta} \left[\alpha_p(\zeta) \, \alpha_q(\zeta) \, (\zeta - \lambda_i)^2 \right] \right\}, \quad p, q = 0, 1, \ldots, n-1. \tag{7.58}$$

With the aid of (7.49) one can obtain an alternative representation for the derivative of the eigenprojections in terms of the eigenprojections themselves. Indeed, substituting the function g in (7.49) by p_i and taking the properties of the latter function (4.51) and (7.45) into account we have

$$\mathbf{P}_i,_{\mathbf{A}} = \sum_{\substack{j=1 \\ j \neq i}}^{s} \frac{\mathbf{P}_i \otimes \mathbf{P}_j + \mathbf{P}_j \otimes \mathbf{P}_i}{\lambda_i - \lambda_j}. \tag{7.59}$$

7.4 Special Case: Three-Dimensional Space

First, we specify the closed-form solutions (7.31) and (7.38) for three-dimensional space ($n = 3$). In this case, the functions $\alpha_k(\zeta)$ $(k = 0, 1, 2)$ (7.30) take the form

$$\alpha_0(\zeta) = \frac{\zeta^2 - \zeta \mathrm{I}_{\zeta\mathbf{I}-\mathbf{A}} + \mathrm{II}_{\zeta\mathbf{I}-\mathbf{A}}}{\mathrm{III}_{\zeta\mathbf{I}-\mathbf{A}}}$$

$$= \frac{\zeta^2 - \zeta(\lambda_1 + \lambda_2 + \lambda_3) + \lambda_1\lambda_2 + \lambda_2\lambda_3 + \lambda_3\lambda_1}{(\zeta - \lambda_1)(\zeta - \lambda_2)(\zeta - \lambda_3)},$$

$$\alpha_1(\zeta) = \frac{\mathrm{I}_{\zeta\mathbf{I}-\mathbf{A}} - 2\zeta}{\mathrm{III}_{\zeta\mathbf{I}-\mathbf{A}}} = \frac{\zeta - \lambda_1 - \lambda_2 - \lambda_3}{(\zeta - \lambda_1)(\zeta - \lambda_2)(\zeta - \lambda_3)},$$

$$\alpha_2(\zeta) = \frac{1}{\mathrm{III}_{\zeta\mathbf{I}-\mathbf{A}}} = \frac{1}{(\zeta - \lambda_1)(\zeta - \lambda_2)(\zeta - \lambda_3)}. \tag{7.60}$$

Inserting these expressions into (7.35) and (7.40) and considering separately cases of distinct and repeated eigenvalues, we obtain the following result [23].

Distinct eigenvalues: $\lambda_1 \neq \lambda_2 \neq \lambda_3 \neq \lambda_1$,

$$\varphi_0 = \sum_{i=1}^{3} \frac{g(\lambda_i)\lambda_j\lambda_k}{D_i},$$

$$\varphi_1 = -\sum_{i=1}^{3} \frac{g(\lambda_i)(\lambda_j + \lambda_k)}{D_i},$$

$$\varphi_2 = \sum_{i=1}^{3} \frac{g(\lambda_i)}{D_i}, \tag{7.61}$$

$$\eta_{00} = \sum_{i=1}^{3} \frac{\lambda_j^2\lambda_k^2 g'(\lambda_i)}{D_i^2} - \sum_{\substack{i,j=1\\ i\neq j}}^{3} \frac{\lambda_i\lambda_j\lambda_k^2 \left[g(\lambda_i) - g(\lambda_j)\right]}{(\lambda_i - \lambda_j)^3 D_k},$$

$$\eta_{01} = \eta_{10} = -\sum_{i=1}^{3} \frac{(\lambda_j + \lambda_k)\lambda_j\lambda_k g'(\lambda_i)}{D_i^2}$$

$$+ \sum_{\substack{i,j=1\\ i\neq j}}^{3} \frac{(\lambda_j + \lambda_k)\lambda_i\lambda_k \left[g(\lambda_i) - g(\lambda_j)\right]}{(\lambda_i - \lambda_j)^3 D_k},$$

$$\eta_{02} = \eta_{20} = \sum_{i=1}^{3} \frac{\lambda_j \lambda_k g'(\lambda_i)}{D_i^2} - \sum_{\substack{i,j=1\\ i\neq j}}^{3} \frac{\lambda_i \lambda_k \left[g(\lambda_i) - g(\lambda_j)\right]}{(\lambda_i - \lambda_j)^3 D_k},$$

$$\eta_{11} = \sum_{i=1}^{3} \frac{(\lambda_j + \lambda_k)^2 g'(\lambda_i)}{D_i^2} - \sum_{\substack{i,j=1\\ i\neq j}}^{3} \frac{(\lambda_j + \lambda_k)(\lambda_i + \lambda_k)\left[g(\lambda_i) - g(\lambda_j)\right]}{(\lambda_i - \lambda_j)^3 D_k},$$

$$\eta_{12} = \eta_{21} = -\sum_{i=1}^{3} \frac{(\lambda_j + \lambda_k) g'(\lambda_i)}{D_i^2} + \sum_{\substack{i,j=1\\ i\neq j}}^{3} \frac{(\lambda_i + \lambda_k)\left[g(\lambda_i) - g(\lambda_j)\right]}{(\lambda_i - \lambda_j)^3 D_k},$$

$$\eta_{22} = \sum_{i=1}^{3} \frac{g'(\lambda_i)}{D_i^2} - \sum_{\substack{i,j=1\\ i\neq j}}^{3} \frac{g(\lambda_i) - g(\lambda_j)}{(\lambda_i - \lambda_j)^3 D_k}, \quad i \neq j \neq k \neq i, \tag{7.62}$$

where

$$D_i = (\lambda_i - \lambda_j)(\lambda_i - \lambda_k), \quad i \neq j \neq k \neq i = 1,2,3. \tag{7.63}$$

Double coalescence of eigenvalues: $\lambda_i \neq \lambda_j = \lambda_k = \lambda,\ j \neq k$,

$$\varphi_0 = \lambda \frac{\lambda g(\lambda_i) - \lambda_i g(\lambda)}{(\lambda_i - \lambda)^2} + \frac{\lambda_i g(\lambda)}{(\lambda_i - \lambda)} - \frac{\lambda \lambda_i g'(\lambda)}{(\lambda_i - \lambda)},$$

$$\varphi_1 = -2\lambda \frac{g(\lambda_i) - g(\lambda)}{(\lambda_i - \lambda)^2} + \frac{g'(\lambda)(\lambda_i + \lambda)}{(\lambda_i - \lambda)},$$

$$\varphi_2 = \frac{g(\lambda_i) - g(\lambda)}{(\lambda_i - \lambda)^2} - \frac{g'(\lambda)}{(\lambda_i - \lambda)}, \tag{7.64}$$

$$\eta_{00} = \frac{\left(2\lambda^2\lambda_i^2 - 6\lambda^3\lambda_i\right)\left[g(\lambda_i) - g(\lambda)\right]}{(\lambda_i - \lambda)^5}$$

$$+ \frac{\lambda^4 g'(\lambda_i) + \left(2\lambda^3\lambda_i + 4\lambda^2\lambda_i^2 - 4\lambda\lambda_i^3 + \lambda_i^4\right) g'(\lambda)}{(\lambda_i - \lambda)^4}$$

$$+ \frac{\left(2\lambda^2\lambda_i^2 - \lambda_i^3\lambda\right) g''(\lambda)}{(\lambda_i - \lambda)^3} + \frac{\lambda^2\lambda_i^2 g'''(\lambda)}{6(\lambda_i - \lambda)^2},$$

$$\eta_{01} = \eta_{10} = \frac{\left(3\lambda^3 + 7\lambda_i\lambda^2 - 2\lambda_i^2\lambda\right)\left[g\left(\lambda_i\right) - g\left(\lambda\right)\right]}{\left(\lambda_i - \lambda\right)^5}$$

$$- \frac{2\lambda^3 g'\left(\lambda_i\right) + \left(\lambda^3 + 7\lambda_i\lambda^2 - 2\lambda_i^2\lambda\right) g'\left(\lambda\right)}{\left(\lambda_i - \lambda\right)^4}$$

$$- \frac{\left(4\lambda^2\lambda_i + \lambda_i^2\lambda - \lambda_i^3\right) g''\left(\lambda\right)}{2\left(\lambda_i - \lambda\right)^3} - \frac{\lambda_i\lambda\left(\lambda_i + \lambda\right) g'''\left(\lambda\right)}{6\left(\lambda_i - \lambda\right)^2},$$

$$\eta_{02} = \eta_{20} = \frac{\left(\lambda_i^2 - 3\lambda_i\lambda - 2\lambda^2\right)\left[g\left(\lambda_i\right) - g\left(\lambda\right)\right]}{\left(\lambda_i - \lambda\right)^5}$$

$$+ \frac{\lambda^2 g'\left(\lambda_i\right) + \left(\lambda^2 + 3\lambda_i\lambda - \lambda_i^2\right) g'\left(\lambda\right)}{\left(\lambda_i - \lambda\right)^4}$$

$$+ \frac{\left(3\lambda\lambda_i - \lambda_i^2\right) g''\left(\lambda\right)}{2\left(\lambda_i - \lambda\right)^3} + \frac{\lambda_i\lambda g'''\left(\lambda\right)}{6\left(\lambda_i - \lambda\right)^2},$$

$$\eta_{11} = -4\frac{\lambda\left(\lambda_i + 3\lambda\right)\left[g\left(\lambda_i\right) - g\left(\lambda\right)\right]}{\left(\lambda_i - \lambda\right)^5} + 4\frac{\lambda^2 g'\left(\lambda_i\right) + \lambda\left(\lambda_i + 2\lambda\right) g'\left(\lambda\right)}{\left(\lambda_i - \lambda\right)^4}$$

$$+ \frac{2\lambda\left(\lambda_i + \lambda\right) g''\left(\lambda\right)}{\left(\lambda_i - \lambda\right)^3} + \frac{\left(\lambda_i + \lambda\right)^2 g'''\left(\lambda\right)}{6\left(\lambda_i - \lambda\right)^2},$$

$$\eta_{12} = \eta_{21} = \frac{\left(\lambda_i + 7\lambda\right)\left[g\left(\lambda_i\right) - g\left(\lambda\right)\right]}{\left(\lambda_i - \lambda\right)^5} - \frac{2\lambda g'\left(\lambda_i\right) + \left(\lambda_i + 5\lambda\right) g'\left(\lambda\right)}{\left(\lambda_i - \lambda\right)^4}$$

$$- \frac{\left(\lambda_i + 3\lambda\right) g''\left(\lambda\right)}{2\left(\lambda_i - \lambda\right)^3} - \frac{\left(\lambda_i + \lambda\right) g'''\left(\lambda\right)}{6\left(\lambda_i - \lambda\right)^2},$$

$$\eta_{22} = -4\frac{g\left(\lambda_i\right) - g\left(\lambda\right)}{\left(\lambda_i - \lambda\right)^5} + \frac{g'\left(\lambda_i\right) + 3g'\left(\lambda\right)}{\left(\lambda_i - \lambda\right)^4} + \frac{g''\left(\lambda\right)}{\left(\lambda_i - \lambda\right)^3} + \frac{g'''\left(\lambda\right)}{6\left(\lambda_i - \lambda\right)^2}. \quad (7.65)$$

Triple coalescence of eigenvalues: $\lambda_1 = \lambda_2 = \lambda_3 = \lambda$,

$$\varphi_0 = g\left(\lambda\right) - \lambda g'\left(\lambda\right) + \frac{1}{2}\lambda^2 g''\left(\lambda\right),$$

$$\varphi_1 = g'(\lambda) - \lambda g''(\lambda),$$
$$\varphi_2 = \frac{1}{2} g''(\lambda), \tag{7.66}$$

$$\begin{aligned}
\eta_{00} &= g'(\lambda) - \lambda g''(\lambda) + \frac{\lambda^2 g'''(\lambda)}{2} - \frac{\lambda^3 g^{IV}(\lambda)}{12} + \frac{\lambda^4 g^{V}(\lambda)}{120}, \\
\eta_{01} &= \eta_{10} = \frac{g''(\lambda)}{2} - \frac{\lambda g'''(\lambda)}{2} + \frac{\lambda^2 g^{IV}(\lambda)}{8} - \frac{\lambda^3 g^{V}(\lambda)}{60}, \\
\eta_{02} &= \eta_{20} = \frac{g'''(\lambda)}{6} - \frac{\lambda g^{IV}(\lambda)}{24} + \frac{\lambda^2 g^{V}(\lambda)}{120}, \\
\eta_{11} &= \frac{g'''(\lambda)}{6} - \frac{\lambda g^{IV}(\lambda)}{6} + \frac{\lambda^2 g^{V}(\lambda)}{30}, \\
\eta_{12} &= \eta_{21} = \frac{g^{IV}(\lambda)}{24} - \frac{\lambda g^{V}(\lambda)}{60}, \\
\eta_{22} &= \frac{g^{V}(\lambda)}{120},
\end{aligned} \tag{7.67}$$

where superposed Roman numerals denote the order of the derivative.

Example 7.2. To illustrate the application of the above closed-form solution we consider the exponential function of the velocity gradient under simple shear. The velocity gradient is defined as the material time derivative of the deformation gradient by $\mathbf{L} = \dot{\mathbf{F}}$. Using the representation of $\mathbf{F}$ in the case of simple shear (2.69) we can write

$$\mathbf{L} = \mathrm{L}^{i}_{\cdot j} \boldsymbol{e}_i \otimes \boldsymbol{e}^j, \quad \text{where} \quad \left[\mathrm{L}^{i}_{\cdot j}\right] = \begin{bmatrix} 0 & \dot{\gamma} & 0 \\ 0 & 0 & 0 \\ 0 & 0 & 0 \end{bmatrix}. \tag{7.68}$$

We observe that $\mathbf{L}$ has a triple ($r_1 = 3$) zero eigenvalue

$$\lambda_1 = \lambda_2 = \lambda_3 = \lambda = 0. \tag{7.69}$$

This eigenvalue is, however, defect since it is associated with only two ($t_1 = 2$) linearly independent (right) eigenvectors

$$\boldsymbol{a}_1 = \boldsymbol{e}_1, \quad \boldsymbol{a}_2 = \boldsymbol{e}_3. \tag{7.70}$$

Therefore, $\mathbf{L}$ (7.68) is not diagonalizable and admits no spectral decomposition in the form (7.1). For this reason, isotropic functions of $\mathbf{L}$ as well as their derivative cannot be obtained on the basis of eigenprojections. Instead, we exploit the closed-form solution (7.31) and (7.38) with the coefficients calculated for the case of triple coalescence of eigenvalues by (7.66) and (7.67). Thus, we can write

$$\exp(\mathbf{L}) = \exp(\lambda)\left[\left(\frac{1}{2}\lambda^2 - \lambda + 1\right)\mathbf{I} + (1-\lambda)\mathbf{L} + \frac{1}{2}\mathbf{L}^2\right], \tag{7.71}$$

$$\begin{aligned}\exp(\mathbf{L})_{,\mathbf{L}} = \exp(\lambda)\Bigg[&\left(1 - \lambda + \frac{\lambda^2}{2} - \frac{\lambda^3}{12} + \frac{\lambda^4}{120}\right)\mathfrak{I} \\ &+ \left(\frac{1}{2} - \frac{\lambda}{2} + \frac{\lambda^2}{8} - \frac{\lambda^3}{60}\right)(\mathbf{L}\otimes\mathbf{I} + \mathbf{I}\otimes\mathbf{L}) \\ &+ \left(\frac{1}{6} - \frac{\lambda}{6} + \frac{\lambda^2}{30}\right)\mathbf{L}\otimes\mathbf{L} \\ &+ \left(\frac{1}{6} - \frac{\lambda}{24} + \frac{\lambda^2}{120}\right)(\mathbf{L}^2\otimes\mathbf{I} + \mathbf{I}\otimes\mathbf{L}^2) \\ &+ \left(\frac{1}{24} - \frac{\lambda}{60}\right)(\mathbf{L}^2\otimes\mathbf{L} + \mathbf{L}\otimes\mathbf{L}^2) + \frac{1}{120}\mathbf{L}^2\otimes\mathbf{L}^2\Bigg].\end{aligned} \tag{7.72}$$

On use of (7.69) this finally leads to the following expressions

$$\exp(\mathbf{L}) = \mathbf{I} + \mathbf{L} + \frac{1}{2}\mathbf{L}^2, \tag{7.73}$$

$$\begin{aligned}\exp(\mathbf{L})_{,\mathbf{L}} = \mathfrak{I} &+ \frac{1}{2}(\mathbf{L}\otimes\mathbf{I} + \mathbf{I}\otimes\mathbf{L}) + \frac{1}{6}\mathbf{L}\otimes\mathbf{L} + \frac{1}{6}(\mathbf{L}^2\otimes\mathbf{I} + \mathbf{I}\otimes\mathbf{L}^2) \\ &+ \frac{1}{24}(\mathbf{L}^2\otimes\mathbf{L} + \mathbf{L}\otimes\mathbf{L}^2) + \frac{1}{120}\mathbf{L}^2\otimes\mathbf{L}^2.\end{aligned} \tag{7.74}$$

Taking into account a special property of $\mathbf{L}$ (7.68):

$$\mathbf{L}^k = \mathbf{0}, \quad k = 2, 3, \ldots \tag{7.75}$$

the same results can also be obtained directly from the power series (1.114) and its derivative. By virtue of (6.134) the latter one can be given by

$$\exp(\mathbf{L})_{,\mathbf{L}} = \sum_{r=1}^{\infty}\frac{1}{r!}\sum_{k=0}^{r-1}\mathbf{L}^{r-1-k}\otimes\mathbf{L}^k. \tag{7.76}$$

For diagonalizable tensor functions the representations (7.31) and (7.38) can be simplified in the cases of repeated eigenvalues where the coefficients φ_p and η_{pq} are given by (7.64)–(7.67). To this end, we use the identities $\mathbf{A}^2 = (\lambda_i + \lambda)\,\mathbf{A} - \lambda_i \lambda \mathbf{I}$ for the case of double coalescence of eigenvalues $\left(\lambda_i \neq \lambda_j = \lambda_k = \lambda\right)$ and $\mathbf{A} = \lambda \mathbf{I}$, $\mathbf{A}^2 = \lambda^2 \mathbf{I}$ for the case of triple coalescence of eigenvalues $(\lambda_1 = \lambda_2 = \lambda_3 = \lambda)$. Thus, we obtain the following result well-known for symmetric isotropic tensor functions [7].

Double coalescence of eigenvalues: $\lambda_i \neq \lambda_j = \lambda_k = \lambda$, $\mathbf{A}^2 = (\lambda_i + \lambda)\,\mathbf{A} - \lambda_i \lambda \mathbf{I}$,

$$\varphi_0 = \frac{\lambda_i g(\lambda) - \lambda g(\lambda_i)}{\lambda_i - \lambda}, \quad \varphi_1 = \frac{g(\lambda_i) - g(\lambda)}{\lambda_i - \lambda}, \quad \varphi_2 = 0, \tag{7.77}$$

$$\begin{aligned}
\eta_{00} &= -2\lambda_i \lambda \frac{g(\lambda_i) - g(\lambda)}{(\lambda_i - \lambda)^3} + \frac{\lambda^2 g'(\lambda_i) + \lambda_i^2 g'(\lambda)}{(\lambda_i - \lambda)^2}, \\
\eta_{01} &= \eta_{10} = (\lambda_i + \lambda) \frac{g(\lambda_i) - g(\lambda)}{(\lambda_i - \lambda)^3} - \frac{\lambda g'(\lambda_i) + \lambda_i g'(\lambda)}{(\lambda_i - \lambda)^2}, \\
\eta_{11} &= -2\frac{g(\lambda_i) - g(\lambda)}{(\lambda_i - \lambda)^3} + \frac{g'(\lambda_i) + g'(\lambda)}{(\lambda_i - \lambda)^2}, \\
\eta_{02} &= \eta_{20} = \eta_{12} = \eta_{21} = \eta_{22} = 0.
\end{aligned} \tag{7.78}$$

Triple coalescence of eigenvalues: $\lambda_1 = \lambda_2 = \lambda_3 = \lambda$, $\mathbf{A} = \lambda \mathbf{I}$, $\mathbf{A}^2 = \lambda^2 \mathbf{I}$,

$$\varphi_0 = g(\lambda), \quad \varphi_1 = \varphi_2 = 0, \tag{7.79}$$

$$\eta_{00} = g'(\lambda), \quad \eta_{01} = \eta_{10} = \eta_{11} = \eta_{02} = \eta_{20} = \eta_{12} = \eta_{21} = \eta_{22} = 0. \tag{7.80}$$

Finally, we specify the representations for eigenprojections (7.43) and their derivative (7.56) for three-dimensional space. The expressions for the functions ρ_{ip} (7.46) and υ_{ipq} (7.58) can be obtained from the representations for φ_p (7.61), (7.77), (7.79) and η_{pq} (7.62), (7.78), respectively. To this end, we set there $g(\lambda_i) = 1$, $g\left(\lambda_j\right) = g(\lambda_k) = g'(\lambda_i) = g'\left(\lambda_j\right) = g'(\lambda_k) = 0$. Accordingly, we obtain the following representations.

Distinct eigenvalues: $\lambda_1 \neq \lambda_2 \neq \lambda_3 \neq \lambda_1$,

$$\rho_{i0} = \frac{\lambda_j \lambda_k}{D_i}, \quad \rho_{i1} = -\frac{\lambda_j + \lambda_k}{D_i}, \quad \rho_{i2} = \frac{1}{D_i}, \tag{7.81}$$

$$\begin{aligned}
\upsilon_{i00} &= -2\lambda_i \lambda_j \lambda_k \left[\frac{\lambda_k}{\left(\lambda_i - \lambda_j\right)^3 D_k} + \frac{\lambda_j}{(\lambda_i - \lambda_k)^3 D_j} \right], \\
\upsilon_{i01} &= \upsilon_{i10} = \lambda_k \frac{\lambda_i \left(\lambda_j + \lambda_k\right) + \lambda_j (\lambda_i + \lambda_k)}{\left(\lambda_i - \lambda_j\right)^3 D_k} + \lambda_j \frac{\lambda_i \left(\lambda_j + \lambda_k\right) + \lambda_k \left(\lambda_i + \lambda_j\right)}{(\lambda_i - \lambda_k)^3 D_j},
\end{aligned}$$

$$v_{i02} = v_{i20} = -\lambda_k \frac{\lambda_i + \lambda_j}{(\lambda_i - \lambda_j)^3 D_k} - \lambda_j \frac{\lambda_i + \lambda_k}{(\lambda_i - \lambda_k)^3 D_j},$$

$$v_{i11} = -2(\lambda_j + \lambda_k)\left[\frac{\lambda_i + \lambda_k}{(\lambda_i - \lambda_j)^3 D_k} + \frac{\lambda_i + \lambda_j}{(\lambda_i - \lambda_k)^3 D_j}\right],$$

$$v_{i12} = v_{i21} = \frac{\lambda_i + \lambda_j + 2\lambda_k}{(\lambda_i - \lambda_j)^3 D_k} + \frac{\lambda_i + 2\lambda_j + \lambda_k}{(\lambda_i - \lambda_k)^3 D_j},$$

$$v_{i22} = -\frac{2}{(\lambda_i - \lambda_j)^3 D_k} - \frac{2}{(\lambda_i - \lambda_k)^3 D_j}, \quad i \neq j \neq k \neq i = 1,2,3. \tag{7.82}$$

Double coalescence of eigenvalues: $\lambda_i \neq \lambda_j = \lambda_k = \lambda,\ j \neq k$,

$$\rho_{i0} = -\frac{\lambda}{\lambda_i - \lambda}, \quad \rho_{i1} = \frac{1}{\lambda_i - \lambda}, \quad \rho_{i2} = 0, \tag{7.83}$$

$$v_{i00} = -\frac{2\lambda\lambda_i}{(\lambda_i - \lambda)^3}, \quad v_{i01} = v_{i10} = \frac{\lambda_i + \lambda}{(\lambda_i - \lambda)^3}, \quad v_{i11} = -\frac{2}{(\lambda_i - \lambda)^3},$$

$$v_{i02} = v_{i20} = v_{i12} = v_{i21} = v_{i22} = 0. \tag{7.84}$$

Triple coalescence of eigenvalues: $\lambda_1 = \lambda_2 = \lambda_3 = \lambda$,

$$\rho_{10} = 1, \quad \rho_{11} = \rho_{12} = 0. \tag{7.85}$$

The functions v_{1pq} $(p, q = 0, 1, 2)$ are in this case undefined since the only eigenprojection $\mathbf{P}_1$ is not differentiable.

7.5 Recurrent Calculation of Tensor Power Series and Their Derivatives

In numerical calculations with a limited number of digits the above presented closed-form solutions especially those ones for the derivative of analytic tensor functions can lead to inexact results if at least two eigenvalues of the tensor argument are close to each other but do not coincide (see [20]). In this case, a numerical calculation of the derivative of an analytic tensor function on the basis of the corresponding power series expansion might be advantageous provided this series converges very fast so that only a relatively small number of terms are sufficient in order to ensure a desired precision. This numerical calculation can be carried out by means of a recurrent procedure presented below.

The recurrent procedure is based on the sequential application of the Cayley-Hamilton equation (4.95). Accordingly, we can write for an arbitrary second-order tensor $\mathbf{A} \in \mathbf{Lin}^n$

$$\mathbf{A}^n = \sum_{k=0}^{n-1} (-1)^{n-k+1} \mathrm{I}_{\mathbf{A}}^{(n-k)} \mathbf{A}^k. \tag{7.86}$$

With the aid of this relation any non-negative integer power of $\mathbf{A}$ can be represented by

$$\mathbf{A}^r = \sum_{k=0}^{n-1} \omega_k^{(r)} \mathbf{A}^k, \quad r = 0, 1, 2, \ldots \tag{7.87}$$

Indeed, for $r \leq n$ one obtains directly from (7.86)

$$\omega_k^{(r)} = \delta_{rk}, \quad \omega_k^{(n)} = (-1)^{n-k+1} \mathrm{I}_{\mathbf{A}}^{(n-k)}, \quad r, k = 0, 1, \ldots, n-1. \tag{7.88}$$

Further powers of $\mathbf{A}$ can be expressed by composing (7.87) with $\mathbf{A}$ and representing $\mathbf{A}^n$ by (7.86) as

$$\mathbf{A}^{r+1} = \sum_{k=0}^{n-1} \omega_k^{(r)} \mathbf{A}^{k+1} = \sum_{k=1}^{n-1} \omega_{k-1}^{(r)} \mathbf{A}^k + \omega_{n-1}^{(r)} \mathbf{A}^n$$

$$= \sum_{k=1}^{n-1} \omega_{k-1}^{(r)} \mathbf{A}^k + \omega_{n-1}^{(r)} \sum_{k=0}^{n-1} (-1)^{n-k-1} \mathrm{I}_{\mathbf{A}}^{(n-k)} \mathbf{A}^k.$$

Comparing with (7.87) we obtain the following recurrent relations (see also [39])

$$\omega_0^{(r+1)} = \omega_{n-1}^{(r)} (-1)^{n-1} \mathrm{I}_{\mathbf{A}}^{(n)},$$

$$\omega_k^{(r+1)} = \omega_{k-1}^{(r)} + \omega_{n-1}^{(r)} (-1)^{n-k-1} \mathrm{I}_{\mathbf{A}}^{(n-k)}, \quad k = 1, 2, \ldots, n-1. \tag{7.89}$$

With the aid of representation (7.87) the infinite power series (7.9) can thus be expressed by (7.31)

$$g(\mathbf{A}) = \sum_{p=0}^{n-1} \varphi_p \mathbf{A}^p, \tag{7.90}$$

where

$$\varphi_p = \sum_{r=0}^{\infty} a_r \omega_p^{(r)}. \tag{7.91}$$

Thus, the infinite power series (7.9) with the coefficients (7.13) results in the same representation as the corresponding analytic tensor function (7.15) provided the infinite series (7.91) converges.

Further, inserting (7.87) into (7.48) we obtain again the representation (7.38)

$$g(\mathbf{A})_{,\mathbf{A}} = \sum_{p,q=0}^{n-1} \eta_{pq} \mathbf{A}^p \otimes \mathbf{A}^q, \tag{7.92}$$

where

$$\eta_{pq} = \eta_{qp} = \sum_{r=1}^{\infty} a_r \sum_{k=0}^{r-1} \omega_p^{(r-1-k)} \omega_q^{(k)}, \quad p, q = 0, 1, \ldots, n-1. \tag{7.93}$$

The procedure computing the coefficients η_{pq} (7.93) can be simplified by means of the following recurrent identity (see also [31])

$$\begin{aligned}
\sum_{k=0}^{r} \mathbf{A}^{r-k} \otimes \mathbf{A}^k &= \mathbf{A}^r \otimes \mathbf{I} + \left[\sum_{k=0}^{r-1} \mathbf{A}^{r-1-k} \otimes \mathbf{A}^k \right] \mathbf{A} \\
&= \mathbf{A} \left[\sum_{k=0}^{r-1} \mathbf{A}^{r-1-k} \otimes \mathbf{A}^k \right] + \mathbf{I} \otimes \mathbf{A}^r, \quad r = 1, 2 \ldots,
\end{aligned} \tag{7.94}$$

where

$$\sum_{k=0}^{r-1} \mathbf{A}^{r-1-k} \otimes \mathbf{A}^k = \sum_{p,q=0}^{n-1} \xi_{pq}^{(r)} \mathbf{A}^p \otimes \mathbf{A}^q, \quad r = 1, 2 \ldots \tag{7.95}$$

Thus, we obtain

$$\eta_{pq} = \sum_{r=1}^{\infty} a_r \xi_{pq}^{(r)}, \tag{7.96}$$

where [20]

$$\begin{aligned}
&\xi_{pq}^{(1)} = \xi_{qp}^{(1)} = \omega_p^{(0)} \omega_q^{(0)} = \delta_{0p}\delta_{0q}, \quad p \le q;\ p, q = 0, 1, \ldots, n-1, \\
&\xi_{00}^{(r)} = \xi_{0\,n-1}^{(r-1)} \omega_0^{(n)} + \omega_0^{(r-1)}, \\
&\xi_{0q}^{(r)} = \xi_{q0}^{(r)} = \xi_{0\,q-1}^{(r-1)} + \xi_{0\,n-1}^{(r-1)} \omega_q^{(n)} = \xi_{n-1\,q}^{(r-1)} \omega_0^{(n)} + \omega_q^{(r-1)}, \\
&\xi_{pq}^{(r)} = \xi_{qp}^{(r)} = \xi_{p\,q-1}^{(r-1)} + \xi_{p\,n-1}^{(r-1)} \omega_q^{(n)} = \xi_{p-1\,q}^{(r-1)} + \xi_{n-1\,q}^{(r-1)} \omega_p^{(n)}, \\
&p \le q;\ p, q = 1, 2, \ldots, n-1, \quad r = 2, 3, \ldots
\end{aligned} \tag{7.97}$$

The calculation of coefficient series (7.89) and (7.97) can be finished as soon as for some r

Table 7.1 Recurrent calculation of the coefficients $\omega_p^{(r)}$ and $\xi_{pq}^{(r)}$

r	$\omega_0^{(r)}$	$\omega_1^{(r)}$	$\omega_2^{(r)}$	$\xi_{00}^{(r)}$	$\xi_{01}^{(r)}$	$\xi_{02}^{(r)}$	$\xi_{11}^{(r)}$	$\xi_{12}^{(r)}$	$\xi_{22}^{(r)}$	a_r
0	1	0	0							1
1	0	1	0	1	0	0	0	0	0	1
2	0	0	1	0	1	0	0	0	0	1/2
3	0	0	0	0	0	1	1	0	0	1/6
4	0	0	0	0	0	0	0	1	0	1/24
5	0	0	0	0	0	0	0	0	1	1/120
6	0	0	0	0	0	0	0	0	0	1/720
$\sum\limits_{r=0} a_r \omega_p^{(r)}$	1	1	$\frac{1}{2}$							
$\sum\limits_{r=1} a_r \xi_{pq}^{(r)}$				1	$\frac{1}{2}$	$\frac{1}{6}$	$\frac{1}{6}$	$\frac{1}{24}$	$\frac{1}{120}$	

$$\left| a_r \omega_p^{(r)} \right| \leq \varepsilon \left| \sum_{t=0}^{r} a_t \omega_p^{(t)} \right|,$$

$$\left| a_r \xi_{pq}^{(r)} \right| \leq \varepsilon \left| \sum_{t=1}^{r} a_t \xi_{pq}^{(t)} \right|, \quad p, q = 0, 1, \ldots, n-1, \tag{7.98}$$

where $\varepsilon > 0$ denotes a precision parameter.

Example 7.3. To illustrate the application of the above recurrent procedure we consider again the exponential function of the velocity gradient under simple shear (7.68). In view of (7.69) we can write

$$\mathrm{I}_{\mathbf{L}}^{(1)} = \mathrm{I}_{\mathbf{L}}^{(2)} = \mathrm{I}_{\mathbf{L}}^{(3)} = 0. \tag{7.99}$$

With this result in hand the coefficients $\omega_p^{(r)}$ and $\xi_{pq}^{(r)}$ $(p, q = 0, 1, 2)$ appearing in the representation of the analytic tensor function (7.90), (7.91) and its derivative (7.92) and (7.96) can easily be calculated by means of the above recurrent formulas (7.88), (7.89) and (7.97). The results of the calculation are summarized in Table 7.1.

Considering these results in (7.90)–(7.92) and (7.96) we obtain the representations (7.73) and (7.74). Note that the recurrent procedure delivers an exact result only in some special cases like this where the argument tensor is characterized by the property (7.75).

Exercises

7.1. Let $\mathbf{R}(\omega)$ be a proper orthogonal tensor describing a rotation about some axis $\boldsymbol{e} \in \mathbb{E}^3$ by the angle ω. Prove that $\mathbf{R}^a(\omega) = \mathbf{R}(a\omega)$ for any real number a.

7.2. Specify the right stretch tensor **U** $(7.5)_1$ for simple shear utilizing the results of Exercise 4.1.

7.3. Prove the properties of analytic tensor functions (7.21).

7.4. Represent the tangen moduli for the Ogden material (6.12) in the case of simple shear by means of (7.49)–(7.50) and by using the result of Exercises 4.13 and 6.9.

7.5. Prove representation (7.54) for eigenprojections of diagonalizable second-order tensors.

7.6. Calculate eigenprojections and their derivatives for the tensor **A** (Exercise 4.14) using representations (7.81)–(7.85).

7.7. Calculate by means of the closed-form solution $\exp(\mathbf{A})$ and $\exp(\mathbf{A})_{,\mathbf{A}}$, where the tensor **A** is defined in Exercise 4.14. Compare the results for $\exp(\mathbf{A})$ with those of Exercise 4.15.

7.8. Compute $\exp(\mathbf{A})$ and $\exp(\mathbf{A})_{,\mathbf{A}}$ by means of the recurrent procedure with the precision parameter $\varepsilon = 1 \cdot 10^{-6}$, where the tensor **A** is defined in Exercise 4.14. Compare the results with those of Exercise 7.7.

Chapter 8
Applications to Continuum Mechanics

8.1 Polar Decomposition of the Deformation Gradient

The deformation gradient $\mathbf{F}$ represents an invertible second-order tensor generally permitting a unique polar decomposition by

$$\mathbf{F} = \mathbf{R}\mathbf{U} = \mathbf{v}\mathbf{R}, \tag{8.1}$$

where $\mathbf{R}$ is an orthogonal tensor while $\mathbf{U}$ and $\mathbf{v}$ are symmetric tensors. In continuum mechanics, $\mathbf{R}$ is called rotation tensor while $\mathbf{U}$ and $\mathbf{v}$ are referred to as the right and left stretch tensor, respectively. The latter ones have already been introduced in Sect. 7.1 in the context of generalized strain measures.

In order to show that the polar decomposition (8.1) always exists and is unique we first consider the so-called right and left Cauchy-Green tensors respectively by

$$\mathbf{C} = \mathbf{F}^{\mathrm{T}}\mathbf{F}, \quad \mathbf{b} = \mathbf{F}\mathbf{F}^{\mathrm{T}}. \tag{8.2}$$

These tensors are symmetric and have principal traces in common. Indeed, in view of (1.151)

$$\mathrm{tr}(\mathbf{C}^k) = \mathrm{tr}\,\underbrace{(\mathbf{F}^{\mathrm{T}}\mathbf{F}\ldots\mathbf{F}^{\mathrm{T}}\mathbf{F})}_{k\ \text{times}} = \mathrm{tr}\,\underbrace{(\mathbf{F}\mathbf{F}^{\mathrm{T}}\ldots\mathbf{F}\mathbf{F}^{\mathrm{T}})}_{k\ \text{times}} = \mathrm{tr}(\mathbf{b}^k). \tag{8.3}$$

For this reason, all scalar-valued isotropic functions of $\mathbf{C}$ and $\mathbf{b}$ such as principal invariants or eigenvalues coincide. Thus, we can write

$$\mathbf{C} = \sum_{i=1}^{s} \Lambda_i \mathbf{P}_i, \quad \mathbf{b} = \sum_{i=1}^{s} \Lambda_i \mathbf{p}_i, \tag{8.4}$$

where eigenvalues Λ_i are positive. Indeed, let $\boldsymbol{a}_i$ be a unit eigenvector associated with the eigenvalue Λ_i. Then, in view of (1.78), (1.104) and (1.115) and by

M. Itskov, *Tensor Algebra and Tensor Analysis for Engineers*, Mathematical Engineering,
DOI 10.1007/978-3-642-30879-6_8, © Springer-Verlag Berlin Heidelberg 2013

Theorem 1.8 one can write

$$\Lambda_i = \boldsymbol{a}_i \cdot (\Lambda_i \boldsymbol{a}_i) = \boldsymbol{a}_i \cdot (\mathbf{C}\boldsymbol{a}_i) = \boldsymbol{a}_i \cdot (\mathbf{F}^{\mathrm{T}}\mathbf{F}\boldsymbol{a}_i)$$
$$= (\boldsymbol{a}_i \mathbf{F}^{\mathrm{T}}) \cdot (\mathbf{F}\boldsymbol{a}_i) = (\mathbf{F}\boldsymbol{a}_i) \cdot (\mathbf{F}\boldsymbol{a}_i) > 0.$$

Thus, square roots of **C** and **b** are unique tensors defined by

$$\mathbf{U} = \sqrt{\mathbf{C}} = \sum_{i=1}^{s} \sqrt{\Lambda_i}\mathbf{P}_i, \quad \mathbf{v} = \sqrt{\mathbf{b}} = \sum_{i=1}^{s} \sqrt{\Lambda_i}\mathbf{p}_i. \tag{8.5}$$

Further, one can show that

$$\mathbf{R} = \mathbf{F}\mathbf{U}^{-1} \tag{8.6}$$

represents an orthogonal tensor. Indeed,

$$\mathbf{R}\mathbf{R}^{\mathrm{T}} = \mathbf{F}\mathbf{U}^{-1}\mathbf{U}^{-1}\mathbf{F}^{\mathrm{T}} = \mathbf{F}\mathbf{U}^{-2}\mathbf{F}^{\mathrm{T}} = \mathbf{F}\mathbf{C}^{-1}\mathbf{F}^{\mathrm{T}}$$
$$= \mathbf{F}(\mathbf{F}^{\mathrm{T}}\mathbf{F})^{-1}\mathbf{F}^{\mathrm{T}} = \mathbf{F}\mathbf{F}^{-1}\mathbf{F}^{-\mathrm{T}}\mathbf{F}^{\mathrm{T}} = \mathbf{I}.$$

Thus, we can write taking (8.6) into account

$$\mathbf{F} = \mathbf{R}\mathbf{U} = \left(\mathbf{R}\mathbf{U}\mathbf{R}^{\mathrm{T}}\right)\mathbf{R}. \tag{8.7}$$

The tensor

$$\mathbf{R}\mathbf{U}\mathbf{R}^{\mathrm{T}} = \mathbf{F}\mathbf{R}^{\mathrm{T}} \tag{8.8}$$

in (8.7) is symmetric due to symmetry of **U** $(8.5)_1$. Thus, one can write

$$(\mathbf{R}\mathbf{U}\mathbf{R}^{\mathrm{T}})^2 = (\mathbf{R}\mathbf{U}\mathbf{R}^{\mathrm{T}})(\mathbf{R}\mathbf{U}\mathbf{R}^{\mathrm{T}})^{\mathrm{T}} = (\mathbf{F}\mathbf{R}^{\mathrm{T}})(\mathbf{F}\mathbf{R}^{\mathrm{T}})^{\mathrm{T}}$$
$$= \mathbf{F}\mathbf{R}^{\mathrm{T}}\mathbf{R}\mathbf{F}^{\mathrm{T}} = \mathbf{F}\mathbf{F}^{\mathrm{T}} = \mathbf{b}. \tag{8.9}$$

In view of $(8.5)_2$ there exists only one real symmetric tensor whose square is **b**. Hence,

$$\mathbf{R}\mathbf{U}\mathbf{R}^{\mathrm{T}} = \mathbf{v}, \tag{8.10}$$

which by virtue of (8.7) results in the polar decomposition (8.1).

8.2 Basis-Free Representations for the Stretch and Rotation Tensor

With the aid of the closed-form representations for analytic tensor functions discussed in Chap. 7 the stretch and rotation tensors can be expressed directly in terms of the deformation gradient and Cauchy-Green tensors without any reference to their eigenprojections. First, we deal with the stretch tensors (8.5). Inserting in (7.61) $g(\Lambda_i) = \sqrt{\Lambda_i} = \lambda_i$ and keeping in mind (7.31) we write

$$\mathbf{U} = \varphi_0 \mathbf{I} + \varphi_1 \mathbf{C} + \varphi_2 \mathbf{C}^2, \quad \mathbf{v} = \varphi_0 \mathbf{I} + \varphi_1 \mathbf{b} + \varphi_2 \mathbf{b}^2, \tag{8.11}$$

where [45]

$$\begin{aligned} \varphi_0 &= \frac{\lambda_1\lambda_2\lambda_3(\lambda_1+\lambda_2+\lambda_3)}{(\lambda_1+\lambda_2)(\lambda_2+\lambda_3)(\lambda_3+\lambda_1)}, \\ \varphi_1 &= \frac{\lambda_1^2+\lambda_2^2+\lambda_3^2+\lambda_1\lambda_2+\lambda_2\lambda_3+\lambda_3\lambda_1}{(\lambda_1+\lambda_2)(\lambda_2+\lambda_3)(\lambda_3+\lambda_1)}, \\ \varphi_2 &= -\frac{1}{(\lambda_1+\lambda_2)(\lambda_2+\lambda_3)(\lambda_3+\lambda_1)}. \end{aligned} \tag{8.12}$$

These representations for φ_i are free of singularities and are therefore generally valid for the case of simple as well as repeated eigenvalues of $\mathbf{C}$ and $\mathbf{b}$.

The rotation tensor results from (8.6) where we can again write

$$\mathbf{U}^{-1} = \varsigma_0 \mathbf{I} + \varsigma_1 \mathbf{C} + \varsigma_2 \mathbf{C}^2. \tag{8.13}$$

The representations for ς_p $(p = 0, 1, 2)$ can be obtained either again by (7.61) where $g(\Lambda_i) = \Lambda_i^{-1/2} = \lambda_i^{-1}$ or by applying the Cayley-Hamilton equation (4.95) leading to

$$\begin{aligned} \mathbf{U}^{-1} &= \mathrm{III}_{\mathbf{U}}^{-1}\left(\mathbf{U}^2 - \mathrm{I}_{\mathbf{U}}\mathbf{U} + \mathrm{II}_{\mathbf{U}}\mathbf{I}\right) \\ &= \mathrm{III}_{\mathbf{U}}^{-1}\left[\left(\mathrm{II}_{\mathbf{U}} - \varphi_0\mathrm{I}_{\mathbf{U}}\right)\mathbf{I} + \left(1 - \varphi_1\mathrm{I}_{\mathbf{U}}\right)\mathbf{C} - \varphi_2\mathrm{I}_{\mathbf{U}}\mathbf{C}^2\right], \end{aligned} \tag{8.14}$$

where

$$\mathrm{I}_{\mathbf{U}} = \lambda_1+\lambda_2+\lambda_3, \quad \mathrm{II}_{\mathbf{U}} = \lambda_1\lambda_2+\lambda_2\lambda_3+\lambda_3\lambda_1, \quad \mathrm{III}_{\mathbf{U}} = \lambda_1\lambda_2\lambda_3. \tag{8.15}$$

Both procedures yield the same representation (8.13) where

$$\begin{aligned} \varsigma_0 &= \frac{\lambda_1\lambda_2+\lambda_2\lambda_3+\lambda_3\lambda_1}{\lambda_1\lambda_2\lambda_3} - \frac{(\lambda_1+\lambda_2+\lambda_3)^2}{(\lambda_1+\lambda_2)(\lambda_2+\lambda_3)(\lambda_3+\lambda_1)}, \\ \varsigma_1 &= \frac{1}{\lambda_1\lambda_2\lambda_3} - \frac{\left(\lambda_1^2+\lambda_2^2+\lambda_3^2+\lambda_1\lambda_2+\lambda_2\lambda_3+\lambda_3\lambda_1\right)(\lambda_1+\lambda_2+\lambda_3)}{\lambda_1\lambda_2\lambda_3(\lambda_1+\lambda_2)(\lambda_2+\lambda_3)(\lambda_3+\lambda_1)}, \\ \varsigma_2 &= \frac{\lambda_1+\lambda_2+\lambda_3}{\lambda_1\lambda_2\lambda_3(\lambda_1+\lambda_2)(\lambda_2+\lambda_3)(\lambda_3+\lambda_1)}. \end{aligned} \tag{8.16}$$

Thus, the rotation tensor (8.6) can be given by

$$\mathbf{R} = \mathbf{F}\left(\varsigma_0\mathbf{I} + \varsigma_1\mathbf{C} + \varsigma_2\mathbf{C}^2\right), \tag{8.17}$$

where the functions ς_i $(i = 0, 1, 2)$ are given by (8.16) in terms of the principal stretches $\lambda_i = \sqrt{\Lambda_i}$, while Λ_i $(i = 1, 2, 3)$ denote the eigenvalues of the right Cauchy-Green tensor $\mathbf{C}$ (8.2).

Example 8.1. Stretch and rotation tensor in the case of simple shear. In this loading case the right and left Cauchy-Green tensors take the form (see Exercise 4.1)

$$\mathbf{C} = \mathrm{C}^i_{\cdot j} \boldsymbol{e}_i \otimes \boldsymbol{e}^j, \quad \left[\mathrm{C}^i_{\cdot j}\right] = \begin{bmatrix} 1 & \gamma & 0 \\ \gamma & 1+\gamma^2 & 0 \\ 0 & 0 & 1 \end{bmatrix}, \tag{8.18}$$

$$\mathbf{b} = \mathrm{b}^i_{\cdot j} \boldsymbol{e}_i \otimes \boldsymbol{e}^j, \quad \left[\mathrm{b}^i_{\cdot j}\right] = \begin{bmatrix} 1+\gamma^2 & \gamma & 0 \\ \gamma & 1 & 0 \\ 0 & 0 & 1 \end{bmatrix} \tag{8.19}$$

with the eigenvalues

$$\Lambda_{1/2} = 1 + \frac{\gamma^2 \pm \sqrt{4\gamma^2 + \gamma^4}}{2} = \left(\frac{\sqrt{4+\gamma^2} \pm \gamma}{2}\right)^2, \quad \Lambda_3 = 1. \tag{8.20}$$

For the principal stretches we thus obtain

$$\lambda_{1/2} = \sqrt{\Lambda_{1/2}} = \frac{\sqrt{4+\gamma^2} \pm \gamma}{2}, \quad \lambda_3 = \sqrt{\Lambda_3} = 1. \tag{8.21}$$

The stretch tensors result from (8.11) where

$$\begin{aligned} \varphi_0 &= \frac{1+\sqrt{\gamma^2+4}}{2\sqrt{\gamma^2+4}+\gamma^2+4}, \\ \varphi_1 &= \frac{1+\sqrt{\gamma^2+4}}{2+\sqrt{\gamma^2+4}}, \\ \varphi_2 &= -\frac{1}{2\sqrt{\gamma^2+4}+\gamma^2+4}. \end{aligned} \tag{8.22}$$

This yields the following result (cf. Exercise 7.2)

$$\mathbf{U} = \mathrm{U}^i_{\cdot j} \boldsymbol{e}_i \otimes \boldsymbol{e}^j, \quad \left[\mathrm{U}^i_{\cdot j}\right] = \begin{bmatrix} \frac{2}{\sqrt{\gamma^2+4}} & \frac{\gamma}{\sqrt{\gamma^2+4}} & 0 \\ \frac{\gamma}{\sqrt{\gamma^2+4}} & \frac{\gamma^2+2}{\sqrt{\gamma^2+4}} & 0 \\ 0 & 0 & 1 \end{bmatrix}, \tag{8.23}$$

$$\mathbf{v} = \mathrm{v}^i_{j} \boldsymbol{e}_i \otimes \boldsymbol{e}^j, \quad \left[\mathrm{v}^i_{j}\right] = \begin{bmatrix} \dfrac{\gamma^2+2}{\sqrt{\gamma^2+4}} & \dfrac{\gamma}{\sqrt{\gamma^2+4}} & 0 \\ \dfrac{\gamma}{\sqrt{\gamma^2+4}} & \dfrac{2}{\sqrt{\gamma^2+4}} & 0 \\ 0 & 0 & 1 \end{bmatrix}. \tag{8.24}$$

The rotation tensor can be calculated by (8.17) where

$$\varsigma_0 = \sqrt{\gamma^2+4} - \frac{1}{2\sqrt{\gamma^2+4}+\gamma^2+4},$$

$$\varsigma_1 = -\frac{3+\sqrt{\gamma^2+4}+\gamma^2}{2+\sqrt{\gamma^2+4}},$$

$$\varsigma_2 = \frac{1+\sqrt{\gamma^2+4}}{2\sqrt{\gamma^2+4}+\gamma^2+4}. \tag{8.25}$$

By this means we obtain

$$\mathbf{R} = \mathrm{R}^i_{\cdot j} \boldsymbol{e}_i \otimes \boldsymbol{e}^j, \quad \left[\mathrm{R}^i_{\cdot j}\right] = \begin{bmatrix} \dfrac{2}{\sqrt{\gamma^2+4}} & \dfrac{\gamma}{\sqrt{\gamma^2+4}} & 0 \\ -\dfrac{\gamma}{\sqrt{\gamma^2+4}} & \dfrac{2}{\sqrt{\gamma^2+4}} & 0 \\ 0 & 0 & 1 \end{bmatrix}. \tag{8.26}$$

8.3 The Derivative of the Stretch and Rotation Tensor with Respect to the Deformation Gradient

In continuum mechanics these derivatives are used for the evaluation of the rate of the stretch and rotation tensor. We begin with a very simple representation in terms of eigenprojections of the right and left Cauchy-Green tensors (8.2). Applying the chain rule of differentiation and using (6.142) we first write

$$\mathbf{U}_{,\mathbf{F}} = \mathbf{C}^{1/2}{}_{,\mathbf{C}} : \mathbf{C}_{,\mathbf{F}} = \mathbf{C}^{1/2}{}_{,\mathbf{C}} : \left[(\mathbf{I} \otimes \mathbf{F})^{\mathrm{t}} + \mathbf{F}^{\mathrm{T}} \otimes \mathbf{I}\right]. \tag{8.27}$$

Further, taking into account the spectral representation of $\mathbf{C}$ $(8.4)_1$ and keeping its symmetry in mind we obtain by virtue of (7.49) and (7.50)

$$\mathbf{C}^{1/2}{}_{,\mathbf{C}} = \sum_{i,j=1}^{s} (\lambda_i + \lambda_j)^{-1} (\mathbf{P}_i \otimes \mathbf{P}_j)^{\mathrm{s}}. \tag{8.28}$$

Inserting this result into (8.27) delivers by means of (5.33), (5.47), $(5.54)_2$ and (5.55)

$$\mathbf{U},_{\mathbf{F}} = \sum_{i,j=1}^{s} (\lambda_i + \lambda_j)^{-1} [(\mathbf{P}_i \otimes \mathbf{F}\mathbf{P}_j)^{\mathrm{t}} + \mathbf{P}_i \mathbf{F}^{\mathrm{T}} \otimes \mathbf{P}_j]. \tag{8.29}$$

The same procedure applied to the left stretch tensor yields by virtue of (6.143)

$$\mathbf{v},_{\mathbf{F}} = \sum_{i,j=1}^{s} (\lambda_i + \lambda_j)^{-1} \left[\mathbf{p}_i \otimes \mathbf{F}^{\mathrm{T}}\mathbf{p}_j + (\mathbf{p}_i \mathbf{F} \otimes \mathbf{p}_j)^{\mathrm{t}}\right]. \tag{8.30}$$

Now, applying the product rule of differentiation (6.140) to (8.6) and taking (6.139) into account we write

$$\begin{aligned} \mathbf{R},_{\mathbf{F}} = (\mathbf{F}\mathbf{U}^{-1}),_{\mathbf{F}} &= \mathbf{I} \otimes \mathbf{U}^{-1} + \mathbf{F}\mathbf{U}^{-1},_{\mathbf{U}} : \mathbf{U},_{\mathbf{F}} \\ &= \mathbf{I} \otimes \mathbf{U}^{-1} - \mathbf{F}(\mathbf{U}^{-1} \otimes \mathbf{U}^{-1})^{\mathrm{s}} : \mathbf{U},_{\mathbf{F}} . \end{aligned} \tag{8.31}$$

With the aid of (7.2) and (8.29) this finally leads to

$$\begin{aligned} \mathbf{R},_{\mathbf{F}} = \mathbf{I} \otimes &\left(\sum_{i=1}^{s} \lambda_i^{-1} \mathbf{P}_i\right) \\ &- \mathbf{F} \sum_{i,j=1}^{s} [(\lambda_i + \lambda_j)\lambda_i\lambda_j]^{-1} [(\mathbf{P}_i \otimes \mathbf{F}\mathbf{P}_j)^{\mathrm{t}} + \mathbf{P}_i \mathbf{F}^{\mathrm{T}} \otimes \mathbf{P}_j]. \end{aligned} \tag{8.32}$$

Note that the eigenprojections $\mathbf{P}_i$ and $\mathbf{p}_i$ $(i = 1, 2, \ldots, s)$ are uniquely defined by the Sylvester formula (4.55) or its alternative form (7.43) in terms of $\mathbf{C}$ and $\mathbf{b}$, respectively. The functions ρ_{ip} appearing in (7.43) are, in turn, expressed in the unique form by (7.81), (7.83) and (7.85) in terms of the eigenvalues $\Lambda_i = \lambda_i^2$ $(i = 1, 2, \ldots, s)$.

In order to avoid the direct reference to the eigenprojections one can obtain the so-called basis-free solutions for $\mathbf{U},_{\mathbf{F}}$, $\mathbf{v},_{\mathbf{F}}$ and $\mathbf{R},_{\mathbf{F}}$ (see, e.g., [8, 14, 18, 38, 47, 49]). As a rule, they are given in terms of the stretch and rotation tensors themselves and require therefore either the explicit polar decomposition of the deformation gradient or a closed-form representation for $\mathbf{U}$, $\mathbf{v}$ and $\mathbf{R}$ like (8.11) and (8.17). In the following we present the basis-free solutions for $\mathbf{U},_{\mathbf{F}}$, $\mathbf{v},_{\mathbf{F}}$ and $\mathbf{R},_{\mathbf{F}}$ in terms of the Cauchy-Green tensors $\mathbf{C}$ and $\mathbf{b}$ (8.2) and the principal stretches $\lambda_i = \sqrt{\Lambda_i}$ $(i = 1, 2, \ldots, s)$. To this end, we apply the representation (7.38) for the derivative of the square root. Thus, we obtain instead of (8.28)

$$\mathbf{C}^{1/2},_{\mathbf{C}} = \sum_{p,q=0}^{2} \eta_{pq} (\mathbf{C}^p \otimes \mathbf{C}^q)^{\mathrm{s}}, \quad \mathbf{b}^{1/2},_{\mathbf{b}} = \sum_{p,q=0}^{2} \eta_{pq} (\mathbf{b}^p \otimes \mathbf{b}^q)^{\mathrm{s}}, \tag{8.33}$$

where the functions η_{pq} result from (7.62) by setting again $g(\Lambda_i) = \sqrt{\Lambda_i} = \lambda_i$. This leads to the following expressions (cf. [18])

$$\eta_{00} = \Delta^{-1}\left[\mathrm{I}_\mathrm{U}^5\mathrm{III}_\mathrm{U}^2 - \mathrm{I}_\mathrm{U}^4\mathrm{II}_\mathrm{U}^2\mathrm{III}_\mathrm{U} + \mathrm{I}_\mathrm{U}^3\mathrm{II}_\mathrm{U}^4\right.$$
$$\left. - \mathrm{I}_\mathrm{U}^2\mathrm{III}_\mathrm{U}\left(3\mathrm{II}_\mathrm{U}^3 - 2\mathrm{III}_\mathrm{U}^2\right) + 3\mathrm{I}_\mathrm{U}\mathrm{II}_\mathrm{U}^2\mathrm{III}_\mathrm{U}^2 - \mathrm{II}_\mathrm{U}\mathrm{III}_\mathrm{U}^3\right],$$

$$\eta_{01} = \eta_{10} = \Delta^{-1}\left[\mathrm{I}_\mathrm{U}^6\mathrm{III}_\mathrm{U} - \mathrm{I}_\mathrm{U}^5\mathrm{II}_\mathrm{U}^2 - \mathrm{I}_\mathrm{U}^4\mathrm{II}_\mathrm{U}\mathrm{III}_\mathrm{U}\right.$$
$$\left. + 2\mathrm{I}_\mathrm{U}^3\left(\mathrm{II}_\mathrm{U}^3 + \mathrm{III}_\mathrm{U}^2\right) - 4\mathrm{I}_\mathrm{U}^2\mathrm{II}_\mathrm{U}^2\mathrm{III}_\mathrm{U} + 2\mathrm{I}_\mathrm{U}\mathrm{II}_\mathrm{U}\mathrm{III}_\mathrm{U}^2 - \mathrm{III}_\mathrm{U}^3\right],$$

$$\eta_{02} = \eta_{20} = \Delta^{-1}\left[-\mathrm{I}_\mathrm{U}^4\mathrm{III}_\mathrm{U} + \mathrm{I}_\mathrm{U}^3\mathrm{II}_\mathrm{U}^2 - \mathrm{I}_\mathrm{U}^2\mathrm{II}_\mathrm{U}\mathrm{III}_\mathrm{U} - \mathrm{I}_\mathrm{U}\mathrm{III}_\mathrm{U}^2\right],$$

$$\eta_{11} = \Delta^{-1}\left[\mathrm{I}_\mathrm{U}^7 - 4\mathrm{I}_\mathrm{U}^5\mathrm{II}_\mathrm{U} + 3\mathrm{I}_\mathrm{U}^4\mathrm{III}_\mathrm{U}\right.$$
$$\left. + 4\mathrm{I}_\mathrm{U}^3\mathrm{II}_\mathrm{U}^2 - 6\mathrm{I}_\mathrm{U}^2\mathrm{II}_\mathrm{U}\mathrm{III}_\mathrm{U} + \mathrm{I}_\mathrm{U}\mathrm{III}_\mathrm{U}^2 + \mathrm{II}_\mathrm{U}^2\mathrm{III}_\mathrm{U}\right],$$
$$\eta_{12} = \eta_{21} = \Delta^{-1}\left[-\mathrm{I}_\mathrm{U}^5 + 2\mathrm{I}_\mathrm{U}^3\mathrm{II}_\mathrm{U} - 2\mathrm{I}_\mathrm{U}^2\mathrm{III}_\mathrm{U} + \mathrm{II}_\mathrm{U}\mathrm{III}_\mathrm{U}\right],$$
$$\eta_{22} = \Delta^{-1}\left[\mathrm{I}_\mathrm{U}^3 + \mathrm{III}_\mathrm{U}\right], \tag{8.34}$$

where

$$\Delta = 2\left(\mathrm{I}_\mathrm{U}\mathrm{II}_\mathrm{U} - \mathrm{III}_\mathrm{U}\right)^3\mathrm{III}_\mathrm{U} \tag{8.35}$$

and the principal invariants I_U, II_U and III_U are given by (8.15).

Finally, substitution of (8.33) into (8.27) yields

$$\mathbf{U}_{,\mathbf{F}} = \sum_{p,q=0}^{2} \eta_{pq}\left[\left(\mathbf{C}^p \otimes \mathbf{F}\mathbf{C}^q\right)^\mathrm{t} + \mathbf{C}^p\mathbf{F}^\mathrm{T} \otimes \mathbf{C}^q\right]. \tag{8.36}$$

Similar we can also write

$$\mathbf{v}_{,\mathbf{F}} = \sum_{p,q=0}^{2} \eta_{pq}\left[\mathbf{b}^p \otimes \mathbf{F}^\mathrm{T}\mathbf{b}^q + \left(\mathbf{b}^p\mathbf{F} \otimes \mathbf{b}^q\right)^\mathrm{t}\right]. \tag{8.37}$$

Inserting further (8.13) and (8.36) into (8.31) we get

$$\mathbf{R}_{,\mathbf{F}} = \mathbf{I} \otimes \sum_{p=0}^{2} \varsigma_p \mathbf{C}^p$$
$$- \mathbf{F} \sum_{p,q,r,t=0}^{2} \varsigma_r\varsigma_t\eta_{pq}\left[\left(\mathbf{C}^{p+r} \otimes \mathbf{F}\mathbf{C}^{q+t}\right)^\mathrm{t} + \mathbf{C}^{p+r}\mathbf{F}^\mathrm{T} \otimes \mathbf{C}^{q+t}\right], \tag{8.38}$$

where ς_p and η_{pq} $(p,q=0,1,2)$ are given by (8.16) and (8.34), respectively. The third and fourth powers of $\mathbf{C}$ in (8.38) can be expressed by means of the Cayley-Hamilton equation (4.95):

$$\mathbf{C}^3 - \mathrm{I_C}\mathbf{C}^2 + \mathrm{II_C}\mathbf{C} - \mathrm{III_C}\mathbf{I} = \mathbf{0}. \tag{8.39}$$

Composing both sides with $\mathbf{C}$ we can also write

$$\mathbf{C}^4 - \mathrm{I_C}\mathbf{C}^3 + \mathrm{II_C}\mathbf{C}^2 - \mathrm{III_C}\mathbf{C} = \mathbf{0}. \tag{8.40}$$

Thus,

$$\begin{aligned}\mathbf{C}^3 &= \mathrm{I_C}\mathbf{C}^2 - \mathrm{II_C}\mathbf{C} + \mathrm{III_C}\mathbf{I},\\ \mathbf{C}^4 &= \left(\mathrm{I_C^2} - \mathrm{II_C}\right)\mathbf{C}^2 + \left(\mathrm{III_C} - \mathrm{I_C}\mathrm{II_C}\right)\mathbf{C} + \mathrm{I_C}\mathrm{III_C}\mathbf{I}.\end{aligned} \tag{8.41}$$

Considering these expressions in (8.38) and taking into account that (see, e.g., [44])

$$\mathrm{I_C} = \mathrm{I_U^2} - 2\mathrm{II_U}, \quad \mathrm{II_C} = \mathrm{II_U^2} - 2\mathrm{I_U}\mathrm{III_U}, \quad \mathrm{III_C} = \mathrm{III_U^2} \tag{8.42}$$

we finally obtain

$$\mathbf{R}_{,\mathbf{F}} = \mathbf{I} \otimes \sum_{p=0}^{2} \varsigma_p \mathbf{C}^p + \mathbf{F} \sum_{p,q=0}^{2} \mu_{pq}\left[\left(\mathbf{C}^p \otimes \mathbf{F}\mathbf{C}^q\right)^{\mathrm{t}} + \mathbf{C}^p\mathbf{F}^{\mathrm{T}} \otimes \mathbf{C}^q\right], \tag{8.43}$$

where

$$\begin{aligned}\mu_{00} = \Upsilon^{-1}\big[&\mathrm{I_U^6}\mathrm{III_U^3} + 2\mathrm{I_U^5}\mathrm{II_U^2}\mathrm{III_U^2} - 3\mathrm{I_U^4}\mathrm{II_U^4}\mathrm{III_U} - 7\mathrm{I_U^4}\mathrm{II_U}\mathrm{III_U^3}\\ &+ \mathrm{I_U^3}\mathrm{II_U^6} + 8\mathrm{I_U^3}\mathrm{II_U^3}\mathrm{III_U^2} + 6\mathrm{I_U^3}\mathrm{III_U^4} - 3\mathrm{I_U^2}\mathrm{II_U^5}\mathrm{III_U}\\ &- 6\mathrm{I_U^2}\mathrm{II_U^2}\mathrm{III_U^3} + 3\mathrm{I_U}\mathrm{II_U^4}\mathrm{III_U^2} - \mathrm{II_U^3}\mathrm{III_U^3} + \mathrm{III_U^5}\big],\end{aligned}$$

$$\begin{aligned}\mu_{01} = \mu_{10} = \Upsilon^{-1}\big[&\mathrm{I_U^7}\mathrm{III_U^2} + \mathrm{I_U^6}\mathrm{II_U^2}\mathrm{III_U} - \mathrm{I_U^5}\mathrm{II_U^4} - 6\mathrm{I_U^5}\mathrm{II_U}\mathrm{III_U^2} + \mathrm{I_U^4}\mathrm{II_U^3}\mathrm{III_U}\\ &+ 5\mathrm{I_U^4}\mathrm{III_U^3} + 2\mathrm{I_U^3}\mathrm{II_U^5} + 4\mathrm{I_U^3}\mathrm{II_U^2}\mathrm{III_U^2} - 6\mathrm{I_U^2}\mathrm{II_U^4}\mathrm{III_U}\\ &- 6\mathrm{I_U^2}\mathrm{II_U}\mathrm{III_U^3} + 6\mathrm{I_U}\mathrm{II_U^3}\mathrm{III_U^2} + \mathrm{I_U}\mathrm{III_U^4} - 2\mathrm{II_U^2}\mathrm{III_U^3}\big],\end{aligned}$$

$$\begin{aligned}\mu_{02} = \mu_{20} = -\Upsilon^{-1}\big[&\mathrm{I_U^5}\mathrm{III_U^2} + \mathrm{I_U^4}\mathrm{II_U^2}\mathrm{III_U} - \mathrm{I_U^3}\mathrm{II_U^4} - 4\mathrm{I_U^3}\mathrm{II_U}\mathrm{III_U^2}\\ &+ 3\mathrm{I_U^2}\mathrm{II_U^3}\mathrm{III_U} + 4\mathrm{I_U^2}\mathrm{III_U^3} - 3\mathrm{I_U}\mathrm{II_U^2}\mathrm{III_U^2} + \mathrm{II_U}\mathrm{III_U^3}\big],\end{aligned}$$

$$\begin{aligned}\mu_{11} = \Upsilon^{-1}\big[&\mathrm{I_U^8}\mathrm{III_U} + \mathrm{I_U^7}\mathrm{II_U^2} - 7\mathrm{I_U^6}\mathrm{II_U}\mathrm{III_U} - 4\mathrm{I_U^5}\mathrm{II_U^3}\\ &+ 5\mathrm{I_U^5}\mathrm{III_U^2} + 16\mathrm{I_U^4}\mathrm{II_U^2}\mathrm{III_U} + 4\mathrm{I_U^3}\mathrm{II_U^4} - 16\mathrm{I_U^3}\mathrm{II_U}\mathrm{III_U^2}\\ &- 12\mathrm{I_U^2}\mathrm{II_U^3}\mathrm{III_U} + 3\mathrm{I_U^2}\mathrm{III_U^3} + 12\mathrm{I_U}\mathrm{II_U^2}\mathrm{III_U^2} - 3\mathrm{II_U}\mathrm{III_U^3}\big],\end{aligned}$$

$$\mu_{12} = \mu_{21} = -\Upsilon^{-1}\left[\mathrm{I}_U^6\mathrm{III}_U + \mathrm{I}_U^5\mathrm{II}_U^2 - 5\mathrm{I}_U^4\mathrm{II}_U\mathrm{III}_U - 2\mathrm{I}_U^3\mathrm{II}_U^3 \right.$$
$$\left. + 4\mathrm{I}_U^3\mathrm{III}_U^2 + 6\mathrm{I}_U^2\mathrm{II}_U^2\mathrm{III}_U - 6\mathrm{I}_U\mathrm{II}_U\mathrm{III}_U^2 + \mathrm{III}_U^3\right],$$
$$\mu_{22} = \Upsilon^{-1}\mathrm{I}_U\left[\mathrm{I}_U^3\mathrm{III}_U + \mathrm{I}_U^2\mathrm{II}_U^2 - 3\mathrm{I}_U\mathrm{II}_U\mathrm{III}_U + 3\mathrm{III}_U^2\right] \tag{8.44}$$

and

$$\Upsilon = -2\left(\mathrm{I}_U\mathrm{II}_U - \mathrm{III}_U\right)^3\mathrm{III}_U^3, \tag{8.45}$$

while the principal invariants I_U, II_U and III_U are given by (8.15).

The same result for $\mathbf{R}_{,\mathbf{F}}$ also follows from

$$\mathbf{R}_{,\mathbf{F}} = \left(\mathbf{F}\mathbf{U}^{-1}\right)_{,\mathbf{F}} = \mathbf{I}\otimes\mathbf{U}^{-1} + \mathbf{F}\mathbf{U}^{-1}{}_{,\mathbf{C}} : \mathbf{C}_{,\mathbf{F}} \tag{8.46}$$

by applying for $\mathbf{U}^{-1}{}_{,\mathbf{C}}$ (7.38) and (7.62) where we set $g(\Lambda_i) = (\Lambda_i)^{-1/2} = \lambda_i^{-1}$. Indeed, this yields

$$\mathbf{C}^{-1/2}{}_{,\mathbf{C}} = \mathbf{U}^{-1}{}_{,\mathbf{C}} = \sum_{p,q=0}^{2}\mu_{pq}\left(\mathbf{C}^p\otimes\mathbf{C}^q\right)^{\mathrm{s}}, \tag{8.47}$$

where μ_{pq} $(p, q = 0, 1, 2)$ are given by (8.44).

8.4 Time Rate of Generalized Strains

Applying the chain rule of differentiation we first write

$$\dot{\mathbf{E}} = \mathbf{E}_{,\mathbf{C}} : \dot{\mathbf{C}}, \tag{8.48}$$

where the superposed dot denotes the so-called material time derivative. The derivative $\mathbf{E}_{,\mathbf{C}}$ can be expressed in a simple form in terms of the eigenprojections of $\mathbf{E}$ and $\mathbf{C}$. To this end, we apply (7.49) and (7.50) taking (7.18) and (8.5) into account which yields

$$\mathbf{E}_{,\mathbf{C}} = \sum_{i,j=1}^{s} f_{ij}\left(\mathbf{P}_i\otimes\mathbf{P}_j\right)^{\mathrm{s}}, \tag{8.49}$$

where

$$f_{ij} = \begin{cases} \dfrac{f'(\lambda_i)}{2\lambda_i} & \text{if } i = j, \\ \dfrac{f(\lambda_i) - f(\lambda_j)}{\lambda_i^2 - \lambda_j^2} & \text{if } i \neq j. \end{cases} \tag{8.50}$$

A basis-free representation for $\mathbf{E}_{,\mathbf{C}}$ can be obtained either from (8.49) by expressing the eigenprojections by (7.43) with (7.81), (7.83) and (7.85) or directly by using the closed-form solution (7.38) with (7.62), (7.78) and (7.80). Both procedures lead to the same result as follows (cf. [22, 48]).

$$\mathbf{E},_{\mathbf{C}} = \sum_{p,q=0}^{2} \eta_{pq} (\mathbf{C}^p \otimes \mathbf{C}^q)^{\mathrm{s}}. \tag{8.51}$$

Distinct eigenvalues: $\lambda_1 \neq \lambda_2 \neq \lambda_3 \neq \lambda_1$,

$$\eta_{00} = \sum_{i=1}^{3} \frac{\lambda_j^4 \lambda_k^4 f'(\lambda_i)}{2\lambda_i \Delta_i^2} - \sum_{\substack{i,j=1 \\ i \neq j}}^{3} \frac{\lambda_i^2 \lambda_j^2 \lambda_k^4 \left[f(\lambda_i) - f(\lambda_j) \right]}{\left(\lambda_i^2 - \lambda_j^2 \right)^3 \Delta_k},$$

$$\eta_{01} = \eta_{10} = -\sum_{i=1}^{3} \frac{\left(\lambda_j^2 + \lambda_k^2 \right) \lambda_j^2 \lambda_k^2 f'(\lambda_i)}{2\lambda_i \Delta_i^2} + \sum_{\substack{i,j=1 \\ i \neq j}}^{3} \frac{\left(\lambda_j^2 + \lambda_k^2 \right) \lambda_i^2 \lambda_k^2 \left[f(\lambda_i) - f(\lambda_j) \right]}{\left(\lambda_i^2 - \lambda_j^2 \right)^3 \Delta_k},$$

$$\eta_{02} = \eta_{20} = \sum_{i=1}^{3} \frac{\lambda_j^2 \lambda_k^2 f'(\lambda_i)}{2\lambda_i \Delta_i^2} - \sum_{\substack{i,j=1 \\ i \neq j}}^{3} \frac{\lambda_i^2 \lambda_k^2 \left[f(\lambda_i) - f(\lambda_j) \right]}{\left(\lambda_i^2 - \lambda_j^2 \right)^3 \Delta_k},$$

$$\eta_{11} = \sum_{i=1}^{3} \frac{\left(\lambda_j^2 + \lambda_k^2 \right)^2 f'(\lambda_i)}{2\lambda_i \Delta_i^2} - \sum_{\substack{i,j=1 \\ i \neq j}}^{3} \frac{\left(\lambda_j^2 + \lambda_k^2 \right) \left(\lambda_i^2 + \lambda_k^2 \right) \left[f(\lambda_i) - f(\lambda_j) \right]}{\left(\lambda_i^2 - \lambda_j^2 \right)^3 \Delta_k},$$

$$\eta_{12} = \eta_{21} = -\sum_{i=1}^{3} \frac{\left(\lambda_j^2 + \lambda_k^2 \right) f'(\lambda_i)}{2\lambda_i \Delta_i^2} + \sum_{\substack{i,j=1 \\ i \neq j}}^{3} \frac{\left(\lambda_i^2 + \lambda_k^2 \right) \left[f(\lambda_i) - f(\lambda_j) \right]}{\left(\lambda_i^2 - \lambda_j^2 \right)^3 \Delta_k},$$

$$\eta_{22} = \sum_{i=1}^{3} \frac{f'(\lambda_i)}{2\lambda_i \Delta_i^2} - \sum_{\substack{i,j=1 \\ i \neq j}}^{3} \frac{f(\lambda_i) - f(\lambda_j)}{(\lambda_i^2 - \lambda_j^2)^3 \Delta_k}, \quad i \neq j \neq k \neq i, \tag{8.52}$$

with

$$\Delta_i = \left(\lambda_i^2 - \lambda_j^2 \right) \left(\lambda_i^2 - \lambda_k^2 \right), \quad i \neq j \neq k \neq i = 1, 2, 3. \tag{8.53}$$

Double coalescence of eigenvalues: $\lambda_i \neq \lambda_j = \lambda_k = \lambda$,

$$\eta_{00} = -2\lambda_i^2 \lambda^2 \frac{f(\lambda_i) - f(\lambda)}{\left(\lambda_i^2 - \lambda^2 \right)^3} + \frac{\lambda^5 f'(\lambda_i) + \lambda_i^5 f'(\lambda)}{2\lambda_i \lambda \left(\lambda_i^2 - \lambda^2 \right)^2},$$

$$\eta_{01} = \eta_{10} = \left(\lambda_i^2 + \lambda^2\right) \frac{f(\lambda_i) - f(\lambda)}{(\lambda_i^2 - \lambda^2)^3} - \frac{\lambda^3 f'(\lambda_i) + \lambda_i^3 f'(\lambda)}{2\lambda_i \lambda (\lambda_i^2 - \lambda^2)^2},$$

$$\eta_{11} = -2\frac{f(\lambda_i) - f(\lambda)}{\left(\lambda_i^2 - \lambda^2\right)^3} + \frac{\lambda f'(\lambda_i) + \lambda_i f'(\lambda)}{2\lambda_i \lambda \left(\lambda_i^2 - \lambda^2\right)^2},$$

$$\eta_{02} = \eta_{20} = \eta_{12} = \eta_{21} = \eta_{22} = 0. \tag{8.54}$$

Triple coalescence of eigenvalues: $\lambda_1 = \lambda_2 = \lambda_3 = \lambda$,

$$\eta_{00} = \frac{f'(\lambda)}{2\lambda}, \quad \eta_{01} = \eta_{10} = \eta_{11} = \eta_{02} = \eta_{20} = \eta_{12} = \eta_{21} = \eta_{22} = 0. \tag{8.55}$$

Insertion of (8.49) or alternatively (8.51) into (8.48) finally yields by $(5.17)_1$ and $(5.48)_1$

$$\dot{\mathbf{E}} = \sum_{i,j=1}^{s} f_{ij} \mathbf{P}_i \dot{\mathbf{C}} \mathbf{P}_j = \sum_{p,q=0}^{2} \eta_{pq} \mathbf{C}^p \dot{\mathbf{C}} \mathbf{C}^q. \tag{8.56}$$

Example 8.2. Material time derivative of the Biot strain tensor $\mathbf{E}^{(1)} = \mathbf{U} - \mathbf{I}$. Insertion of $f(\lambda) = \lambda - 1$ into (8.50) and $(8.56)_1$ yields

$$\dot{\mathbf{E}}^{(1)} = \sum_{i,j=1}^{s} \frac{1}{\lambda_i + \lambda_j} \mathbf{P}_i \dot{\mathbf{C}} \mathbf{P}_j. \tag{8.57}$$

Keeping (8.33) in mind and applying the chain rule of differentiation we can also write

$$\dot{\mathbf{E}}^{(1)} = \dot{\mathbf{U}} = \mathbf{C}^{1/2},_{\mathbf{C}} : \dot{\mathbf{C}} = \sum_{p,q=0}^{2} \eta_{pq} \mathbf{C}^p \dot{\mathbf{C}} \mathbf{C}^q, \tag{8.58}$$

where the coefficients η_{pq} $(p, q = 0, 1, 2)$ are given by (8.34) in terms of the principal invariants of $\mathbf{U}$ (8.15).

8.5 Stress Conjugate to a Generalized Strain

Let $\mathbf{E}$ be an arbitrary Lagrangian strain $(7.6)_1$. Assume existence of the so-called strain energy function $\psi(\mathbf{E})$ differentiable with respect to $\mathbf{E}$. The symmetric tensor

$$\mathbf{T} = \psi(\mathbf{E}),_{\mathbf{E}} \tag{8.59}$$

is referred to as stress conjugate to $\mathbf{E}$. With the aid of the chain rule it can be represented by

$$\mathbf{T} = \psi(\mathbf{E}),_{\mathbf{C}} : \mathbf{C},_{\mathbf{E}} = \frac{1}{2}\mathbf{S} : \mathbf{C},_{\mathbf{E}}, \tag{8.60}$$

where $\mathbf{S} = 2\psi(\mathbf{E})_{,\mathbf{C}}$ denotes the second Piola-Kirchhoff stress tensor. The latter one is defined in terms of the Cauchy stress $\boldsymbol{\sigma}$ by (see, e.g., [46])

$$\mathbf{S} = \det(\mathbf{F})\mathbf{F}^{-1}\boldsymbol{\sigma}\mathbf{F}^{-\mathrm{T}}. \tag{8.61}$$

Using (8.59) and (7.7) one can also write

$$\dot{\psi} = \mathbf{T} : \dot{\mathbf{E}} = \mathbf{S} : \frac{1}{2}\dot{\mathbf{C}} = \mathbf{S} : \dot{\mathbf{E}}^{(2)}. \tag{8.62}$$

The fourth-order tensor $\mathbf{C}_{,\mathbf{E}}$ appearing in (8.60) can be expressed in terms of the right Cauchy-Green tensor $\mathbf{C}$ by means of the relation

$$\mathcal{J}^{\mathrm{s}} = \mathbf{E}_{,\mathbf{E}} = \mathbf{E}_{,\mathbf{C}} : \mathbf{C}_{,\mathbf{E}}\,, \tag{8.63}$$

where the derivative $\mathbf{E}_{,\mathbf{C}}$ is given by (8.49) and (8.50). The basis tensors of the latter representation are

$$\mathcal{P}_{ij} = \begin{cases} (\mathbf{P}_i \otimes \mathbf{P}_i)^{\mathrm{s}} & \text{if } i = j, \\ \left(\mathbf{P}_i \otimes \mathbf{P}_j + \mathbf{P}_j \otimes \mathbf{P}_i\right)^{\mathrm{s}} & \text{if } i \neq j. \end{cases} \tag{8.64}$$

In view of (4.44), (5.33) and (5.55) they are pairwise orthogonal (see Exercise 8.2) such that (cf. [48])

$$\mathcal{P}_{ij} : \mathcal{P}_{kl} = \begin{cases} \mathcal{P}_{ij} & \text{if } i = k \text{ and } j = l \text{ or } i = l \text{ and } j = k, \\ \mathcal{O} & \text{otherwise.} \end{cases} \tag{8.65}$$

By means of (4.46) and (5.86) we can also write

$$\sum_{\substack{i,j=1 \\ j \geq i}}^{s} \mathcal{P}_{ij} = \left[\left(\sum_{i=1}^{s} \mathbf{P}_i\right) \otimes \left(\sum_{j=1}^{s} \mathbf{P}_j\right)\right]^{\mathrm{s}} = (\mathbf{I} \otimes \mathbf{I})^{\mathrm{s}} = \mathcal{J}^{\mathrm{s}}. \tag{8.66}$$

Using these properties we thus obtain

$$\mathbf{C}_{,\mathbf{E}} = \sum_{i,j=1}^{s} f_{ij}^{-1} (\mathbf{P}_i \otimes \mathbf{P}_j)^{\mathrm{s}}, \tag{8.67}$$

where f_{ij} $(i, j = 1, 2, \ldots, s)$ are given by (8.50). Substituting this result into (8.60) and taking $(5.22)_1$, (5.46) and (5.47) into account yields [19]

$$\mathbf{T} = \frac{1}{2} \sum_{i,j=1}^{s} f_{ij}^{-1} \mathbf{P}_i \mathbf{S} \mathbf{P}_j\,. \tag{8.68}$$

In order to avoid any reference to eigenprojections we can again express them by (7.43) with (7.81), (7.83) and (7.85) or alternatively use the closed-form solution (7.38) with (7.62), (7.78) and (7.80). Both procedures lead to the following result (cf. [48]).

$$\mathbf{T} = \sum_{p,q=0}^{2} \eta_{pq} \mathbf{C}^p \mathbf{S} \mathbf{C}^q. \tag{8.69}$$

Distinct eigenvalues: $\lambda_1 \neq \lambda_2 \neq \lambda_3 \neq \lambda_1$,

$$\eta_{00} = \sum_{i=1}^{3} \frac{\lambda_j^4 \lambda_k^4 \lambda_i}{f'(\lambda_i)\Delta_i^2} - \sum_{\substack{i,j=1 \\ i \neq j}}^{3} \frac{\lambda_i^2 \lambda_j^2 \lambda_k^4}{2\left(\lambda_i^2 - \lambda_j^2\right)\left[f(\lambda_i) - f(\lambda_j)\right]\Delta_k},$$

$$\eta_{01} = \eta_{10} = -\sum_{i=1}^{3} \frac{\left(\lambda_j^2 + \lambda_k^2\right)\lambda_j^2 \lambda_k^2 \lambda_i}{f'(\lambda_i)\Delta_i^2} + \sum_{\substack{i,j=1 \\ i \neq j}}^{3} \frac{\left(\lambda_j^2 + \lambda_k^2\right)\lambda_i^2 \lambda_k^2}{2\left(\lambda_i^2 - \lambda_j^2\right)\left[f(\lambda_i) - f(\lambda_j)\right]\Delta_k},$$

$$\eta_{02} = \eta_{20} = \sum_{i=1}^{3} \frac{\lambda_j^2 \lambda_k^2 \lambda_i}{f'(\lambda_i)\Delta_i^2} - \sum_{\substack{i,j=1 \\ i \neq j}}^{3} \frac{\lambda_i^2 \lambda_k^2}{2\left(\lambda_i^2 - \lambda_j^2\right)\left[f(\lambda_i) - f(\lambda_j)\right]\Delta_k},$$

$$\eta_{11} = \sum_{i=1}^{3} \frac{\left(\lambda_j^2 + \lambda_k^2\right)^2 \lambda_i}{f'(\lambda_i)\Delta_i^2} - \sum_{\substack{i,j=1 \\ i \neq j}}^{3} \frac{\left(\lambda_j^2 + \lambda_k^2\right)\left(\lambda_i^2 + \lambda_k^2\right)}{2\left(\lambda_i^2 - \lambda_j^2\right)\left[f(\lambda_i) - f(\lambda_j)\right]\Delta_k},$$

$$\eta_{12} = \eta_{21} = -\sum_{i=1}^{3} \frac{\left(\lambda_j^2 + \lambda_k^2\right)\lambda_i}{f'(\lambda_i)\Delta_i^2} + \sum_{\substack{i,j=1 \\ i \neq j}}^{3} \frac{\lambda_i^2 + \lambda_k^2}{2\left(\lambda_i^2 - \lambda_j^2\right)\left[f(\lambda_i) - f(\lambda_j)\right]\Delta_k},$$

$$\eta_{22} = \sum_{i=1}^{3} \frac{\lambda_i}{f'(\lambda_i)\Delta_i^2} - \sum_{\substack{i,j=1 \\ i \neq j}}^{3} \frac{1}{2\left(\lambda_i^2 - \lambda_j^2\right)\left[f(\lambda_i) - f(\lambda_j)\right]\Delta_k}, \tag{8.70}$$

where $i \neq j \neq k \neq i$ and Δ_i are given by (8.53).

Double coalescence of eigenvalues: $\lambda_i \neq \lambda_j = \lambda_k = \lambda$,

$$\eta_{00} = -\frac{\lambda_i^2 \lambda^2}{\left(\lambda_i^2 - \lambda^2\right)\left[f(\lambda_i) - f(\lambda)\right]} + \frac{\lambda_i \lambda}{\left(\lambda_i^2 - \lambda^2\right)^2}\left[\frac{\lambda^3}{f'(\lambda_i)} + \frac{\lambda_i^3}{f'(\lambda)}\right],$$

$$\eta_{01} = \eta_{10} = \frac{\lambda_i^2 + \lambda^2}{2\left(\lambda_i^2 - \lambda^2\right)\left[f(\lambda_i) - f(\lambda)\right]} - \frac{\lambda_i \lambda}{\left(\lambda_i^2 - \lambda^2\right)^2}\left[\frac{\lambda}{f'(\lambda_i)} + \frac{\lambda_i}{f'(\lambda)}\right],$$

$$\eta_{11} = -\frac{1}{\left(\lambda_i^2 - \lambda^2\right)\left[f(\lambda_i) - f(\lambda)\right]} + \frac{1}{\left(\lambda_i^2 - \lambda^2\right)^2}\left[\frac{\lambda_i}{f'(\lambda_i)} + \frac{\lambda}{f'(\lambda)}\right],$$

$$\eta_{02} = \eta_{20} = \eta_{12} = \eta_{21} = \eta_{22} = 0. \tag{8.71}$$

Triple coalescence of eigenvalues: $\lambda_1 = \lambda_2 = \lambda_3 = \lambda$,

$$\eta_{00} = \frac{\lambda}{f'(\lambda)}, \quad \eta_{01} = \eta_{10} = \eta_{11} = \eta_{02} = \eta_{20} = \eta_{12} = \eta_{21} = \eta_{22} = 0. \tag{8.72}$$

8.6 Finite Plasticity Based on the Additive Decomposition of Generalized Strains

Keeping in mind the above results regarding generalized strains we are concerned in this section with a thermodynamically based plasticity theory. The basic kinematic assumption of this theory is the additive decomposition of generalized strains (7.6) into an elastic part $\mathbf{E}_e$ and a plastic part $\mathbf{E}_p$ as

$$\mathbf{E} = \mathbf{E}_e + \mathbf{E}_p. \tag{8.73}$$

The derivation of evolution equations for the plastic strain is based on the second law of thermodynamics and the principle of maximum plastic dissipation. The second law of thermodynamics can be written in the Clausius-Planck form as (see, e.g. [46])

$$\mathcal{D} = \mathbf{T} : \dot{\mathbf{E}} - \dot{\psi} \geq 0, \tag{8.74}$$

where $\mathcal{D}$ denotes the dissipation and $\mathbf{T}$ is again the stress tensor work conjugate to $\mathbf{E}$. Inserting (8.73) into (8.74) we further write

$$\mathcal{D} = \left(\mathbf{T} - \frac{\partial \psi}{\partial \mathbf{E}_e}\right) : \dot{\mathbf{E}}_e + \mathbf{T} : \dot{\mathbf{E}}_p \geq 0, \tag{8.75}$$

where the strain energy is assumed to be a function of the elastic strain as $\psi = \hat{\psi}(\mathbf{E}_e)$. The first term in the expression of the dissipation (8.75) depends solely on the elastic strain rate $\dot{\mathbf{E}}_e$, while the second one on the plastic strain rate $\dot{\mathbf{E}}_p$. Since the elastic and plastic strain rates are independent of each other the dissipation inequality (8.75) requires that

$$\mathbf{T} = \frac{\partial \psi}{\partial \mathbf{E}_e}. \tag{8.76}$$

This leads to the so-called reduced dissipation inequality

$$\mathcal{D} = \mathbf{T} : \dot{\mathbf{E}}_p \geq 0. \tag{8.77}$$

Among all admissible processes the real one maximizes the dissipation (8.77). This statement is based on the postulate of maximum plastic dissipation (see, e.g., [29]). According to the converse Kuhn-Tucker theorem (see, e.g., [6]) the sufficient conditions of this maximum are written as

$$\dot{\mathbf{E}}_p = \dot{\zeta}\frac{\partial \Phi}{\partial \mathbf{T}}, \quad \dot{\zeta} \geq 0, \quad \dot{\zeta}\Phi = 0, \quad \Phi \leq 0, \tag{8.78}$$

where Φ represents a convex yield function and $\dot{\zeta}$ denotes a consistency parameter. In the following, we will deal with an ideal-plastic isotropic material described by a von Mises-type yield criterion. Written in terms of the stress tensor $\mathbf{T}$ the von Mises yield function takes the form [32]

$$\Phi = \|\mathrm{dev}\mathbf{T}\| - \sqrt{\frac{2}{3}}\sigma_Y, \tag{8.79}$$

where σ_Y denotes the normal yield stress. With the aid of (6.51) and (6.147) the evolution equation $(8.78)_1$ can thus be given by

$$\begin{aligned}\dot{\mathbf{E}}_p &= \dot{\zeta}\|\mathrm{dev}\mathbf{T}\|,_{\mathbf{T}} \\ &= \dot{\zeta}\|\mathrm{dev}\mathbf{T}\|,_{\mathrm{dev}\mathbf{T}} : \mathrm{dev}\mathbf{T},_{\mathbf{T}} = \dot{\zeta}\frac{\mathrm{dev}\mathbf{T}}{\|\mathrm{dev}\mathbf{T}\|} : \mathcal{P}_{\mathrm{dev}} = \dot{\zeta}\frac{\mathrm{dev}\mathbf{T}}{\|\mathrm{dev}\mathbf{T}\|}.\end{aligned} \tag{8.80}$$

Taking the quadratic norm on both the right and left hand side of this identity delivers the consistency parameter as $\dot{\zeta} = ||\dot{\mathbf{E}}_p||$. In view of the yield condition $\Phi = 0$ we thus obtain

$$\mathrm{dev}\mathbf{T} = \sqrt{\frac{2}{3}}\sigma_Y\frac{\dot{\mathbf{E}}_p}{\|\dot{\mathbf{E}}_p\|}, \tag{8.81}$$

which immediately requires that (see Exercise 1.49)

$$\mathrm{tr}\dot{\mathbf{E}}_p = 0. \tag{8.82}$$

In the following, we assume small elastic but large plastic strains and specify the above plasticity model for finite simple shear. In this case all three principal stretches (8.21) are distinct so that we can write by virtue of (7.6)

$$\dot{\mathbf{E}}_p = \dot{\mathbf{E}} = \sum_{i=1}^{3} f'(\lambda_i)\dot{\lambda}_i\mathbf{P}_i + \sum_{i=1}^{3} f(\lambda_i)\dot{\mathbf{P}}_i. \tag{8.83}$$

By means of the identities $\mathrm{tr}\mathbf{P}_i = 1$ and $\mathrm{tr}\dot{\mathbf{P}}_i = 0$ following from (4.62) and (4.63) where $r_i = 1$ $(i = 1, 2, 3)$ the condition (8.82) requires that

$$\sum_{i=1}^{3} f'(\lambda_i)\dot{\lambda}_i = 0. \tag{8.84}$$

In view of (8.21) it leads to the equation

$$f'(\lambda) - f'(\lambda^{-1})\lambda^{-2} = 0, \quad \forall \lambda > 0, \tag{8.85}$$

where we set $\lambda_1 = \lambda$ and consequently $\lambda_2 = \lambda^{-1}$. Solutions of this equations can be given by [22]

$$f_a(\lambda) = \begin{cases} \dfrac{1}{2a}(\lambda^a - \lambda^{-a}) & \text{for } a \neq 0, \\ \ln \lambda & \text{for } a = 0. \end{cases} \tag{8.86}$$

By means of $(7.6)_1$ or $(7.18)_1$ the functions f_a (8.86) yield a set of new generalized strain measures

$$\mathbf{E}^{\langle a \rangle} = \begin{cases} \dfrac{1}{2a}(\mathbf{U}^a - \mathbf{U}^{-a}) = \dfrac{1}{2a}(\mathbf{C}^{a/2} - \mathbf{C}^{-a/2}) & \text{for } a \neq 0, \\ \ln \mathbf{U} = \dfrac{1}{2}\ln \mathbf{C} & \text{for } a = 0, \end{cases} \tag{8.87}$$

among which only the logarithmic one $(a = 0)$ belongs to Seth's family (7.7). Henceforth, we will deal only with the generalized strains (8.87) as able to provide the traceless deformation rate (8.82). For these strains Eq. (8.81) takes the form

$$\mathrm{dev}\mathbf{T}^{\langle a \rangle} = \sqrt{\frac{2}{3}}\sigma_Y \frac{\dot{\mathbf{E}}^{\langle a \rangle}}{\|\dot{\mathbf{E}}^{\langle a \rangle}\|}, \tag{8.88}$$

where $\mathbf{T}^{\langle a \rangle}$ denotes the stress tensor work conjugate to $\mathbf{E}^{\langle a \rangle}$. $\mathbf{T}^{\langle a \rangle}$ itself has no physical meaning and should be transformed to the Cauchy stresses. With the aid of (8.60), (8.61) and (8.63) we can write

$$\boldsymbol{\sigma} = \frac{1}{\det\mathbf{F}}\mathbf{F}\mathbf{S}\mathbf{F}^{\mathrm{T}} = \frac{1}{\det\mathbf{F}}\mathbf{F}\left(\mathbf{T}^{\langle a \rangle} : \mathcal{P}_a\right)\mathbf{F}^{\mathrm{T}}, \tag{8.89}$$

where

$$\mathcal{P}_a = 2\mathbf{E}^{\langle a \rangle},_{\mathbf{C}} \tag{8.90}$$

can be expressed either by (8.49) and (8.50) or by (8.51)–(8.55). It is seen that this fourth-order tensor is super-symmetric (see Exercise 5.11), so that $\mathbf{T}^{\langle a \rangle} : \mathcal{P}_a = \mathcal{P}_a : \mathbf{T}^{\langle a \rangle}$. Thus, by virtue of (1.162) and (1.163) representation (8.89) can be rewritten as

$$\begin{aligned}\boldsymbol{\sigma} &= \frac{1}{\det\mathbf{F}}\mathbf{F}(\mathcal{P}_a : \mathbf{T}^{\langle a\rangle})\mathbf{F}^{\mathrm{T}} \\ &= \frac{1}{\det\mathbf{F}}\mathbf{F}\left[\mathcal{P}_a : \mathrm{dev}\mathbf{T}^{\langle a\rangle} + \frac{1}{3}\mathrm{tr}\mathbf{T}^{\langle a\rangle}(\mathcal{P}_a : \mathbf{I})\right]\mathbf{F}^{\mathrm{T}}. \end{aligned} \tag{8.91}$$

With the aid of the relation

$$\begin{aligned}\mathcal{P}_a : \mathbf{I} &= 2\,\frac{\mathrm{d}}{\mathrm{d}t}\mathbf{E}^{\langle a\rangle}(\mathbf{C} + t\mathbf{I})\Big|_{t=0} \\ &= 2\frac{\mathrm{d}}{\mathrm{d}t}\sum_{i=1}^{3} f_a\left(\sqrt{\lambda_i^2 + t}\right)\mathbf{P}_i\Big|_{t=0} = \sum_{i=1}^{3} f_a'(\lambda_i)\lambda_i^{-1}\mathbf{P}_i \end{aligned} \tag{8.92}$$

following from (6.128) and taking (8.86) into account one obtains

$$\mathbf{F}(\mathcal{P}_a : \mathbf{I})\mathbf{F}^{\mathrm{T}} = \frac{1}{2}\mathbf{F}\left(\mathbf{C}^{a/2-1} + \mathbf{C}^{-a/2-1}\right)\mathbf{F}^{\mathrm{T}} = \frac{1}{2}(\mathbf{b}^{a/2} + \mathbf{b}^{-a/2}).$$

Inserting this result into (8.91) yields

$$\boldsymbol{\sigma} = \frac{1}{\det\mathbf{F}}\mathbf{F}\left(\mathcal{P}_a : \mathrm{dev}\mathbf{T}^{\langle a\rangle}\right)\mathbf{F}^{\mathrm{T}} + \hat{\boldsymbol{\sigma}} \tag{8.93}$$

with the abbreviation

$$\hat{\boldsymbol{\sigma}} = \frac{\mathrm{tr}\mathbf{T}^{\langle a\rangle}}{6\det\mathbf{F}}(\mathbf{b}^{a/2} + \mathbf{b}^{-a/2}). \tag{8.94}$$

Using the spectral decomposition of $\mathbf{b}$ by (8.4) and taking into account that in the case of simple shear $\det\mathbf{F} = 1$ we can further write

$$\hat{\boldsymbol{\sigma}} = \frac{1}{6}\mathrm{tr}\mathbf{T}^{\langle a\rangle}\left[(\lambda^a + \lambda^{-a})(\mathbf{p}_1 + \mathbf{p}_2) + 2\mathbf{p}_3\right], \tag{8.95}$$

where λ is given by (8.21). Thus, in the 1–2 shear plane the stress tensor $\hat{\boldsymbol{\sigma}}$ has the double eigenvalue $\frac{1}{6}\mathrm{tr}\mathbf{T}^{\langle a\rangle}(\lambda^a + \lambda^{-a})$ and causes equibiaxial tension or compression. Hence, in this plane the component $\hat{\boldsymbol{\sigma}}$ (8.94) is shear free and does not influence the shear stress response. Inserting (8.88) into (8.93) and taking (8.18) and (8.48) into account we finally obtain

$$\boldsymbol{\sigma} = \sqrt{\frac{2}{3}}\sigma_Y\mathbf{F}\left[\frac{\mathcal{P}_a : \mathcal{P}_a : \mathbf{A}}{\|\mathcal{P}_a : \mathbf{A}\|}\right]\mathbf{F}^{\mathrm{T}} + \hat{\boldsymbol{\sigma}}, \tag{8.96}$$

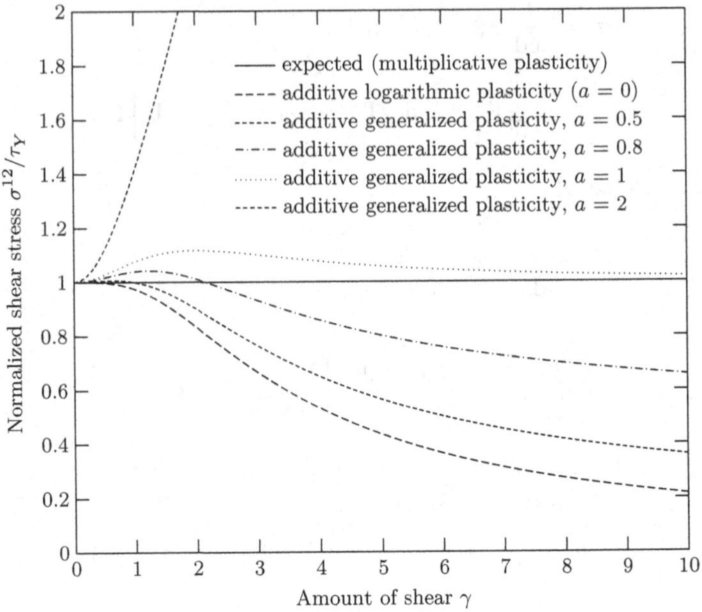

Fig. 8.1 Simple shear of an ideal-plastic material: shear stress responses based on the additive decomposition of generalized strains

where

$$\mathbf{A} = \frac{1}{2\dot{\gamma}}\dot{\mathbf{C}} = \begin{bmatrix} 0 & 1/2 & 0 \\ 1/2 & \gamma & 0 \\ 0 & 0 & 0 \end{bmatrix} \boldsymbol{e}_i \otimes \boldsymbol{e}_j. \tag{8.97}$$

Of particular interest is the shear stress σ^{12} as a function of the amount of shear γ. Inserting (8.51), (8.52) and (8.90) into (8.96) we obtain after some algebraic manipulations

$$\frac{\sigma^{12}}{\tau_Y} = \frac{2\sqrt{(4+\gamma^2)\,\Gamma^2 f_a'^2(\Gamma) + 4f_a^2(\Gamma)}}{4+\gamma^2}, \tag{8.98}$$

where

$$\Gamma = \frac{\gamma}{2} + \frac{\sqrt{4+\gamma^2}}{2} \tag{8.99}$$

and $\tau_Y = \sigma_Y/\sqrt{3}$ denotes the shear yield stress. Equation (8.98) is illustrated graphically in Fig. 8.1 for several values of the parameter a. Since the presented plasticity model considers neither softening nor hardening and is restricted to small elastic strains a constant shear stress response even at large plastic deformations is expected. It is also predicted by a plasticity model based on the multiplicative decomposition of the deformation gradient (see, e.g., [22] for more details). The plasticity model based on the additive decomposition of generalized strains exhibits, however, a non-constant shear stress for all examined values of a. This restricts the applicability of this model to moderate plastic shears. Indeed, in the vicinity of the point $\gamma = 0$ the power series expansion of (8.98) takes the form

$$\frac{\sigma^{12}}{\tau_Y} = 1 + \frac{1}{4}a^2\gamma^2 + \left(\frac{1}{16}a^4 - \frac{3}{4}a^2 - 1\right)\gamma^4 + O(\gamma^6). \tag{8.100}$$

Thus, in the case of simple shear the amount of shear is limited for the logarithmic strain ($a = 0$) by $\gamma^4 \ll 1$ and for other generalized strain measures by $\gamma^2 \ll 1$.

Exercises

8.1. The deformation gradient is given by $\mathbf{F} = \mathrm{F}^i_{\cdot j}\boldsymbol{e}_i \otimes \boldsymbol{e}^j$, where

$$\left[\mathrm{F}^i_{\cdot j}\right] = \begin{bmatrix} 1 & 2 & 0 \\ -2 & 2 & 0 \\ 0 & 0 & 1 \end{bmatrix}.$$

Evaluate the stretch tensors $\mathbf{U}$ and $\mathbf{v}$ and the rotation tensor $\mathbf{R}$ using (8.11), (8.12) and (8.16), (8.17).

8.2. Prove the orthogonality (8.65) of the basis tensors (8.64) using (4.44), (5.33) and (5.55).

8.3. Express the time derivative of the logarithmic strain $\mathbf{E}^{(0)}$ by means of the relations (8.48)–(8.50).

Chapter 9
Solutions

9.1 Exercises of Chap. 1

1.1

(a) (A.4) and (A.3):

$$\boldsymbol{0} = \boldsymbol{0} + (-\boldsymbol{0}) = -\boldsymbol{0}.$$

(b) (A.1)–(A.4) and (B.3):

$$\alpha\boldsymbol{0} = \boldsymbol{0} + \alpha\boldsymbol{0} = \alpha\boldsymbol{x} + (-\alpha\boldsymbol{x}) + \alpha\boldsymbol{0}$$
$$= \alpha(\boldsymbol{0} + \boldsymbol{x}) + (-\alpha\boldsymbol{x}) = \alpha\boldsymbol{x} + (-\alpha\boldsymbol{x}) = \boldsymbol{0}.$$

(c) (A.2)–(A.4) and (B.4):

$$0\boldsymbol{x} = 0\boldsymbol{x} + \boldsymbol{0} = 0\boldsymbol{x} + 0\boldsymbol{x} + (-0\boldsymbol{x}) = 0\boldsymbol{x} + (-0\boldsymbol{x}) = \boldsymbol{0}, \quad \forall \boldsymbol{x} \in \mathbb{V}.$$

(d) (A.2)–(A.4), (B.2), (B.4) and (c):

$$(-1)\boldsymbol{x} = (-1)\boldsymbol{x} + \boldsymbol{0} = (-1)\boldsymbol{x} + \boldsymbol{x} + (-\boldsymbol{x})$$
$$= (-1+1)\boldsymbol{x} + (-\boldsymbol{x}) = 0\boldsymbol{x} + (-\boldsymbol{x}) = \boldsymbol{0} + (-\boldsymbol{x}) = -\boldsymbol{x}, \quad \forall \boldsymbol{x} \in \mathbb{V}.$$

(e) If, on the contrary, $\alpha \neq 0$ and $\boldsymbol{x} \neq \boldsymbol{0}$, then according to (b), (B.1) and (B.2):

$$\boldsymbol{0} = \alpha^{-1}\boldsymbol{0} = \alpha^{-1}(\alpha\boldsymbol{x}) = \boldsymbol{x}.$$

1.2 Let, on the contrary, $\boldsymbol{x}_k = \boldsymbol{0}$ for some k. Then, $\sum_{i=1}^{n} \alpha_i \boldsymbol{x}_i = \boldsymbol{0}$, where $\alpha_k = 1$, $\alpha_i = 0, i = 1, \ldots, k-1, k+1, \ldots, n$.

1.3 If, on the contrary, for some $k < n$: $\sum_{i=1}^{k} \alpha_i \boldsymbol{x}_i = \boldsymbol{0}$, where not all α_i, $(i = 1, 2, \ldots, k)$ are zero, then we can also write: $\sum_{i=1}^{n} \alpha_i \boldsymbol{x}_i = \boldsymbol{0}$, where $\alpha_i = 0$, for $i = k+1, \ldots, n$.

M. Itskov, *Tensor Algebra and Tensor Analysis for Engineers*, Mathematical Engineering, DOI 10.1007/978-3-642-30879-6_9,

1.4

(a) $\delta^i_j a^j = \delta^i_1 a^1 + \delta^i_2 a^2 + \delta^i_3 a^3 = a^i$,
(b) $\delta_{ij} x^i x^j = \delta_{11} x^1 x^1 + \delta_{12} x^1 x^2 + \ldots + \delta_{33} x^3 x^3 = x^1 x^1 + x^2 x^2 + x^3 x^3$,
(c) $\delta^i_i = \delta^1_1 + \delta^2_2 + \delta^3_3 = 3$,
(d) $\dfrac{\partial f_i}{\partial x^j} dx^j = \dfrac{\partial f_i}{\partial x^1} dx^1 + \dfrac{\partial f_i}{\partial x^2} dx^2 + \dfrac{\partial f_i}{\partial x^3} dx^3$.

1.5 (A.4), (C.2), (C.3) and Exercises 1.1(d):

$$\boldsymbol{0} \cdot \boldsymbol{x} = [\boldsymbol{x} + (-\boldsymbol{x})] \cdot \boldsymbol{x} = [\boldsymbol{x} + (-1)\boldsymbol{x}] \cdot \boldsymbol{x} = \boldsymbol{x} \cdot \boldsymbol{x} - \boldsymbol{x} \cdot \boldsymbol{x} = 0.$$

1.6 Let on the contrary $\sum_{i=1}^m \alpha_i \boldsymbol{g}_i = \boldsymbol{0}$, where not all α_i $(i = 1, 2, \ldots, m)$ are zero. Multiplying scalarly by $\boldsymbol{g}_j$ we obtain: $0 = \boldsymbol{g}_j \cdot \left(\sum_{i=1}^m \alpha_i \boldsymbol{g}_i\right)$. Since $\boldsymbol{g}_i \cdot \boldsymbol{g}_j = 0$ for $i \neq j$, we can write: $\alpha_j \boldsymbol{g}_j \cdot \boldsymbol{g}_j = 0\,(j = 1, 2, \ldots, m)$. The fact that the vectors $\boldsymbol{g}_j$ are non-zero leads in view of (C.4) to the conclusion that $\alpha_j = 0\,(j = 1, 2, \ldots, m)$ which contradicts the earlier assumption.

1.7 Equation (1.6), (C.1) and (C.2):

$$\begin{aligned} \|\boldsymbol{x} + \boldsymbol{y}\|^2 &= (\boldsymbol{x} + \boldsymbol{y}) \cdot (\boldsymbol{x} + \boldsymbol{y}) \\ &= \boldsymbol{x} \cdot \boldsymbol{x} + \boldsymbol{x} \cdot \boldsymbol{y} + \boldsymbol{y} \cdot \boldsymbol{x} + \boldsymbol{y} \cdot \boldsymbol{y} = \|\boldsymbol{x}\|^2 + 2\boldsymbol{x} \cdot \boldsymbol{y} + \|\boldsymbol{y}\|^2. \end{aligned}$$

1.8 Since $\mathcal{G} = \{\boldsymbol{g}_1, \boldsymbol{g}_2, \ldots, \boldsymbol{g}_n\}$ is a basis we can write $\boldsymbol{a} = a^i \boldsymbol{g}_i$. Then, $\boldsymbol{a} \cdot \boldsymbol{a} = a^i (\boldsymbol{g}_i \cdot \boldsymbol{a})$. Thus, if $\boldsymbol{a} \cdot \boldsymbol{g}_i = 0\,(i = 1, 2, \ldots, n)$, then $\boldsymbol{a} \cdot \boldsymbol{a} = \boldsymbol{0}$ and according to (C.4) $\boldsymbol{a} = \boldsymbol{0}$ (sufficiency). Conversely, if $\boldsymbol{a} = \boldsymbol{0}$, then (see Exercise 1.5) $\boldsymbol{a} \cdot \boldsymbol{g}_i = 0\,(i = 1, 2, \ldots, n)$ (necessity).

1.9 Necessity. (C.2): $\boldsymbol{a} \cdot \boldsymbol{x} = \boldsymbol{b} \cdot \boldsymbol{x} \Rightarrow \boldsymbol{a} \cdot \boldsymbol{x} - \boldsymbol{b} \cdot \boldsymbol{x} = (\boldsymbol{a} - \boldsymbol{b}) \cdot \boldsymbol{x} = 0,\ \forall \boldsymbol{x} \in \mathbb{E}^n$. Let $\boldsymbol{x} = \boldsymbol{a} - \boldsymbol{b}$, then $(\boldsymbol{a} - \boldsymbol{b}) \cdot (\boldsymbol{a} - \boldsymbol{b}) = 0$ and according to (C.4) $\boldsymbol{a} - \boldsymbol{b} = \boldsymbol{0}$. This implies that $\boldsymbol{a} = \boldsymbol{b}$. The sufficiency is evident.

1.10

(a) Orthonormal vectors $\boldsymbol{e}_1$, $\boldsymbol{e}_2$ and $\boldsymbol{e}_3$ can be calculated by means of the Gram-Schmidt procedure (1.10)–(1.12) as follows

$$\boldsymbol{e}_1 = \frac{\boldsymbol{g}_1}{\|\boldsymbol{g}_1\|} = \begin{Bmatrix} \sqrt{2}/2 \\ \sqrt{2}/2 \\ 0 \end{Bmatrix},$$

$$\boldsymbol{e}'_2 = \boldsymbol{g}_2 - (\boldsymbol{g}_2 \cdot \boldsymbol{e}_1)\boldsymbol{e}_1 = \begin{Bmatrix} 1/2 \\ -1/2 \\ -2 \end{Bmatrix}, \quad \boldsymbol{e}_2 = \frac{\boldsymbol{e}'_2}{\|\boldsymbol{e}'_2\|} = \begin{Bmatrix} \sqrt{2}/6 \\ -\sqrt{2}/6 \\ -2\sqrt{2}/3 \end{Bmatrix},$$

$$\boldsymbol{e}'_3 = \boldsymbol{g}_3 - (\boldsymbol{g}_3 \cdot \boldsymbol{e}_2)\boldsymbol{e}_2 - (\boldsymbol{g}_3 \cdot \boldsymbol{e}_1)\boldsymbol{e}_1 = \begin{Bmatrix} 10/9 \\ -10/9 \\ 5/9 \end{Bmatrix}, \ \boldsymbol{e}_3 = \frac{\boldsymbol{e}'_3}{\|\boldsymbol{e}'_3\|} = \begin{Bmatrix} 2/3 \\ -2/3 \\ 1/3 \end{Bmatrix}.$$

(b) According to $(1.16)_2$ the matrix $[\beta_i^j]$ is composed from the components of the vectors $\boldsymbol{g}_i$ $(i = 1, 2, 3)$ as follows:

$$[\beta_i^j] = \begin{bmatrix} 1 & 1 & 0 \\ 2 & 1 & -2 \\ 4 & 2 & 1 \end{bmatrix}.$$

In view of (1.18)

$$[\alpha_i^j] = [\beta_i^j]^{-1} = \begin{bmatrix} -1 & \frac{1}{5} & \frac{2}{5} \\ 2 & -\frac{1}{5} & -\frac{2}{5} \\ 0 & -\frac{2}{5} & \frac{1}{5} \end{bmatrix}.$$

By (1.19) the columns of this matrix represent components of the dual vectors so that

$$\boldsymbol{g}^1 = \begin{Bmatrix} -1 \\ 2 \\ 0 \end{Bmatrix}, \quad \boldsymbol{g}^2 = \begin{Bmatrix} 1/5 \\ -1/5 \\ -2/5 \end{Bmatrix}, \quad \boldsymbol{g}^3 = \begin{Bmatrix} 2/5 \\ -2/5 \\ 1/5 \end{Bmatrix}.$$

(c) First, we calculate the matrices $[g_{ij}]$ and $\left[g^{ij}\right]$ by $(1.25)_2$ and (1.24)

$$[g_{ij}] = [\boldsymbol{g}_i \cdot \boldsymbol{g}_j] = \begin{bmatrix} 2 & 3 & 6 \\ 3 & 9 & 8 \\ 6 & 8 & 21 \end{bmatrix}, \quad [g^{ij}] = [g_{ij}]^{-1} = \begin{bmatrix} 5 & -\frac{3}{5} & -\frac{6}{5} \\ -\frac{3}{5} & \frac{6}{25} & \frac{2}{25} \\ -\frac{6}{5} & \frac{2}{25} & \frac{9}{25} \end{bmatrix},$$

With the aid of (1.21) we thus obtain

$$\boldsymbol{g}^1 = g^{11}\boldsymbol{g}_1 + g^{12}\boldsymbol{g}_2 + g^{13}\boldsymbol{g}_3 = \begin{Bmatrix} -1 \\ 2 \\ 0 \end{Bmatrix},$$

$$\boldsymbol{g}^2 = g^{21}\boldsymbol{g}_1 + g^{22}\boldsymbol{g}_2 + g^{23}\boldsymbol{g}_3 = \begin{Bmatrix} 1/5 \\ -1/5 \\ -2/5 \end{Bmatrix},$$

$$\boldsymbol{g}^3 = g^{31}\boldsymbol{g}_1 + g^{32}\boldsymbol{g}_2 + g^{33}\boldsymbol{g}_3 = \begin{Bmatrix} 2/5 \\ -2/5 \\ 1/5 \end{Bmatrix}.$$

(d) By virtue of (1.35) we write

$$g = \left|\beta^i_{j}\right| = \begin{vmatrix} 1 & 1 & 0 \\ 2 & 1 & -2 \\ 4 & 2 & 1 \end{vmatrix} = -5.$$

Applying (1.33) we further obtain with the aid of (1.47)

$$\boldsymbol{g}^1 = g^{-1}\boldsymbol{g}_2 \times \boldsymbol{g}_3 = -\frac{1}{5}\begin{vmatrix} 2 & 1 & -2 \\ 4 & 2 & 1 \\ \boldsymbol{a}_1 & \boldsymbol{a}_2 & \boldsymbol{a}_3 \end{vmatrix} = -\boldsymbol{a}_1 + 2\boldsymbol{a}_2,$$

$$\boldsymbol{g}^2 = g^{-1}\boldsymbol{g}_3 \times \boldsymbol{g}_1 = -\frac{1}{5}\begin{vmatrix} 4 & 2 & 1 \\ 1 & 1 & 0 \\ \boldsymbol{a}_1 & \boldsymbol{a}_2 & \boldsymbol{a}_3 \end{vmatrix} = \frac{1}{5}\left(\boldsymbol{a}_1 - \boldsymbol{a}_2 - 2\boldsymbol{a}_3\right),$$

$$\boldsymbol{g}^3 = g^{-1}\boldsymbol{g}_1 \times \boldsymbol{g}_2 = -\frac{1}{5}\begin{vmatrix} 1 & 1 & 0 \\ 2 & 1 & -2 \\ \boldsymbol{a}_1 & \boldsymbol{a}_2 & \boldsymbol{a}_3 \end{vmatrix} = \frac{1}{5}(2\boldsymbol{a}_1 - 2\boldsymbol{a}_2 + \boldsymbol{a}_3),$$

where $\boldsymbol{a}_i$ denote the orthonormal basis the components of the original vectors $\boldsymbol{g}_i$ $(i = 1, 2, 3)$ are related to.

1.11 Let on the contrary $\alpha_i \boldsymbol{g}^i = \boldsymbol{0}$, where not all α_i are zero. Multiplying scalarly by $\boldsymbol{g}_j$ we obtain by virtue of (1.15): $0 = \boldsymbol{g}_j \cdot (\alpha_i \boldsymbol{g}^i) = \alpha_i \delta^i_j = \alpha_j$ $(j = 1, 2, 3)$.

1.12 Similarly to (1.35) we write using also (1.18), (1.19) and (1.36)

$$\left[\boldsymbol{g}^1\boldsymbol{g}^2\boldsymbol{g}^3\right] = \left[\alpha^1_i\boldsymbol{e}^i\alpha^2_j\boldsymbol{e}^j\alpha^3_k\boldsymbol{e}^k\right] = \alpha^1_i\alpha^2_j\alpha^3_k\left[\boldsymbol{e}^i\boldsymbol{e}^j\boldsymbol{e}^k\right]$$

$$= \alpha^1_i\alpha^2_j\alpha^3_k e^{ijk} = \left|\alpha^i_j\right| = \left|\beta^i_j\right|^{-1} = g^{-1}.$$

Equation (1.42) immediately follows from (1.24) and (1.34).

1.13 The components of the vector $\boldsymbol{g}_i \times \boldsymbol{g}_j$ with respect to the basis $\boldsymbol{g}^k$ $(k = 1, 2, 3)$ result from $(1.28)_2$ and (1.39) as

$$(\boldsymbol{g}_i \times \boldsymbol{g}_j) \cdot \boldsymbol{g}_k = [\boldsymbol{g}_i\boldsymbol{g}_j\boldsymbol{g}_k] = e_{ijk}\, g, \quad i, j, k = 1, 2, 3,$$

which immediately implies (1.40). In the same manner one also proves (1.44) using (1.43).

1.14

(a) $\delta^{ij} e_{ijk} = \delta^{11} e_{11k} + \delta^{12} e_{12k} + \ldots + \delta^{33} e_{33k} = 0.$

(b) Writing out the term $e^{ikm}e_{jkm}$ we first obtain

$$e^{ikm}e_{jkm} = e^{i11}e_{j11} + e^{i12}e_{j12} + \ldots + e^{i33}e_{j33}$$
$$= e^{i12}e_{j12} + e^{i21}e_{j21} + e^{i13}e_{j13} + e^{i31}e_{j31} + e^{i32}e_{j32} + e^{i23}e_{j23}.$$

For $i \neq j$ each term in this sum is equal to zero. Let further $i = j = 1$. Then we obtain $e^{i12}e_{j12} + e^{i21}e_{j21} + e^{i13}e_{j13} + e^{i31}e_{j31} + e^{i32}e_{j32} + e^{i23}e_{j23} = e^{132}e_{132} + e^{123}e_{123} = (-1)(-1) + 1 \cdot 1 = 2$. The same result also holds for the cases $i = j = 2$ and $i = j = 3$. Thus, we can write $e^{ikm}e_{jkm} = 2\delta^i_j$.

(c) By means of the previous result (b) we can write: $e^{ijk}e_{ijk} = 2\delta^i_i = 2(\delta^1_1 + \delta^2_2 + \delta^3_3) = 6$. This can also be shown directly by

$$e^{ijk}e_{ijk} = e^{123}e_{123} + e^{132}e_{132} + e^{213}e_{213} + e^{231}e_{231} + e^{312}e_{312} + e^{321}e_{321}$$
$$= 1 \cdot 1 + (-1) \cdot (-1) + (-1) \cdot (-1) + 1 \cdot 1 + 1 \cdot 1 + (-1) \cdot (-1) = 6.$$

(d) $e^{ijm}e_{klm} = e^{ij1}e_{kl1} + e^{ij2}e_{kl2} + e^{ij3}e_{kl3}$. It is seen that in the case $i = j$ or $k = l$ this sum as well as the right hand side $\delta^i_k\delta^j_l - \delta^i_l\delta^j_k$ are both zero. Let further $i \neq j$. Then, only one term in the above sum is non-zero if $k = i$ and $l = j$ or vice versa $l = i$ and $k = j$. In the first case the left hand side is 1 and in the last case -1. The same holds also for the right side $\delta^i_k\delta^j_l - \delta^i_l\delta^j_k$. Indeed, we obtain for $k = i \neq l = j$: $\delta^i_k\delta^j_l - \delta^i_l\delta^j_k = 1 \cdot 1 - 0 = 1$ and for $l = i \neq k = j$: $\delta^i_k\delta^j_l - \delta^i_l\delta^j_k = 0 - 1 \cdot 1 = -1$.

1.15 Using the representations $\boldsymbol{a} = a^i\boldsymbol{g}_i$, $\boldsymbol{b} = b^j\boldsymbol{g}_j$ and $\boldsymbol{c} = c_l\boldsymbol{g}^l$ we can write by virtue of (1.40) and (1.44)

$$(\boldsymbol{a} \times \boldsymbol{b}) \times \boldsymbol{c} = \left[\left(a^i\boldsymbol{g}_i\right) \times \left(b^j\boldsymbol{g}_j\right)\right] \times \boldsymbol{c} = \left(a^ib^je_{ijk}g\boldsymbol{g}^k\right) \times \left(c_l\boldsymbol{g}^l\right)$$
$$= a^ib^jc_le_{ijk}e^{klm}\boldsymbol{g}_m = a^ib^jc_le_{ijk}e^{lmk}\boldsymbol{g}_m.$$

With the aid of the identity $e^{ijm}e_{klm} = \delta^i_k\delta^j_l - \delta^i_l\delta^j_k$ (Exercise 1.14) we finally obtain

$$(\boldsymbol{a} \times \boldsymbol{b}) \times \boldsymbol{c} = a^ib^jc_l\left(\delta^l_i\delta^m_j - \delta^m_i\delta^l_j\right)\boldsymbol{g}_m = a^ib^jc_l\delta^l_i\delta^m_j\boldsymbol{g}_m - a^ib^jc_l\delta^m_i\delta^l_j\boldsymbol{g}_m$$
$$= a^ib^jc_i\boldsymbol{g}_j - a^ib^jc_j\boldsymbol{g}_i = (\boldsymbol{a} \cdot \boldsymbol{c})\,\boldsymbol{b} - (\boldsymbol{b} \cdot \boldsymbol{c})\,\boldsymbol{a}.$$

Relation (1.169) immediately follows from (1.168) taking into account the definition of the operator $\hat{(\bullet)}$ (1.66) and the tensor product (1.80).

1.16 Equation $(1.64)_1$ results immediately from (1.32) and (C.3). Equation $(1.64)_2$ can further be proved by using the representations $\boldsymbol{w} = w_i\boldsymbol{g}^i$, $\boldsymbol{x} = x_i\boldsymbol{g}^i$ and $\boldsymbol{y} = y_i\boldsymbol{g}^i$ and by means of (1.45) as follows:

$$\boldsymbol{w} \times (\boldsymbol{x} + \boldsymbol{y}) = \left(w_i \boldsymbol{g}^i\right) \times \left(x_j + y_j\right) \boldsymbol{g}^j = w_i \left(x_j + y_j\right) e^{ijk} g^{-1} \boldsymbol{g}_k$$
$$= w_i x_j e^{ijk} g^{-1} \boldsymbol{g}_k + w_i y_j e^{ijk} g^{-1} \boldsymbol{g}_k = \boldsymbol{w} \times \boldsymbol{x} + \boldsymbol{w} \times \boldsymbol{y}.$$

1.17 (A.2)–(A.4) and (1.49):

$$\boldsymbol{0} = \mathbf{A}\boldsymbol{x} + (-\mathbf{A}\boldsymbol{x}) = \mathbf{A}(\boldsymbol{x} + \boldsymbol{0}) + (-\mathbf{A}\boldsymbol{x}) = \mathbf{A}\boldsymbol{x} + \mathbf{A}\boldsymbol{0} + (-\mathbf{A}\boldsymbol{x}) = \mathbf{A}\boldsymbol{0}.$$

1.18 Equation (1.50), Exercises 1.1(c) and 1.17: $(0\mathbf{A})\boldsymbol{x} = \mathbf{A}(0\boldsymbol{x}) = \mathbf{A}\boldsymbol{0} = \boldsymbol{0}$, $\forall \boldsymbol{x} \in \mathbb{E}^n$.

1.19 Equation (1.62) and Exercise 1.18: $\mathbf{A} + (-\mathbf{A}) = \mathbf{A} + (-1)\mathbf{A} = (1-1)\mathbf{A} = 0\mathbf{A} = \mathbf{0}$.

1.20 We show that this is not possible, for example, for the identity tensor. Let, on the contrary, $\mathbf{I} = \boldsymbol{a} \otimes \boldsymbol{b}$. Clearly, $\boldsymbol{a} \neq \boldsymbol{0}$, since otherwise $(\boldsymbol{a} \otimes \boldsymbol{b})\,\boldsymbol{x} = \boldsymbol{0}\,\forall \boldsymbol{x} \in \mathbb{E}^n$. Let further $\boldsymbol{x}$ be a vector linearly independent of $\boldsymbol{a}$. Such a vector can be obtained for example by completing $\boldsymbol{a}$ to a basis of $\mathbb{E}^n$. Then, mapping of $\boldsymbol{x}$ by $\mathbf{I}$ leads to the contradiction: $\boldsymbol{x} = (\boldsymbol{b} \cdot \boldsymbol{x})\,\boldsymbol{a}$.

1.21 Indeed, a scalar product of the right-hand side of (1.85) with an arbitrary vector $\boldsymbol{x}$ yields $[(\boldsymbol{y} \cdot \boldsymbol{a})\,\boldsymbol{b}] \cdot \boldsymbol{x} = (\boldsymbol{y} \cdot \boldsymbol{a})\,(\boldsymbol{b} \cdot \boldsymbol{x})$. The same result follows also from $\boldsymbol{y} \cdot [(\boldsymbol{a} \otimes \boldsymbol{b})\,\boldsymbol{x}] = (\boldsymbol{y} \cdot \boldsymbol{a})\,(\boldsymbol{b} \cdot \boldsymbol{x})$, $\forall \boldsymbol{x}, \boldsymbol{y} \in \mathbb{E}^n$ for the left-hand side. This implies that the identity (1.85) is true (see Exercise 1.9).

1.22 For $(1.88)_1$ we have for example

$$\boldsymbol{g}^i \mathbf{A} \boldsymbol{g}^j = \boldsymbol{g}^i \left(\mathrm{A}^{kl} \boldsymbol{g}_k \otimes \boldsymbol{g}_l\right) \boldsymbol{g}^j = \mathrm{A}^{kl} \left(\boldsymbol{g}^i \cdot \boldsymbol{g}_k\right) \left(\boldsymbol{g}_l \cdot \boldsymbol{g}^j\right) = \mathrm{A}^{kl} \delta^i_k \delta^j_l = \mathrm{A}^{ij}.$$

1.23 For an arbitrary vector $\boldsymbol{x} = x^i \boldsymbol{g}_i \in \mathbb{E}^3$ we can write using (1.28), (1.40) and (1.80)

$$\mathbf{W}\boldsymbol{x} = \boldsymbol{w} \times \boldsymbol{x} = \left(w^i \boldsymbol{g}_i\right) \times \left(x^j \boldsymbol{g}_j\right)$$
$$= e_{ijk} g w^i x^j \boldsymbol{g}^k = e_{ijk} g w^i \left(\boldsymbol{x} \cdot \boldsymbol{g}^j\right) \boldsymbol{g}^k = e_{ijk} g w^i \left(\boldsymbol{g}^k \otimes \boldsymbol{g}^j\right) \boldsymbol{x}.$$

Comparing the left and right hand side of this equality we obtain

$$\mathbf{W} = e_{ijk} g w^i \boldsymbol{g}^k \otimes \boldsymbol{g}^j, \tag{9.1}$$

so that the components of $\mathbf{W} = \mathrm{W}_{kj} \boldsymbol{g}^k \otimes \boldsymbol{g}^j$ can be given by $\mathrm{W}_{kj} = e_{ijk} g w^i$ or in the matrix form as

$$\left[\mathrm{W}_{kj}\right] = g \left[e_{ijk} w^i\right] = g \begin{bmatrix} 0 & -w^3 & w^2 \\ w^3 & 0 & -w^1 \\ -w^2 & w^1 & 0 \end{bmatrix}.$$

This yields also an alternative representation for $\mathbf{W}\boldsymbol{x}$ as follows

$$\mathbf{W}\boldsymbol{x} = g\left[\left(w^2x^3 - w^3x^2\right)\boldsymbol{g}^1 + \left(w^3x^1 - w^1x^3\right)\boldsymbol{g}^2 + \left(w^1x^2 - w^2x^1\right)\boldsymbol{g}^3\right].$$

It is seen that the tensor $\mathbf{W}$ is skew-symmetric because $\mathbf{W}^{\mathrm{T}} = -\mathbf{W}$.

1.24 According to (1.73) we can write

$$\mathbf{R} = \cos\alpha\mathbf{I} + \sin\alpha\hat{\boldsymbol{e}}_3 + (1-\cos\alpha)\left(\boldsymbol{e}_3\otimes\boldsymbol{e}_3\right).$$

Thus, an arbitrary vector $\boldsymbol{a} = a^i\boldsymbol{e}_i$ in $\mathbb{E}^3$ is rotated to $\mathbf{R}\boldsymbol{a} = \cos\alpha\left(a^i\boldsymbol{e}_i\right) + \sin\alpha\boldsymbol{e}_3\times\left(a^i\boldsymbol{e}_i\right) + (1-\cos\alpha)\,a^3\boldsymbol{e}_3$. By virtue of (1.46) we can further write

$$\begin{aligned}\mathbf{R}\boldsymbol{a} &= \cos\alpha\left(a^i\boldsymbol{e}_i\right) + \sin\alpha\left(a^1\boldsymbol{e}_2 - a^2\boldsymbol{e}_1\right) + (1-\cos\alpha)\,a^3\boldsymbol{e}_3\\ &= \left(a^1\cos\alpha - a^2\sin\alpha\right)\boldsymbol{e}_1 + \left(a^1\sin\alpha + a^2\cos\alpha\right)\boldsymbol{e}_2 + a^3\boldsymbol{e}_3.\end{aligned}$$

Thus, the rotation tensor can be given by $\mathbf{R} = \mathrm{R}^{ij}\boldsymbol{e}_i\otimes\boldsymbol{e}_j$, where

$$\left[\mathrm{R}^{ij}\right] = \begin{bmatrix}\cos\alpha & -\sin\alpha & 0\\ \sin\alpha & \cos\alpha & 0\\ 0 & 0 & 1\end{bmatrix}.$$

1.25 With the aid of (1.88) and (1.97) we obtain

$$\left[\mathrm{A}^i_{\cdot j}\right] = \left[\mathrm{A}^{ik}g_{kj}\right] = \left[\mathrm{A}^{ik}\right]\left[g_{kj}\right] = \begin{bmatrix}0 & -1 & 0\\ 0 & 0 & 0\\ 1 & 0 & 0\end{bmatrix}\begin{bmatrix}2 & 3 & 6\\ 3 & 9 & 8\\ 6 & 8 & 21\end{bmatrix} = \begin{bmatrix}-3 & -9 & -8\\ 0 & 0 & 0\\ 2 & 3 & 6\end{bmatrix},$$

$$\left[\mathrm{A}_{i\cdot}^{\ j}\right] = \left[g_{ik}\mathrm{A}^{kj}\right] = \left[g_{ik}\right]\left[\mathrm{A}^{kj}\right] = \begin{bmatrix}2 & 3 & 6\\ 3 & 9 & 8\\ 6 & 8 & 21\end{bmatrix}\begin{bmatrix}0 & -1 & 0\\ 0 & 0 & 0\\ 1 & 0 & 0\end{bmatrix} = \begin{bmatrix}6 & -2 & 0\\ 8 & -3 & 0\\ 21 & -6 & 0\end{bmatrix},$$

$$\left[\mathrm{A}_{ij}\right] = \left[g_{ik}\mathrm{A}^k_{\cdot j}\right] = \left[g_{ik}\right]\left[\mathrm{A}^k_{\cdot j}\right] = \left[\mathrm{A}_{i\cdot}^{\ k}g_{kj}\right] = \left[\mathrm{A}_{i\cdot}^{\ k}\right]\left[g_{kj}\right]$$

$$= \begin{bmatrix}2 & 3 & 6\\ 3 & 9 & 8\\ 6 & 8 & 21\end{bmatrix}\begin{bmatrix}-3 & -9 & -8\\ 0 & 0 & 0\\ 2 & 3 & 6\end{bmatrix} = \begin{bmatrix}6 & -2 & 0\\ 8 & -3 & 0\\ 21 & -6 & 0\end{bmatrix}\begin{bmatrix}2 & 3 & 6\\ 3 & 9 & 8\\ 6 & 8 & 21\end{bmatrix} = \begin{bmatrix}6 & 0 & 20\\ 7 & -3 & 24\\ 24 & 9 & 78\end{bmatrix}.$$

1.26 By means of (1.54), (1.89), (1.103) and Exercise 1.17 we can write

$$(\mathbf{A0})\,\boldsymbol{x} = \mathbf{A}\,(\mathbf{0}\boldsymbol{x}) = \mathbf{A}\boldsymbol{0} = \boldsymbol{0},\quad (\mathbf{0A})\,\boldsymbol{x} = \mathbf{0}\,(\mathbf{A}\boldsymbol{x}) = \boldsymbol{0},$$

$$(\mathbf{AI})\,\boldsymbol{x} = \mathbf{A}\,(\mathbf{I}\boldsymbol{x}) = \mathbf{A}\boldsymbol{x},\quad (\mathbf{IA})\,\boldsymbol{x} = \mathbf{I}\,(\mathbf{A}\boldsymbol{x}) = \mathbf{A}\boldsymbol{x},$$

$$\mathbf{A}\,(\mathbf{BC})\,\boldsymbol{x} = \mathbf{A}\left[\mathbf{B}\,(\mathbf{C}\boldsymbol{x})\right] = (\mathbf{AB})\,(\mathbf{C}\boldsymbol{x}) = \left[(\mathbf{AB})\,\mathbf{C}\right]\boldsymbol{x},\quad \forall\boldsymbol{x}\in\mathbb{E}^n.$$

1.27 To check the commutativeness of the tensors **A** and **B** we compute the components of the tensor **AB** − **BA**:

$$\left[(\mathbf{AB}-\mathbf{BA})^{i}_{\cdot j}\right]=\left[\mathrm{A}^{i}_{\cdot k}\mathrm{B}^{k}_{\cdot j}-\mathrm{B}^{i}_{\cdot k}\mathrm{A}^{k}_{\cdot j}\right]=\left[\mathrm{A}^{i}_{\cdot k}\right]\left[\mathrm{B}^{k}_{\cdot j}\right]-\left[\mathrm{B}^{i}_{\cdot k}\right]\left[\mathrm{A}^{k}_{\cdot j}\right]$$

$$=\begin{bmatrix}0&2&0\\0&0&0\\0&0&0\end{bmatrix}\begin{bmatrix}0&0&0\\0&0&0\\0&0&1\end{bmatrix}-\begin{bmatrix}0&0&0\\0&0&0\\0&0&1\end{bmatrix}\begin{bmatrix}0&2&0\\0&0&0\\0&0&0\end{bmatrix}=\begin{bmatrix}0&0&0\\0&0&0\\0&0&0\end{bmatrix}.$$

Similar we also obtain

$$\left[(\mathbf{AC}-\mathbf{CA})^{i}_{\cdot j}\right]=\begin{bmatrix}0&-2&0\\0&0&0\\0&0&0\end{bmatrix},\quad\left[(\mathbf{AD}-\mathbf{DA})^{i}_{\cdot j}\right]=\begin{bmatrix}0&-1&0\\0&0&0\\0&0&0\end{bmatrix},$$

$$\left[(\mathbf{BC}-\mathbf{CB})^{i}_{\cdot j}\right]=\begin{bmatrix}0&0&-3\\0&0&0\\0&1&0\end{bmatrix},\quad\left[(\mathbf{BD}-\mathbf{DB})^{i}_{\cdot j}\right]=\begin{bmatrix}0&0&0\\0&0&0\\0&0&0\end{bmatrix},$$

$$\left[(\mathbf{CD}-\mathbf{DC})^{i}_{\cdot j}\right]=\begin{bmatrix}0&-1&27\\0&0&0\\0&-19/2&0\end{bmatrix}.$$

Thus, **A** commutes with **B** while **B** also commutes with **D**.

1.28 Taking into account commutativeness of **A** and **B** we obtain for example for $k=2$

$$(\mathbf{A}+\mathbf{B})^{2}=(\mathbf{A}+\mathbf{B})(\mathbf{A}+\mathbf{B})=\mathbf{A}^{2}+\mathbf{AB}+\mathbf{BA}+\mathbf{B}^{2}=\mathbf{A}^{2}+2\mathbf{AB}+\mathbf{B}^{2}.$$

Generalizing this result for $k=2,3,\ldots$ we obtain using the Newton formula

$$(\mathbf{A}+\mathbf{B})^{k}=\sum_{i=0}^{k}\binom{k}{i}\mathbf{A}^{k-i}\mathbf{B}^{i},\quad\text{where}\quad\binom{k}{i}=\frac{k!}{i!\,(k-i)!}.\tag{9.2}$$

1.29 Using the result of the previous Exercise we first write out the left hand side of (1.170) by

$$\exp(\mathbf{A}+\mathbf{B})=\sum_{k=0}^{\infty}\frac{(\mathbf{A}+\mathbf{B})^{k}}{k!}=\sum_{k=0}^{\infty}\frac{1}{k!}\sum_{i=0}^{k}\binom{k}{i}\mathbf{A}^{k-i}\mathbf{B}^{i}$$

$$=\sum_{k=0}^{\infty}\frac{1}{k!}\sum_{i=0}^{k}\frac{k!}{i!\,(k-i)\,!}\mathbf{A}^{k-i}\mathbf{B}^{i}=\sum_{k=0}^{\infty}\sum_{i=0}^{k}\frac{\mathbf{A}^{k-i}\mathbf{B}^{i}}{i\,!(k-i)!}.$$

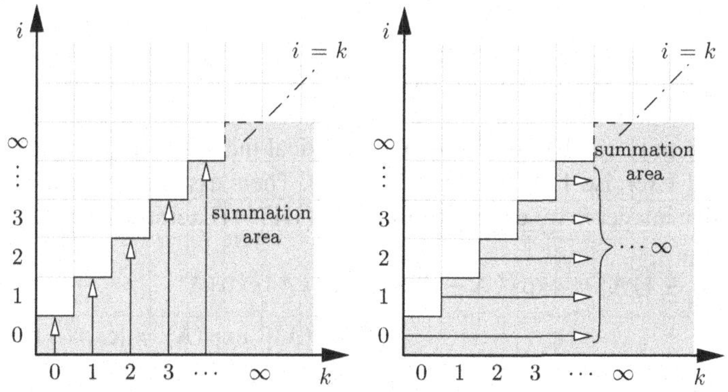

Fig. 9.1 Geometric illustration of the summation area and the summation order

Changing the summation order as shown in Fig. 9.1 we further obtain

$$\exp(\mathbf{A}+\mathbf{B}) = \sum_{i=0}^{\infty}\sum_{k=i}^{\infty} \frac{\mathbf{A}^{k-i}\mathbf{B}^{i}}{i!\,(k-i)!}.$$

By means of the abbreviation $l = k - i$ it yields

$$\exp(\mathbf{A}+\mathbf{B}) = \sum_{i=0}^{\infty}\sum_{l=0}^{\infty} \frac{\mathbf{A}^{l}\mathbf{B}^{i}}{i!l!}.$$

The same expression can alternatively be obtained by applying formally the Cauchy product of infinite series (see e.g. [24]). For the right hand side of (1.170) we finally get the same result as above:

$$\exp(\mathbf{A})\exp(\mathbf{B}) = \left(\sum_{l=0}^{\infty}\frac{\mathbf{A}^{l}}{l!}\right)\left(\sum_{i=0}^{\infty}\frac{\mathbf{B}^{i}}{i!}\right) = \sum_{l=0}^{\infty}\sum_{i=0}^{\infty}\frac{\mathbf{A}^{l}\mathbf{B}^{i}}{l!i!}.$$

1.30 Using the definition of the exponential tensor function (1.114) we get

$$\exp(\mathbf{0}) = \sum_{k=0}^{\infty}\frac{\mathbf{0}^{k}}{k!} = \mathbf{I}+\mathbf{0}+\mathbf{0}+\ldots = \mathbf{I},$$

$$\exp(\mathbf{I}) = \sum_{k=0}^{\infty}\frac{\mathbf{I}^{k}}{k!} = \sum_{k=0}^{\infty}\frac{\mathbf{I}}{k!} = \mathbf{I}\sum_{k=0}^{\infty}\frac{1}{k!} = \exp(1)\,\mathbf{I} = e\mathbf{I}.$$

1.31 Since the tensors $\mathbf{A}$ and $-\mathbf{A}$ commute we can write

$$\exp(\mathbf{A})\exp(-\mathbf{A}) = \exp(-\mathbf{A})\exp(\mathbf{A}) = \exp[\mathbf{A}+(-\mathbf{A})] = \exp(\mathbf{0}) = \mathbf{I}.$$

Accordingly

$$\exp(-\mathbf{A}) = \exp(\mathbf{A})^{-1}. \tag{9.3}$$

1.32 This identity can be proved by mathematical induction. Indeed, according to the results of Exercise 1.30 it holds for $k = 0$. Then, assuming that it is valid for some positive integer k we can write applying (1.170) (Exercise 1.29)

$$\begin{aligned}\exp[(k+1)\mathbf{A}] &= \exp(k\mathbf{A}+\mathbf{A}) = \exp(k\mathbf{A})\exp(\mathbf{A}) \\ &= [\exp(\mathbf{A})]^k \exp(\mathbf{A}) = [\exp(\mathbf{A})]^{k+1}.\end{aligned}$$

For negative integer k we proceed in a similar way using (9.3):

$$\begin{aligned}\exp[(k-1)\mathbf{A}] &= \exp(k\mathbf{A}-\mathbf{A}) = \exp(k\mathbf{A})\exp(-\mathbf{A}) \\ &= [\exp(\mathbf{A})]^k \exp(\mathbf{A})^{-1} = [\exp(\mathbf{A})]^{k-1}.\end{aligned}$$

1.33 Equations (1.114) and (9.2):

$$\begin{aligned}\exp(\mathbf{A}+\mathbf{B}) &= \sum_{k=0}^{\infty}\frac{(\mathbf{A}+\mathbf{B})^k}{k!} = \mathbf{I}+\sum_{k=1}^{\infty}\frac{(\mathbf{A}+\mathbf{B})^k}{k!} \\ &= \mathbf{I}+\sum_{k=1}^{\infty}\frac{\mathbf{A}^k+\mathbf{B}^k}{k!} = \exp(\mathbf{A})+\exp(\mathbf{B})-\mathbf{I}.\end{aligned}$$

1.34 Equations (1.114) and (1.135):

$$\begin{aligned}\exp\left(\mathbf{QAQ}^{\mathrm{T}}\right) &= \sum_{k=0}^{\infty}\frac{1}{k!}\left(\mathbf{QAQ}^{\mathrm{T}}\right)^k = \sum_{k=0}^{\infty}\frac{1}{k!}\underbrace{\mathbf{QAQ}^{\mathrm{T}}\mathbf{QAQ}^{\mathrm{T}}\ldots\mathbf{QAQ}^{\mathrm{T}}}_{k\ \text{times}} \\ &= \sum_{k=0}^{\infty}\frac{1}{k!}\mathbf{QA}^k\mathbf{Q}^{\mathrm{T}} = \mathbf{Q}\left(\sum_{k=0}^{\infty}\frac{1}{k!}\mathbf{A}^k\right)\mathbf{Q}^{\mathrm{T}} = \mathbf{Q}\exp(\mathbf{A})\mathbf{Q}^{\mathrm{T}}.\end{aligned}$$

1.35 We begin with the power of the tensor $\mathbf{D}$.

$$\begin{aligned}\mathbf{D}^2 = \mathbf{DD} &= \left(\mathrm{D}^{i}_{\cdot j}\boldsymbol{g}_i\otimes\boldsymbol{g}^j\right)\left(\mathrm{D}^{k}_{\cdot l}\boldsymbol{g}_k\otimes\boldsymbol{g}^l\right) \\ &= \mathrm{D}^{i}_{\cdot j}\mathrm{D}^{k}_{\cdot l}\delta^j_k\boldsymbol{g}_i\otimes\boldsymbol{g}^l = \mathrm{D}^{i}_{\cdot j}\mathrm{D}^{j}_{\cdot l}\boldsymbol{g}_i\otimes\boldsymbol{g}^l = \left(\mathbf{D}^2\right)^{i}_{\cdot j}\boldsymbol{g}_i\otimes\boldsymbol{g}^j,\end{aligned}$$

where $\left[\left(\mathbf{D}^2\right)^{i}_{\cdot j}\right] = \left[\mathrm{D}^{i}_{\cdot j}\right]\left[\mathrm{D}^{i}_{\cdot j}\right]$. Generalizing this results for an arbitrary integer exponent yields

$$\left[(\mathbf{D}^m)^{i}_{\cdot j} \right] = \underbrace{\left[\mathrm{D}^{i}_{\cdot j} \right] \ldots \left[\mathrm{D}^{i}_{\cdot j} \right]}_{m \text{ times}} = \begin{bmatrix} 2^m & 0 & 0 \\ 0 & 3^m & 0 \\ 0 & 0 & 1^m \end{bmatrix}.$$

We observe that the composition of tensors represented by mixed components related to the same mixed basis can be expressed in terms of the product of the component matrices. With this result in hand we thus obtain

$$\exp(\mathbf{D}) = \sum_{m=0}^{\infty} \frac{\mathbf{D}^m}{m!} = \exp(\mathbf{D})^{i}_{\cdot j}\, \boldsymbol{g}_i \otimes \boldsymbol{g}^j,$$

where

$$\left[\exp(\mathbf{D})^{i}_{\cdot j} \right] = \begin{bmatrix} \sum_{m=0}^{\infty} \frac{2^m}{m!} & 0 & 0 \\ 0 & \sum_{m=0}^{\infty} \frac{3^m}{m!} & 0 \\ 0 & 0 & \sum_{m=0}^{\infty} \frac{1^m}{m!} \end{bmatrix} = \begin{bmatrix} \mathrm{e}^2 & 0 & 0 \\ 0 & \mathrm{e}^3 & 0 \\ 0 & 0 & \mathrm{e} \end{bmatrix}.$$

For the powers of the tensor $\mathbf{E}$ we further obtain

$$\mathbf{E}^k = \mathbf{0}, \; k = 2, 3 \ldots$$

Hence,

$$\exp(\mathbf{E}) = \sum_{m=0}^{\infty} \frac{\mathbf{E}^m}{m!} = \mathbf{I} + \mathbf{E} + \mathbf{0} + \mathbf{0} + \ldots = \mathbf{I} + \mathbf{E},$$

so that

$$\left[\exp(\mathbf{E})^{i}_{\cdot j} \right] = \begin{bmatrix} 1 & 1 & 0 \\ 0 & 1 & 0 \\ 0 & 0 & 1 \end{bmatrix}.$$

To express the exponential of the tensor $\mathbf{F}$ we first decompose it by $\mathbf{F} = \mathbf{X} + \mathbf{Y}$, where

$$\left[\mathrm{X}^{i}_{\cdot j} \right] = \begin{bmatrix} 0 & 0 & 0 \\ 0 & 0 & 0 \\ 0 & 0 & 1 \end{bmatrix}, \quad \left[\mathrm{Y}^{i}_{\cdot j} \right] = \begin{bmatrix} 0 & 2 & 0 \\ 0 & 0 & 0 \\ 0 & 0 & 0 \end{bmatrix}.$$

$\mathbf{X}$ and $\mathbf{Y}$ are commutative since $\mathbf{XY} = \mathbf{YX} = \mathbf{0}$. Hence,

$$\exp(\mathbf{F}) = \exp(\mathbf{X} + \mathbf{Y}) = \exp(\mathbf{X}) \exp(\mathbf{Y}).$$

Noticing that $\mathbf{X}$ has the form of $\mathbf{D}$ and $\mathbf{Y}$ that of $\mathbf{E}$ we can write

$$\left[\exp(\mathbf{X})^{i}_{.j}\right] = \begin{bmatrix} 1 & 0 & 0 \\ 0 & 1 & 0 \\ 0 & 0 & e \end{bmatrix}, \quad \left[\exp(\mathbf{Y})^{i}_{.j}\right] = \begin{bmatrix} 1 & 2 & 0 \\ 0 & 1 & 0 \\ 0 & 0 & 1 \end{bmatrix}.$$

Finally, we obtain

$$\left[\exp(\mathbf{F})^{i}_{.j}\right] = \left[\exp(\mathbf{X})^{i}_{.j}\right]\left[\exp(\mathbf{Y})^{i}_{.j}\right] = \begin{bmatrix} 1 & 0 & 0 \\ 0 & 1 & 0 \\ 0 & 0 & e \end{bmatrix}\begin{bmatrix} 1 & 2 & 0 \\ 0 & 1 & 0 \\ 0 & 0 & 1 \end{bmatrix} = \begin{bmatrix} 1 & 2 & 0 \\ 0 & 1 & 0 \\ 0 & 0 & e \end{bmatrix}.$$

1.36 Equation (1.120): $(\mathbf{ABCD})^{\mathrm{T}} = (\mathbf{CD})^{\mathrm{T}}(\mathbf{AB})^{\mathrm{T}} = \mathbf{D}^{\mathrm{T}}\mathbf{C}^{\mathrm{T}}\mathbf{B}^{\mathrm{T}}\mathbf{A}^{\mathrm{T}}$.

1.37 Using the result of the previous Exercise we can write

$$(\underbrace{\mathbf{AA}\ldots\mathbf{A}}_{k \text{ times}})^{\mathrm{T}} = \underbrace{\mathbf{A}^{\mathrm{T}}\mathbf{A}^{\mathrm{T}}\ldots\mathbf{A}^{\mathrm{T}}}_{k \text{ times}} = \left(\mathbf{A}^{\mathrm{T}}\right)^{k}.$$

1.38 According to (1.124) and (1.125) $\mathrm{B}^{ij} = \mathrm{A}^{ji}$, $\mathrm{B}_{ij} = \mathrm{A}_{ji}$, $\mathrm{B}_{i.}^{\ j} = \mathrm{A}_{.i}^{j}$ and $\mathrm{B}^{i}_{.j} = \mathrm{A}_{j.}^{\ i}$ so that (see Exercise 1.25)

$$\left[\mathrm{B}^{ij}\right] = \left[\mathrm{A}^{ij}\right]^{\mathrm{T}} = \begin{bmatrix} 0 & 0 & 1 \\ -1 & 0 & 0 \\ 0 & 0 & 0 \end{bmatrix}, \quad \left[\mathrm{B}_{ij}\right] = \left[\mathrm{A}_{ij}\right]^{\mathrm{T}} = \begin{bmatrix} 6 & 7 & 24 \\ 0 & -3 & 9 \\ 20 & 24 & 78 \end{bmatrix},$$

$$\left[\mathrm{B}_{i.}^{\ j}\right] = \left[\mathrm{A}^{i}_{.j}\right]^{\mathrm{T}} = \begin{bmatrix} -3 & 0 & 2 \\ -9 & 0 & 3 \\ -8 & 0 & 6 \end{bmatrix}, \quad \left[\mathrm{B}^{i}_{.j}\right] = \left[\mathrm{A}_{i.}^{\ j}\right]^{\mathrm{T}} = \begin{bmatrix} 6 & 8 & 21 \\ -2 & -3 & -6 \\ 0 & 0 & 0 \end{bmatrix}.$$

1.39 Equations (1.120), (1.126) and (1.131):

$$\mathbf{I} = \mathbf{I}^{\mathrm{T}} = \left(\mathbf{AA}^{-1}\right)^{\mathrm{T}} = \left(\mathbf{A}^{-1}\right)^{\mathrm{T}}\mathbf{A}^{\mathrm{T}}.$$

1.40 $\left(\mathbf{A}^{k}\right)^{-1}$ is the tensor satisfying the identity $\left(\mathbf{A}^{k}\right)^{-1}\mathbf{A}^{k} = \mathbf{I}$. On the other hand, $\left(\mathbf{A}^{-1}\right)^{k}\mathbf{A}^{k} = \underbrace{\mathbf{A}^{-1}\mathbf{A}^{-1}\ldots\mathbf{A}^{-1}}_{k \text{ times}}\underbrace{\mathbf{AA}\ldots\mathbf{A}}_{k \text{ times}} = \mathbf{I}$. Thus, $\left(\mathbf{A}^{-1}\right)^{k} = \left(\mathbf{A}^{k}\right)^{-1}$.

1.41 An arbitrary tensor $\mathbf{A} \in \mathbf{Lin}^{n}$ can be represented with respect to a basis for example by $\mathbf{A} = \mathrm{A}^{ij}\boldsymbol{g}_i \otimes \boldsymbol{g}_j$. Thus, by virtue of (1.141) we obtain:

$$\begin{aligned}\boldsymbol{c} \otimes \boldsymbol{d} : \mathbf{A} &= \boldsymbol{c} \otimes \boldsymbol{d} : \left(\mathrm{A}^{ij}\boldsymbol{g}_i \otimes \boldsymbol{g}_j\right) = \mathrm{A}^{ij}\left(\boldsymbol{c} \cdot \boldsymbol{g}_i\right)\left(\boldsymbol{g}_j \cdot \boldsymbol{d}\right) \\ &= \boldsymbol{c}\left(\mathrm{A}^{ij}\boldsymbol{g}_i \otimes \boldsymbol{g}_j\right)\boldsymbol{d} = \boldsymbol{c}\mathbf{A}\boldsymbol{d} = \boldsymbol{d}\mathbf{A}^{\mathrm{T}}\boldsymbol{c}.\end{aligned}$$

1.42 The properties (D.1) and (D.3) directly follow from (1.141) and (1.143). Further, for three arbitrary tensors $\mathbf{A}$, $\mathbf{B} = \mathrm{B}^{ij}\boldsymbol{g}_i \otimes \boldsymbol{g}_j$ and $\mathbf{C} = \mathrm{C}^{ij}\boldsymbol{g}_i \otimes \boldsymbol{g}_j$ we have with the aid of (1.142)

$$\begin{aligned}\mathbf{A} : (\mathbf{B} + \mathbf{C}) &= \mathbf{A} : \left[\left(\mathrm{B}^{ij} + \mathrm{C}^{ij}\right)\left(\boldsymbol{g}_i \otimes \boldsymbol{g}_j\right)\right] = \left(\mathrm{B}^{ij} + \mathrm{C}^{ij}\right)\left(\boldsymbol{g}_i \mathbf{A} \boldsymbol{g}_j\right) \\ &= \mathrm{B}^{ij}\left(\boldsymbol{g}_i \mathbf{A} \boldsymbol{g}_j\right) + \mathrm{C}^{ij}\left(\boldsymbol{g}_i \mathbf{A} \boldsymbol{g}_j\right) \\ &= \mathbf{A} : \left(\mathrm{B}^{ij}\boldsymbol{g}_i \otimes \boldsymbol{g}_j\right) + \mathbf{A} : \left(\mathrm{C}^{ij}\boldsymbol{g}_i \otimes \boldsymbol{g}_j\right) = \mathbf{A} : \mathbf{B} + \mathbf{A} : \mathbf{C},\end{aligned}$$

which implies (D.2).

1.43 By virtue of (1.108), (1.89) and (1.142) we obtain

$$\begin{aligned}\left[(\boldsymbol{a} \otimes \boldsymbol{b})(\boldsymbol{c} \otimes \boldsymbol{d})\right] : \mathbf{I} &= \left[(\boldsymbol{b} \cdot \boldsymbol{c})(\boldsymbol{a} \otimes \boldsymbol{d})\right] : \mathbf{I} \\ &= (\boldsymbol{b} \cdot \boldsymbol{c})(\boldsymbol{a}\mathbf{I}\boldsymbol{d}) = (\boldsymbol{a} \cdot \boldsymbol{d})(\boldsymbol{b} \cdot \boldsymbol{c}).\end{aligned}$$

1.44 By virtue of (1.15), (1.25) and (1.149) we can write

$$\begin{aligned}\mathrm{tr}\mathbf{A} &= \mathrm{tr}\left(\mathrm{A}^{ij}\boldsymbol{g}_i \otimes \boldsymbol{g}_j\right) = \mathrm{A}^{ij}\left(\boldsymbol{g}_i \cdot \boldsymbol{g}_j\right) = \mathrm{A}^{ij} g_{ij} \\ &= \mathrm{tr}\left(\mathrm{A}_{ij}\boldsymbol{g}^i \otimes \boldsymbol{g}^j\right) = \mathrm{A}_{ij}\left(\boldsymbol{g}^i \cdot \boldsymbol{g}^j\right) = \mathrm{A}_{ij} g^{ij} \\ &= \mathrm{tr}\left(\mathrm{A}^i_{\cdot j}\boldsymbol{g}_i \otimes \boldsymbol{g}^j\right) = \mathrm{A}^i_{\cdot j}\left(\boldsymbol{g}_i \cdot \boldsymbol{g}^j\right) = \mathrm{A}^i_{\cdot j}\delta^j_i = \mathrm{A}^i_{\cdot i}.\end{aligned}$$

1.45 Using the results of Exercise 1.10 (c) and by means of (1.159) we obtain

$$\begin{aligned}\boldsymbol{w} &= g\left(\mathrm{W}^{32}\boldsymbol{g}^1 + \mathrm{W}^{13}\boldsymbol{g}^2 + \mathrm{W}^{21}\boldsymbol{g}^3\right) \\ &= -5\left[-(-\boldsymbol{a}_1 + 2\boldsymbol{a}_2) - 3\frac{1}{5}(\boldsymbol{a}_1 - \boldsymbol{a}_2 - 2\boldsymbol{a}_3) + \frac{1}{5}(2\boldsymbol{a}_1 - 2\boldsymbol{a}_2 + \boldsymbol{a}_3)\right] \\ &= -4\boldsymbol{a}_1 + 9\boldsymbol{a}_2 - 7\boldsymbol{a}_3.\end{aligned}$$

1.46 Equation (1.147): $\mathbf{M} : \mathbf{W} = \mathbf{M}^{\mathrm{T}} : \mathbf{W}^{\mathrm{T}} = \mathbf{M} : (-\mathbf{W}) = -(\mathbf{M} : \mathbf{W}) = 0$.

1.47 $\mathbf{W}^k$ is skew-symmetric for odd k. Indeed, $\left(\mathbf{W}^k\right)^{\mathrm{T}} = \left(\mathbf{W}^{\mathrm{T}}\right)^k = (-\mathbf{W})^k = (-1)^k\,\mathbf{W}^k = -\mathbf{W}^k$. Thus, using the result of the previous Exercise we can write: $\mathrm{tr}\mathbf{W}^k = \mathbf{W}^k : \mathbf{I} = 0$.

1.48 By means of the definition (1.153) we obtain

$$\begin{aligned}\mathrm{sym}\left(\mathrm{skew}\mathbf{A}\right) &= \frac{1}{2}\left[\mathrm{skew}\mathbf{A} + \left(\mathrm{skew}\mathbf{A}\right)^{\mathrm{T}}\right] \\ &= \frac{1}{2}\left[\frac{1}{2}\left(\mathbf{A} - \mathbf{A}^{\mathrm{T}}\right) + \frac{1}{2}\left(\mathbf{A} - \mathbf{A}^{\mathrm{T}}\right)^{\mathrm{T}}\right]\end{aligned}$$

$$= \frac{1}{2}\left[\frac{1}{2}\mathbf{A} - \frac{1}{2}\mathbf{A}^{\mathrm{T}} + \frac{1}{2}\mathbf{A}^{\mathrm{T}} - \frac{1}{2}\mathbf{A}\right] = \mathbf{0}.$$

The same procedure leads to the identity skew (sym$\mathbf{A}$) $= \mathbf{0}$.

1.49 On use of (1.163) we can write

$$\mathrm{sph}\,(\mathrm{dev}\mathbf{A}) = \mathrm{sph}\left[\mathbf{A} - \frac{1}{n}\mathrm{tr}\,(\mathbf{A})\,\mathbf{I}\right] = \frac{1}{n}\mathrm{tr}\left[\mathbf{A} - \frac{1}{n}\mathrm{tr}\,(\mathbf{A})\,\mathbf{I}\right]\mathbf{I} = \mathbf{0},$$

where we take into account that tr$\mathbf{I} = n$. In the same way, one proves that dev (sph$\mathbf{A}$) $= \mathbf{0}$.

9.2 Exercises of Chap. 2

2.1 The tangent vectors take the form:

$$\begin{aligned}
\boldsymbol{g}_1 &= \frac{\partial \boldsymbol{r}}{\partial \varphi} = r\cos\varphi\sin\phi\,\boldsymbol{e}_1 - r\sin\varphi\sin\phi\,\boldsymbol{e}_3,\\
\boldsymbol{g}_2 &= \frac{\partial \boldsymbol{r}}{\partial \phi} = r\sin\varphi\cos\phi\,\boldsymbol{e}_1 - r\sin\phi\,\boldsymbol{e}_2 + r\cos\varphi\cos\phi\,\boldsymbol{e}_3,\\
\boldsymbol{g}_3 &= \frac{\partial \boldsymbol{r}}{\partial r} = \sin\varphi\sin\phi\,\boldsymbol{e}_1 + \cos\phi\,\boldsymbol{e}_2 + \cos\varphi\sin\phi\,\boldsymbol{e}_3.
\end{aligned} \tag{9.4}$$

For the metrics coefficients we can further write:

$$\begin{aligned}
\boldsymbol{g}_1\cdot\boldsymbol{g}_1 = &\left(r\cos\varphi\sin\phi\,\boldsymbol{e}_1 - r\sin\varphi\sin\phi\,\boldsymbol{e}_3\right)\\
&\cdot\left(r\cos\varphi\sin\phi\,\boldsymbol{e}_1 - r\sin\varphi\sin\phi\,\boldsymbol{e}_3\right) = r^2\sin^2\phi,
\end{aligned}$$

$$\begin{aligned}
\boldsymbol{g}_1\cdot\boldsymbol{g}_2 = &\left(r\cos\varphi\sin\phi\,\boldsymbol{e}_1 - r\sin\varphi\sin\phi\,\boldsymbol{e}_3\right)\\
&\cdot\left(r\sin\varphi\cos\phi\,\boldsymbol{e}_1 - r\sin\phi\,\boldsymbol{e}_2 + r\cos\varphi\cos\phi\,\boldsymbol{e}_3\right)\\
= &\ r^2\left(\sin\varphi\cos\varphi\sin\phi\cos\phi - \sin\varphi\cos\varphi\sin\phi\cos\phi\right) = 0,
\end{aligned}$$

$$\begin{aligned}
\boldsymbol{g}_1\cdot\boldsymbol{g}_3 = &\left(r\cos\varphi\sin\phi\,\boldsymbol{e}_1 - r\sin\varphi\sin\phi\,\boldsymbol{e}_3\right)\\
&\cdot\left(\sin\varphi\sin\phi\,\boldsymbol{e}_1 + \cos\phi\,\boldsymbol{e}_2 + \cos\varphi\sin\phi\,\boldsymbol{e}_3\right)\\
= &\ r\left(\sin\varphi\cos\varphi\sin^2\phi - \sin\varphi\cos\varphi\sin^2\phi\right) = 0,
\end{aligned}$$

$$\begin{aligned}
\boldsymbol{g}_2\cdot\boldsymbol{g}_2 = &\left(r\sin\varphi\cos\phi\,\boldsymbol{e}_1 - r\sin\phi\,\boldsymbol{e}_2 + r\cos\varphi\cos\phi\,\boldsymbol{e}_3\right)\\
&\cdot\left(r\sin\varphi\cos\phi\,\boldsymbol{e}_1 - r\sin\phi\,\boldsymbol{e}_2 + r\cos\varphi\cos\phi\,\boldsymbol{e}_3\right)\\
= &\ r^2\left(\sin^2\varphi\cos^2\phi + \sin^2\phi + \cos^2\varphi\cos^2\phi\right) = r^2,
\end{aligned}$$

$$\begin{aligned} \boldsymbol{g}_2 \cdot \boldsymbol{g}_3 = &\ (r \sin\varphi \cos\phi \boldsymbol{e}_1 - r \sin\phi \boldsymbol{e}_2 + r \cos\varphi \cos\phi \boldsymbol{e}_3) \\ &\cdot (\sin\varphi \sin\phi \boldsymbol{e}_1 + \cos\phi \boldsymbol{e}_2 + \cos\varphi \sin\phi \boldsymbol{e}_3) \\ = &\ r \left(\sin^2\varphi \sin\phi \cos\phi - \sin\phi \cos\phi + \cos^2\varphi \sin\phi \cos\phi\right) = 0, \end{aligned}$$

$$\begin{aligned} \boldsymbol{g}_3 \cdot \boldsymbol{g}_3 = &\ (\sin\varphi \sin\phi \boldsymbol{e}_1 + \cos\phi \boldsymbol{e}_2 + \cos\varphi \sin\phi \boldsymbol{e}_3) \\ &\cdot (\sin\varphi \sin\phi \boldsymbol{e}_1 + \cos\phi \boldsymbol{e}_2 + \cos\varphi \sin\phi \boldsymbol{e}_3) \\ = &\ \sin^2\varphi \sin^2\phi + \cos^2\phi + \cos^2\varphi \sin^2\phi = 1. \end{aligned}$$

Thus,

$$\left[g_{ij}\right] = \left[\boldsymbol{g}_i \cdot \boldsymbol{g}_j\right] = \begin{bmatrix} r^2 \sin^2\phi & 0 & 0 \\ 0 & r^2 & 0 \\ 0 & 0 & 1 \end{bmatrix}$$

and consequently

$$\left[g^{ij}\right] = \left[g_{ij}\right]^{-1} = \begin{bmatrix} \dfrac{1}{r^2 \sin^2\phi} & 0 & 0 \\ 0 & \dfrac{1}{r^2} & 0 \\ 0 & 0 & 1 \end{bmatrix}. \tag{9.5}$$

Finally, we calculate the dual basis by $(1.21)_1$:

$$\boldsymbol{g}^1 = \frac{1}{r^2 \sin^2\phi} \boldsymbol{g}_1 = r^{-1} \frac{\cos\varphi}{\sin\phi} \boldsymbol{e}_1 - r^{-1} \frac{\sin\varphi}{\sin\phi} \boldsymbol{e}_3,$$

$$\boldsymbol{g}^2 = \frac{1}{r^2} \boldsymbol{g}_2 = r^{-1} \sin\varphi \cos\phi \boldsymbol{e}_1 - r^{-1} \sin\phi \boldsymbol{e}_2 + r^{-1} \cos\varphi \cos\phi \boldsymbol{e}_3,$$

$$\boldsymbol{g}^3 = \boldsymbol{g}_3 = \sin\varphi \sin\phi \boldsymbol{e}_1 + \cos\phi \boldsymbol{e}_2 + \cos\varphi \sin\phi \boldsymbol{e}_3.$$

2.2 The connection between the linear and spherical coordinates (2.157) can be expressed by

$$x^1 = r \sin\varphi \sin\phi, \quad x^2 = r \cos\phi, \quad x^3 = r \cos\varphi \sin\phi.$$

Thus, we obtain

$$\begin{aligned} &\frac{\partial x^1}{\partial \varphi} = r \cos\varphi \sin\phi, && \frac{\partial x^1}{\partial \phi} = r \sin\varphi \cos\phi, && \frac{\partial x^1}{\partial r} = \sin\varphi \sin\phi, \\ &\frac{\partial x^2}{\partial \varphi} = 0, && \frac{\partial x^2}{\partial \phi} = -r \sin\phi, && \frac{\partial x^2}{\partial r} = \cos\phi, \\ &\frac{\partial x^3}{\partial \varphi} = -r \sin\varphi \sin\phi, && \frac{\partial x^3}{\partial \phi} = r \cos\varphi \cos\phi, && \frac{\partial x^3}{\partial r} = \cos\varphi \sin\phi. \end{aligned}$$

Inverting the so-constructed matrix $\left[\dfrac{\partial x^i}{\partial \varphi}\ \dfrac{\partial x^i}{\partial \phi}\ \dfrac{\partial x^i}{\partial r}\right]$ further yields according to (2.23)

$$\frac{\partial \varphi}{\partial x^1} = \frac{\cos\varphi}{r\sin\phi}, \qquad \frac{\partial \varphi}{\partial x^2} = 0, \qquad \frac{\partial \varphi}{\partial x^3} = -\frac{\sin\varphi}{r\sin\phi},$$

$$\frac{\partial \phi}{\partial x^1} = \frac{\sin\varphi\cos\phi}{r}, \qquad \frac{\partial \phi}{\partial x^2} = -\frac{\sin\phi}{r}, \qquad \frac{\partial \phi}{\partial x^3} = \frac{\cos\varphi\cos\phi}{r},$$

$$\frac{\partial r}{\partial x^1} = \sin\varphi\sin\phi, \qquad \frac{\partial r}{\partial x^2} = \cos\phi, \qquad \frac{\partial r}{\partial x^3} = \cos\varphi\sin\phi.$$

2.3 Applying the directional derivative we have

$$(\text{a}):\ \frac{\mathrm{d}}{\mathrm{d}s}\left\|\boldsymbol{r}+s\boldsymbol{a}\right\|^{-1}\Big|_{s=0} = \frac{\mathrm{d}}{\mathrm{d}s}\left[(\boldsymbol{r}+s\boldsymbol{a})\cdot(\boldsymbol{r}+s\boldsymbol{a})\right]^{-1/2}\Big|_{s=0}$$

$$= \frac{\mathrm{d}}{\mathrm{d}s}\left[\boldsymbol{r}\cdot\boldsymbol{r}+2s\boldsymbol{r}\cdot\boldsymbol{a}+s^2\boldsymbol{a}\cdot\boldsymbol{a}\right]^{-1/2}\Big|_{s=0}$$

$$= -\frac{1}{2}\frac{2\boldsymbol{r}\cdot\boldsymbol{a}+2s\boldsymbol{a}\cdot\boldsymbol{a}}{\left[(\boldsymbol{r}+s\boldsymbol{a})\cdot(\boldsymbol{r}+s\boldsymbol{a})\right]^{3/2}}\Big|_{s=0} = -\frac{\boldsymbol{r}\cdot\boldsymbol{a}}{\|\boldsymbol{r}\|^3}.$$

Comparing with (2.54) finally yields

$$\operatorname{grad}\|\boldsymbol{r}\|^{-1} = -\frac{\boldsymbol{r}}{\|\boldsymbol{r}\|^3}.$$

$$(\text{b}):\ \frac{\mathrm{d}}{\mathrm{d}s}(\boldsymbol{r}+s\boldsymbol{a})\cdot\boldsymbol{w}\Big|_{s=0} = \frac{\mathrm{d}}{\mathrm{d}s}(\boldsymbol{r}\cdot\boldsymbol{w}+s\boldsymbol{a}\cdot\boldsymbol{w})\Big|_{s=0} = \boldsymbol{a}\cdot\boldsymbol{w}.$$

Hence, $\operatorname{grad}(\boldsymbol{r}\cdot\boldsymbol{w}) = \boldsymbol{w}$.

$$(\text{c}):\ \frac{\mathrm{d}}{\mathrm{d}s}(\boldsymbol{r}+s\boldsymbol{a})\,\mathbf{A}\,(\boldsymbol{r}+s\boldsymbol{a})\Big|_{s=0} = \frac{\mathrm{d}}{\mathrm{d}s}\left(\boldsymbol{r}\mathbf{A}\boldsymbol{r}+s\boldsymbol{a}\mathbf{A}\boldsymbol{r}+s\boldsymbol{r}\mathbf{A}\boldsymbol{a}+s^2\boldsymbol{a}\mathbf{A}\boldsymbol{a}\right)\Big|_{s=0}$$

$$= \boldsymbol{a}\mathbf{A}\boldsymbol{r}+\boldsymbol{r}\mathbf{A}\boldsymbol{a} = (\mathbf{A}\boldsymbol{r})\cdot\boldsymbol{a}+(\boldsymbol{r}\mathbf{A})\cdot\boldsymbol{a} = (\mathbf{A}\boldsymbol{r}+\boldsymbol{r}\mathbf{A})\cdot\boldsymbol{a},$$

Thus, applying (1.115) and $(1.153)_1$ we can write

$$\operatorname{grad}(\boldsymbol{r}\mathbf{A}\boldsymbol{r}) = \mathbf{A}\boldsymbol{r}+\boldsymbol{r}\mathbf{A} = \left(\mathbf{A}+\mathbf{A}^{\mathrm{T}}\right)\boldsymbol{r} = 2\,(\operatorname{sym}\mathbf{A})\,\boldsymbol{r}.$$

$$(\text{d}):\ \frac{\mathrm{d}}{\mathrm{d}s}\mathbf{A}\,(\boldsymbol{r}+s\boldsymbol{a})\Big|_{s=0} = \frac{\mathrm{d}}{\mathrm{d}s}(\mathbf{A}\boldsymbol{r}+s\mathbf{A}\boldsymbol{a})\Big|_{s=0} = \mathbf{A}\boldsymbol{a}.$$

Comparing with (2.57) we then have

$$\operatorname{grad}(\mathbf{A}\boldsymbol{r}) = \mathbf{A}.$$

(e): In view of (1.65) and using the results of (d) we obtain

$$\operatorname{grad}(\boldsymbol{w} \times \boldsymbol{r}) = \operatorname{grad}(\mathbf{W}\boldsymbol{r}) = \mathbf{W}.$$

With the aid of the representation $\boldsymbol{w} = w^i \boldsymbol{g}_i$ we can further write (see Exercise 1.23)

$$\mathbf{W} = \mathrm{W}_{ij}\boldsymbol{g}^i \otimes \boldsymbol{g}^j, \quad \left[\mathrm{W}_{ij}\right] = g \begin{bmatrix} 0 & -w^3 & w^2 \\ w^3 & 0 & -w^1 \\ -w^2 & w^1 & 0 \end{bmatrix}.$$

2.4 We begin with the derivative of the metrics coefficients obtained in Exercise 2.1:

$$\left[g_{ij},_1\right] = \left[\frac{\partial g_{ij}}{\partial \varphi}\right] = \begin{bmatrix} 0 & 0 & 0 \\ 0 & 0 & 0 \\ 0 & 0 & 0 \end{bmatrix}, \quad \left[g_{ij},_2\right] = \left[\frac{\partial g_{ij}}{\partial \phi}\right] = \begin{bmatrix} 2r^2 \sin\phi\cos\phi & 0 & 0 \\ 0 & 0 & 0 \\ 0 & 0 & 0 \end{bmatrix},$$

$$\left[g_{ij},_3\right] = \left[\frac{\partial g_{ij}}{\partial r}\right] = \begin{bmatrix} 2r\sin^2\phi & 0 & 0 \\ 0 & 2r & 0 \\ 0 & 0 & 0 \end{bmatrix}.$$

Thus, according to (2.84)

$$\left[\Gamma_{ij1}\right] = \left[\frac{1}{2}\left(g_{1i},_j + g_{1j},_i - g_{ij},_1\right)\right]$$

$$= \begin{bmatrix} 0 & r^2\sin\phi\cos\phi & r\sin^2\phi \\ r^2\sin\phi\cos\phi & 0 & 0 \\ r\sin^2\phi & 0 & 0 \end{bmatrix},$$

$$\left[\Gamma_{ij2}\right] = \left[\frac{1}{2}\left(g_{2i},_j + g_{2j},_i - g_{ij},_2\right)\right] = \begin{bmatrix} -r^2\sin\phi\cos\phi & 0 & 0 \\ 0 & 0 & r \\ 0 & r & 0 \end{bmatrix},$$

$$\left[\Gamma_{ij3}\right] = \left[\frac{1}{2}\left(g_{3i},_j + g_{3j},_i - g_{ij},_3\right)\right] = \begin{bmatrix} -r\sin^2\phi & 0 & 0 \\ 0 & -r & 0 \\ 0 & 0 & 0 \end{bmatrix}.$$

With the aid of (2.77) we further obtain

$$\Gamma^1_{ij} = g^{1l}\Gamma_{ijl} = g^{11}\Gamma_{ij1} + g^{12}\Gamma_{ij2} + g^{13}\Gamma_{ij3} = \frac{\Gamma_{ij1}}{r^2 \sin^2\phi}, \quad i,j = 1,2,3,$$

$$\left[\Gamma^1_{ij}\right] = \begin{bmatrix} 0 & \cot\phi & r^{-1} \\ \cot\phi & 0 & 0 \\ r^{-1} & 0 & 0 \end{bmatrix}, \tag{9.6}$$

$$\Gamma^2_{ij} = g^{2l}\Gamma_{ijl} = g^{21}\Gamma_{ij1} + g^{22}\Gamma_{ij2} + g^{23}\Gamma_{ij3} = \frac{\Gamma_{ij2}}{r^2}, \quad i,j = 1,2,3,$$

$$\left[\Gamma^2_{ij}\right] = \begin{bmatrix} -\sin\phi\cos\phi & 0 & 0 \\ 0 & 0 & r^{-1} \\ 0 & r^{-1} & 0 \end{bmatrix}, \tag{9.7}$$

$$\Gamma^3_{ij} = g^{3l}\Gamma_{ijl} = g^{31}\Gamma_{ij1} + g^{32}\Gamma_{ij2} + g^{33}\Gamma_{ij3} = \Gamma_{ij3}, \quad i,j = 1,2,3,$$

$$\left[\Gamma^3_{ij}\right] = \begin{bmatrix} -r\sin^2\phi & 0 & 0 \\ 0 & -r & 0 \\ 0 & 0 & 0 \end{bmatrix}. \tag{9.8}$$

2.5 Relations (2.96) can be obtained in the same manner as (2.94). Indeed, using the representation $\mathbf{A} = \mathrm{A}_{ij}\, \boldsymbol{g}^i \otimes \boldsymbol{g}^j$ and by virtue of (2.82) we have for example for $(2.96)_1$:

$$\begin{aligned} \mathbf{A},_k &= \left(\mathrm{A}_{ij}\, \boldsymbol{g}^i \otimes \boldsymbol{g}^j\right),_k \\ &= \mathrm{A}_{ij},_k\, \boldsymbol{g}^i \otimes \boldsymbol{g}^j + \mathrm{A}_{ij}\, \boldsymbol{g}^i,_k \otimes \boldsymbol{g}^j + \mathrm{A}_{ij}\, \boldsymbol{g}^i \otimes \boldsymbol{g}^j,_k \\ &= \mathrm{A}_{ij},_k\, \boldsymbol{g}^i \otimes \boldsymbol{g}^j + \mathrm{A}_{ij}\left(-\Gamma^i_{lk}\boldsymbol{g}^l\right) \otimes \boldsymbol{g}^j + \mathrm{A}_{ij}\, \boldsymbol{g}^i \otimes \left(-\Gamma^j_{lk}\boldsymbol{g}^l\right) \\ &= \left(\mathrm{A}_{ij},_k - \mathrm{A}_{lj}\Gamma^l_{ik} - \mathrm{A}_{il}\Gamma^l_{jk}\right) \boldsymbol{g}^i \otimes \boldsymbol{g}^j. \end{aligned}$$

2.6 Equations (1.91) and $(2.72)_2$:

$$\mathbf{0} = \mathbf{I},_k = \left(g_{ij}\, \boldsymbol{g}^i \otimes \boldsymbol{g}^j\right),_k = g_{ij}|_k\, \boldsymbol{g}^i \otimes \boldsymbol{g}^j = \left(g^{ij}\, \boldsymbol{g}_i \otimes \boldsymbol{g}_j\right),_k = g^{ij}|_k\, \boldsymbol{g}_i \otimes \boldsymbol{g}_j.$$

2.7 Using $(2.96)_1$ we write for example for the left hand side of (2.101)

$$\mathrm{A}_{ij}|_k = \mathrm{A}_{ij},_k - \mathrm{A}_{lj}\Gamma^l_{ik} - \mathrm{A}_{il}\Gamma^l_{jk}.$$

In view of $(2.93)_2$ the same result holds for the right hand side of (2.101) as well. Indeed,

$$a_i|_k\, b_j + a_i b_j|_k = \left(a_i{,_k} - a_l \Gamma^l_{ik}\right) b_j + a_i \left(b_j{,_k} - b_l \Gamma^l_{jk}\right)$$

$$= a_i{,_k}\, b_j + a_i b_j{,_k} - a_l b_j \Gamma^l_{ik} - a_i b_l \Gamma^l_{jk}$$

$$= \mathrm{A}_{ij}{,_k} - \mathrm{A}_{lj}\Gamma^l_{ik} - \mathrm{A}_{il}\Gamma^l_{jk}.$$

2.8 By analogy with (9.1)

$$\hat{\boldsymbol{t}} = e^{ijk} g^{-1} t_i \boldsymbol{g}_k \otimes \boldsymbol{g}_j.$$

Inserting this expression into (2.125) and taking (2.112) into account we further write

$$\mathrm{curl}\boldsymbol{t} = -\mathrm{div}\hat{\boldsymbol{t}} = -\left(e^{ijk} g^{-1} t_i \boldsymbol{g}_k \otimes \boldsymbol{g}_j\right){,_l}\, \boldsymbol{g}^l.$$

With the aid of the identities $\left(g^{-1}\boldsymbol{g}_j\right){,_l} \cdot \boldsymbol{g}^l = 0\ (j = 1, 2, 3)$ following from (2.76) and (2.107) and applying the product rule of differentiation we finally obtain

$$\mathrm{curl}\boldsymbol{t} = -e^{ijk} g^{-1} t_i{,_j}\, \boldsymbol{g}_k - e^{ijk} g^{-1} t_i \boldsymbol{g}_k{,_j}$$

$$= -e^{ijk} g^{-1} t_i{,_j}\, \boldsymbol{g}_k = -e^{ijk} g^{-1} t_i|_j\, \boldsymbol{g}_k = e^{jik} g^{-1} t_i|_j\, \boldsymbol{g}_k$$

keeping (1.36), (2.78) and $(2.93)_2$ in mind.

2.9 We begin with the covariant derivative of the Cauchy stress components (2.118). Using the results of Exercise 2.4 concerning the Christoffel symbols for the spherical coordinates we get

$$\sigma^{1j}|_j = \sigma^{1j}{,_j} + \sigma^{lj}\Gamma^1_{lj} + \sigma^{1l}\Gamma^j_{lj} = \sigma^{11}{,_1} + \sigma^{12}{,_2} + \sigma^{13}{,_3} + 3\sigma^{12}\cot\phi + 4\frac{\sigma^{13}}{r},$$

$$\sigma^{2j}|_j = \sigma^{2j}{,_j} + \sigma^{lj}\Gamma^2_{lj} + \sigma^{2l}\Gamma^j_{lj}$$

$$= \sigma^{21}{,_1} + \sigma^{22}{,_2} + \sigma^{23}{,_3} - \sigma^{11}\sin\phi\cos\phi + \sigma^{22}\cot\phi + 4\frac{\sigma^{23}}{r},$$

$$\sigma^{3j}|_j = \sigma^{3j}{,_j} + \sigma^{lj}\Gamma^3_{lj} + \sigma^{3l}\Gamma^j_{lj}$$

$$= \sigma^{31}{,_1} + \sigma^{32}{,_2} + \sigma^{33}{,_3} - \sigma^{11} r\sin^2\phi - \sigma^{22} r + \sigma^{32}\cot\phi + 2\frac{\sigma^{33}}{r}.$$

The balance equations (2.116) take thus the form

$$\rho a^1 = \sigma^{11}{}_{,1} + \sigma^{12}{}_{,2} + \sigma^{13}{}_{,3} + 3\sigma^{12}\cot\phi + 4\frac{\sigma^{13}}{r} + f^1,$$

$$\rho a^2 = \sigma^{21}{}_{,1} + \sigma^{22}{}_{,2} + \sigma^{23}{}_{,3} - \sigma^{11}\sin\phi\cos\phi + \sigma^{22}\cot\phi + 4\frac{\sigma^{23}}{r} + f^2,$$

$$\rho a^3 = \sigma^{31}{}_{,1} + \sigma^{32}{}_{,2} + \sigma^{33}{}_{,3} - \sigma^{11} r\sin^2\phi - \sigma^{22} r + \sigma^{32}\cot\phi + 2\frac{\sigma^{33}}{r} + f^3.$$

2.10 The tangent vectors take the form:

$$\boldsymbol{g}_1 = \frac{\partial \boldsymbol{r}}{\partial r} = \left(\cos\frac{s}{r} + \frac{s}{r}\sin\frac{s}{r}\right)\boldsymbol{e}_1 + \left(\sin\frac{s}{r} - \frac{s}{r}\cos\frac{s}{r}\right)\boldsymbol{e}_2,$$

$$\boldsymbol{g}_2 = \frac{\partial \boldsymbol{r}}{\partial s} = -\sin\frac{s}{r}\boldsymbol{e}_1 + \cos\frac{s}{r}\boldsymbol{e}_2, \quad \boldsymbol{g}_3 = \frac{\partial \boldsymbol{r}}{\partial z} = \boldsymbol{e}_3.$$

The metrics coefficients can further be written by

$$[g_{ij}] = [\boldsymbol{g}_i \cdot \boldsymbol{g}_j] = \begin{bmatrix} 1 + \frac{s^2}{r^2} & -\frac{s}{r} & 0 \\ -\frac{s}{r} & 1 & 0 \\ 0 & 0 & 1 \end{bmatrix}, \quad [g^{ij}] = [g_{ij}]^{-1} = \begin{bmatrix} 1 & \frac{s}{r} & 0 \\ \frac{s}{r} & 1 + \frac{s^2}{r^2} & 0 \\ 0 & 0 & 1 \end{bmatrix}.$$

For the dual basis we use $(1.21)_1$:

$$\boldsymbol{g}^1 = \boldsymbol{g}_1 + \frac{s}{r}\boldsymbol{g}_2 = \cos\frac{s}{r}\boldsymbol{e}_1 + \sin\frac{s}{r}\boldsymbol{e}_2,$$

$$\boldsymbol{g}^2 = \frac{s}{r}\boldsymbol{g}_1 + \left(1 + \frac{s^2}{r^2}\right)\boldsymbol{g}_2$$

$$= \left(-\sin\frac{s}{r} + \frac{s}{r}\cos\frac{s}{r}\right)\boldsymbol{e}_1 + \left(\cos\frac{s}{r} + \frac{s}{r}\sin\frac{s}{r}\right)\boldsymbol{e}_2,$$

$$\boldsymbol{g}_3 = \boldsymbol{g}^3 = \boldsymbol{e}_3.$$

The derivatives of the metrics coefficients become

$$[g_{ij},_1] = \begin{bmatrix} -2\frac{s^2}{r^3} & \frac{s}{r^2} & 0 \\ \frac{s}{r^2} & 0 & 0 \\ 0 & 0 & 0 \end{bmatrix}, \quad [g_{ij},_2] = \begin{bmatrix} \frac{2s}{r^2} & -\frac{1}{r} & 0 \\ -\frac{1}{r} & 0 & 0 \\ 0 & 0 & 0 \end{bmatrix}, \quad [g_{ij},_3] = \begin{bmatrix} 0 & 0 & 0 \\ 0 & 0 & 0 \\ 0 & 0 & 0 \end{bmatrix}.$$

For the Christoffel symbols we thus obtain by means of (2.84) and (2.77):

$$\left[\Gamma_{ij1}\right] = \begin{bmatrix} -\frac{s^2}{r^3} & \frac{s}{r^2} & 0 \\ \frac{s}{r^2} & -\frac{1}{r} & 0 \\ 0 & 0 & 0 \end{bmatrix}, \quad \left[\Gamma_{ij2}\right] = \begin{bmatrix} 0&0&0 \\ 0&0&0 \\ 0&0&0 \end{bmatrix}, \quad \left[\Gamma_{ij3}\right] = \begin{bmatrix} 0&0&0 \\ 0&0&0 \\ 0&0&0 \end{bmatrix},$$

$$\left[\Gamma^1_{ij}\right] = \begin{bmatrix} -\frac{s^2}{r^3} & \frac{s}{r^2} & 0 \\ \frac{s}{r^2} & -\frac{1}{r} & 0 \\ 0 & 0 & 0 \end{bmatrix}, \quad \left[\Gamma^2_{ij}\right] = \begin{bmatrix} -\frac{s^3}{r^4} & \frac{s^2}{r^3} & 0 \\ \frac{s^2}{r^3} & -\frac{s}{r^2} & 0 \\ 0 & 0 & 0 \end{bmatrix}, \quad \left[\Gamma^3_{ij}\right] = \begin{bmatrix} 0&0&0 \\ 0&0&0 \\ 0&0&0 \end{bmatrix}.$$

2.11 First, we express the covariant derivative of the Cauchy stress components by (2.118) using the results of the previous exercise:

$$\begin{aligned} \sigma^{1j}|_j &= \sigma^{11},_r + \sigma^{12},_s + \sigma^{13},_z - \sigma^{11}\frac{s^2}{r^3} - \frac{\sigma^{22}}{r} + 2\sigma^{12}\frac{s}{r^2}, \\ \Sigma^{2j}|_j &= \sigma^{21},_r + \sigma^{22},_s + \sigma^{23},_z - \sigma^{11}\frac{s^3}{r^4} - \sigma^{22}\frac{s}{r^2} + 2\sigma^{12}\frac{s^2}{r^3}, \\ \sigma^{3j}|_j &= \sigma^{31},_r + \sigma^{32},_s + \sigma^{33},_z . \end{aligned}$$

The balance equations (2.116) become

$$\begin{aligned} \rho a^1 &= \sigma^{11},_r + \sigma^{12},_s + \sigma^{13},_z - \sigma^{11}\frac{s^2}{r^3} - \frac{\sigma^{22}}{r} + 2\sigma^{12}\frac{s}{r^2} + f^1, \\ \rho a^2 &= \sigma^{21},_r + \sigma^{22},_s + \sigma^{23},_z - \sigma^{11}\frac{s^3}{r^4} - \sigma^{22}\frac{s}{r^2} + 2\sigma^{12}\frac{s^2}{r^3} + f^2, \\ \rho a^3 &= \sigma^{31},_r + \sigma^{32},_s + \sigma^{33},_z + f^3. \end{aligned}$$

2.12 Equations (2.126), (2.128), (1.32) and (2.82):

$$\begin{aligned} \operatorname{div}\operatorname{curl}\boldsymbol{t} &= \left(\boldsymbol{g}^i \times \boldsymbol{t},_i\right),_j \cdot \boldsymbol{g}^j = \left(-\Gamma^i_{kj}\boldsymbol{g}^k \times \boldsymbol{t},_i + \boldsymbol{g}^i \times \boldsymbol{t},_{ij}\right) \cdot \boldsymbol{g}^j \\ &= -\left(\Gamma^i_{kj}\boldsymbol{g}^j \times \boldsymbol{g}^k\right) \cdot \boldsymbol{t},_i + \left(\boldsymbol{g}^j \times \boldsymbol{g}^i\right) \cdot \boldsymbol{t},_{ij} = 0, \end{aligned}$$

where we take into consideration that $\boldsymbol{t},_{ij} = \boldsymbol{t},_{ji}$, $\Gamma^l_{ij} = \Gamma^l_{ji}$ and $\boldsymbol{g}^i \times \boldsymbol{g}^j = -\boldsymbol{g}^j \times \boldsymbol{g}^i$ $(i \neq j,\ i, j = 1, 2, 3)$.

Equations (2.126), (2.128) and (1.32):

$$\operatorname{div}(\boldsymbol{u}\times\boldsymbol{v}) = (\boldsymbol{u}\times\boldsymbol{v})_{,i}\cdot\boldsymbol{g}^i = (\boldsymbol{u}_{,i}\times\boldsymbol{v}+\boldsymbol{u}\times\boldsymbol{v}_{,i})\cdot\boldsymbol{g}^i$$

$$= \left(\boldsymbol{g}^i\times\boldsymbol{u}_{,i}\right)\cdot\boldsymbol{v} + \left(\boldsymbol{v}_{,i}\times\boldsymbol{g}^i\right)\cdot\boldsymbol{u} = \boldsymbol{v}\cdot\operatorname{curl}\boldsymbol{u} - \boldsymbol{u}\cdot\operatorname{curl}\boldsymbol{v}.$$

Equations (2.6), $(2.75)_1$ and (2.126):

$$\operatorname{grad}\operatorname{div}\boldsymbol{t} = \left(\boldsymbol{t}_{,i}\cdot\boldsymbol{g}^i\right)_{,j}\boldsymbol{g}^j = \left(\boldsymbol{t}_{,ij}\cdot\boldsymbol{g}^i\right)\boldsymbol{g}^j + \left(\boldsymbol{t}_{,i}\cdot\boldsymbol{g}^i{}_{,j}\right)\boldsymbol{g}^j.$$

Using the relation

$$\left(\boldsymbol{t}_{,i}\cdot\boldsymbol{g}^i{}_{,j}\right)\boldsymbol{g}^j = \left[\boldsymbol{t}_{,i}\cdot\left(-\Gamma^i_{jk}\boldsymbol{g}^k\right)\right]\boldsymbol{g}^j$$

$$= \left(\boldsymbol{t}_{,i}\cdot\boldsymbol{g}^k\right)\left(-\Gamma^i_{jk}\boldsymbol{g}^j\right) = \left(\boldsymbol{t}_{,i}\cdot\boldsymbol{g}^k\right)\boldsymbol{g}^i{}_{,k} = \left(\boldsymbol{t}_{,i}\cdot\boldsymbol{g}^j\right)\boldsymbol{g}^i{}_{,j} \qquad (9.9)$$

following from (2.82) we thus write

$$\operatorname{grad}\operatorname{div}\boldsymbol{t} = \left(\boldsymbol{t}_{,ij}\cdot\boldsymbol{g}^i\right)\boldsymbol{g}^j + \left(\boldsymbol{t}_{,i}\cdot\boldsymbol{g}^j\right)\boldsymbol{g}^i{}_{,j}.$$

Equations (2.128) and (1.168):

$$\operatorname{curl}\operatorname{curl}\boldsymbol{t} = \boldsymbol{g}^j\times\left(\boldsymbol{g}^i\times\boldsymbol{t}_{,i}\right)_{,j} = \boldsymbol{g}^j\times\left(\boldsymbol{g}^i{}_{,j}\times\boldsymbol{t}_{,i}\right) + \boldsymbol{g}^j\times\left(\boldsymbol{g}^i\times\boldsymbol{t}_{,ij}\right)$$

$$= \left(\boldsymbol{g}^j\cdot\boldsymbol{t}_{,i}\right)\boldsymbol{g}^i{}_{,j} - \left(\boldsymbol{g}^j\cdot\boldsymbol{g}^i{}_{,j}\right)\boldsymbol{t}_{,i} + \left(\boldsymbol{g}^j\cdot\boldsymbol{t}_{,ij}\right)\boldsymbol{g}^i - g^{ij}\boldsymbol{t}_{,ij}.$$

Equations (2.8), $(2.64)_1$, (2.126), (1.121) and (9.9):

$$\operatorname{div}\operatorname{grad}\boldsymbol{t} = \left(\boldsymbol{t}_{,i}\otimes\boldsymbol{g}^i\right)_{,j}\boldsymbol{g}^j = \left(\boldsymbol{t}_{,ij}\otimes\boldsymbol{g}^i\right)\boldsymbol{g}^j + \left(\boldsymbol{t}_{,i}\otimes\boldsymbol{g}^i{}_{,j}\right)\boldsymbol{g}^j$$

$$= g^{ij}\boldsymbol{t}_{,ij} + \left(\boldsymbol{g}^i{}_{,j}\cdot\boldsymbol{g}^j\right)\boldsymbol{t}_{,i}. \qquad (9.10)$$

$$\operatorname{div}(\operatorname{grad}\boldsymbol{t})^{\mathrm{T}} = \left(\boldsymbol{t}_{,i}\otimes\boldsymbol{g}^i\right)^{\mathrm{T}}{}_{,j}\boldsymbol{g}^j = \left(\boldsymbol{g}^i\otimes\boldsymbol{t}_{,ij}\right)\boldsymbol{g}^j + \left(\boldsymbol{g}^i{}_{,j}\otimes\boldsymbol{t}_{,i}\right)\boldsymbol{g}^j$$

$$= \left(\boldsymbol{t}_{,ij}\cdot\boldsymbol{g}^j\right)\boldsymbol{g}^i + \left(\boldsymbol{t}_{,i}\cdot\boldsymbol{g}^j\right)\boldsymbol{g}^i{}_{,j}.$$

The latter four relations immediately imply (2.135) and (2.136).

Equations $(1.153)_2$, (1.169), $(2.64)_1$ and (2.128):

$$\operatorname{skew}(\operatorname{grad}\boldsymbol{t}) = \frac{1}{2}\left(\boldsymbol{t}_{,i}\otimes\boldsymbol{g}^i - \boldsymbol{g}^i\otimes\boldsymbol{t}_{,i}\right) = \frac{1}{2}\widehat{\boldsymbol{g}^i\times\boldsymbol{t}_{,i}} = \frac{1}{2}\widehat{\operatorname{curl}\boldsymbol{t}}.$$

Equations (2.5), (1.142), (2.112), (2.126) and $(2.64)_1$:

$$\operatorname{div}(\boldsymbol{t}\mathbf{A}) = (\boldsymbol{t}\mathbf{A})_{,i} \cdot \boldsymbol{g}^i = (\boldsymbol{t}_{,i}\,\mathbf{A}) \cdot \boldsymbol{g}^i + (\boldsymbol{t}\mathbf{A}_{,i}) \cdot \boldsymbol{g}^i$$

$$= \mathbf{A} : \boldsymbol{t}_{,i} \otimes \boldsymbol{g}^i + \boldsymbol{t} \cdot \left(\mathbf{A}_{,i}\,\boldsymbol{g}^i\right) = \mathbf{A} : \operatorname{grad}\boldsymbol{t} + \boldsymbol{t} \cdot \operatorname{div}\mathbf{A}.$$

Equations (2.3), (2.63) and (2.126):

$$\operatorname{div}(\Phi\boldsymbol{t}) = (\Phi\boldsymbol{t})_{,i} \cdot \boldsymbol{g}^i = (\Phi_{,i}\,\boldsymbol{t}) \cdot \boldsymbol{g}^i + (\Phi\boldsymbol{t}_{,i}) \cdot \boldsymbol{g}^i$$

$$= \boldsymbol{t} \cdot \left(\Phi_{,i}\,\boldsymbol{g}^i\right) + \Phi\left(\boldsymbol{t}_{,i} \cdot \boldsymbol{g}^i\right) = \boldsymbol{t} \cdot \operatorname{grad}\Phi + \Phi\operatorname{div}\boldsymbol{t}.$$

Equations (2.4), (2.63) and (2.112):

$$\operatorname{div}(\Phi\mathbf{A}) = (\Phi\mathbf{A})_{,i}\,\boldsymbol{g}^i = (\Phi_{,i}\,\mathbf{A})\,\boldsymbol{g}^i + (\Phi\mathbf{A}_{,i})\,\boldsymbol{g}^i$$

$$= \mathbf{A}\left(\Phi_{,i}\,\boldsymbol{g}^i\right) + \Phi\left(\mathbf{A}_{,i}\,\boldsymbol{g}^i\right) = \mathbf{A}\operatorname{grad}\Phi + \Phi\operatorname{div}\mathbf{A}.$$

2.13 Cylindrical coordinates, $(2.75)_2$, (2.93) and (2.90):

$$\operatorname{grad}\boldsymbol{t} = t_i|_j\,\boldsymbol{g}^i \otimes \boldsymbol{g}^j = \left(t_{i,j} - t_k\Gamma^k_{ij}\right)\boldsymbol{g}^i \otimes \boldsymbol{g}^j$$

$$= t_{i,j}\,\boldsymbol{g}^i \otimes \boldsymbol{g}^j + rt_3\boldsymbol{g}^1 \otimes \boldsymbol{g}^1 - r^{-1}t_1\left(\boldsymbol{g}^1 \otimes \boldsymbol{g}^3 + \boldsymbol{g}^3 \otimes \boldsymbol{g}^1\right),$$

or alternatively

$$\operatorname{grad}\boldsymbol{t} = t^i|_j\,\boldsymbol{g}_i \otimes \boldsymbol{g}^j = \left(t^i{}_{,j} + t^k\Gamma^i_{kj}\right)\boldsymbol{g}_i \otimes \boldsymbol{g}^j$$

$$= t^i{}_{,j}\,\boldsymbol{g}_i \otimes \boldsymbol{g}^j + r^{-1}t^3\boldsymbol{g}_1 \otimes \boldsymbol{g}^1 + t^1\left(r^{-1}\boldsymbol{g}_1 \otimes \boldsymbol{g}^3 - r\boldsymbol{g}_3 \otimes \boldsymbol{g}^1\right).$$

Equations (2.30) and (2.127):

$$\operatorname{div}\boldsymbol{t} = \operatorname{tr}\operatorname{grad}\boldsymbol{t} = t_{i,j}\,g^{ij} + rt_3g^{11} - 2r^{-1}t_1g^{13}$$

$$= r^{-2}t_{1,1} + t_{2,2} + t_{3,3} + r^{-1}t_3,$$

or alternatively

$$\operatorname{div}\boldsymbol{t} = \operatorname{tr}\operatorname{grad}\boldsymbol{t} = t^i{}_{,i} + r^{-1}t^3 = t^i{}_{,i} + r^{-1}t_3 = t^1{}_{,1} + t^2{}_{,2} + t^3{}_{,3} + r^{-1}t_3.$$

Equations (2.93) and (2.129):

$$\operatorname{curl}\boldsymbol{t} = e^{jik}\frac{1}{g}t_i|_j\,\boldsymbol{g}_k$$

$$= g^{-1}\left[(t_3|_2 - t_2|_3)\, \boldsymbol{g}_1 + (t_1|_3 - t_3|_1)\, \boldsymbol{g}_2 + (t_2|_1 - t_1|_2)\, \boldsymbol{g}_3\right]$$

$$= r^{-1}\left[(t_{3,2} - t_{2,3})\, \boldsymbol{g}_1 + (t_{1,3} - t_{3,1})\, \boldsymbol{g}_2 + (t_{2,1} - t_{1,2})\, \boldsymbol{g}_3\right].$$

Spherical coordinates, (9.6)–(9.8):

$$\mathrm{grad}\boldsymbol{t} = \left(t_{i,j} - t_k \Gamma^k_{ij}\right) \boldsymbol{g}^i \otimes \boldsymbol{g}^j$$

$$= \left(t_{1,1} + t_2 \sin\phi\cos\phi + t_3 r \sin^2\phi\right) \boldsymbol{g}^1 \otimes \boldsymbol{g}^1 + \left(t_{2,2} + t_3 r\right) \boldsymbol{g}^2 \otimes \boldsymbol{g}^2$$

$$+ t_{3,3}\, \boldsymbol{g}^3 \otimes \boldsymbol{g}^3 + \left(t_{1,2} - t_1 \cot\phi\right) \boldsymbol{g}^1 \otimes \boldsymbol{g}^2 + \left(t_{2,1} - t_1 \cot\phi\right) \boldsymbol{g}^2 \otimes \boldsymbol{g}^1$$

$$+ \left(t_{1,3} - t_1 r^{-1}\right) \boldsymbol{g}^1 \otimes \boldsymbol{g}^3 + \left(t_{3,1} - t_1 r^{-1}\right) \boldsymbol{g}^3 \otimes \boldsymbol{g}^1$$

$$+ \left(t_{2,3} - t_2 r^{-1}\right) \boldsymbol{g}^2 \otimes \boldsymbol{g}^3 + \left(t_{3,2} - t_2 r^{-1}\right) \boldsymbol{g}^3 \otimes \boldsymbol{g}^2,$$

or alternatively

$$\mathrm{grad}\boldsymbol{t} = \left(t^i{}_{,j} + t^k \Gamma^i_{kj}\right) \boldsymbol{g}_i \otimes \boldsymbol{g}^j = \left(t^1{}_{,1} + t^2 \cot\phi + t^3 r^{-1}\right) \boldsymbol{g}_1 \otimes \boldsymbol{g}^1$$

$$+ \left(t^2{}_{,2} + t^3 r^{-1}\right) \boldsymbol{g}_2 \otimes \boldsymbol{g}^2 + t^3{}_{,3}\, \boldsymbol{g}_3 \otimes \boldsymbol{g}^3$$

$$+ \left(t^1{}_{,2} + t^1 \cot\phi\right) \boldsymbol{g}_1 \otimes \boldsymbol{g}^2 + \left(t^2{}_{,1} - t^1 \sin\phi\cos\phi\right) \boldsymbol{g}_2 \otimes \boldsymbol{g}^1$$

$$+ \left(t^1{}_{,3} + t^1 r^{-1}\right) \boldsymbol{g}_1 \otimes \boldsymbol{g}^3 + \left(t^3{}_{,1} - t^1 r \sin^2\phi\right) \boldsymbol{g}_3 \otimes \boldsymbol{g}^1$$

$$+ \left(t^2{}_{,3} + t^2 r^{-1}\right) \boldsymbol{g}_2 \otimes \boldsymbol{g}^3 + \left(t^3{}_{,2} - t^2 r\right) \boldsymbol{g}_3 \otimes \boldsymbol{g}^2,$$

(2.93), (2.127), (2.129), (9.4)–(9.8):

$$\mathrm{div}\boldsymbol{t} = \left(t_{i,j} - t_k \Gamma^k_{ij}\right) g^{ij} = \frac{t_{1,1}}{r^2 \sin^2\phi} + r^{-2} t_{2,2} + t_{3,3} + r^{-2} \cot\phi\, t_2 + 2r^{-1} t_3$$

$$= t^i{}_{,i} + t^k \Gamma^i_{ki} = t^1{}_{,1} + t^2{}_{,2} + t^3{}_{,3} + \cot\phi\, t^2 + 2r^{-1} t^3,$$

$$\mathrm{curl}\boldsymbol{t} = g^{-1}\left[(t_3|_2 - t_2|_3)\, \boldsymbol{g}_1 + (t_1|_3 - t_3|_1)\, \boldsymbol{g}_2 + (t_2|_1 - t_1|_2)\, \boldsymbol{g}_3\right]$$

$$= -\frac{1}{r^2 \sin\phi}\left[(t_{3,2} - t_{2,3})\, \boldsymbol{g}_1 + (t_{1,3} - t_{3,1})\, \boldsymbol{g}_2 + (t_{2,1} - t_{1,2})\, \boldsymbol{g}_3\right].$$

2.14 According to the result (9.10) of Exercise 2.12

$$\Delta\boldsymbol{t} = \mathrm{div}\,\mathrm{grad}\boldsymbol{t} = g^{ij}\, \boldsymbol{t}_{,ij} + \left(\boldsymbol{g}^i{}_{,j} \cdot \boldsymbol{g}^j\right) \boldsymbol{t}_{,i}\,.$$

By virtue of (2.73), (2.82) and $(2.93)_2$ we further obtain

$$\Delta \boldsymbol{t} = g^{ij} \boldsymbol{t},_{ij} - \Gamma^k_{ij} g^{ij} \boldsymbol{t},_k = g^{ij} \left(\boldsymbol{t},_{ij} - \Gamma^k_{ij} \boldsymbol{t},_k \right) = g^{ij} \boldsymbol{t},_i|_j = \boldsymbol{t},_i|^i .$$

In Cartesian coordinates it leads to the well-known relation

$$\operatorname{div} \operatorname{grad} \boldsymbol{t} = \boldsymbol{t},_{11} + \boldsymbol{t},_{22} + \boldsymbol{t},_{33} .$$

2.15 Specifying the result of Exercise 2.14 to scalar functions we can write

$$\Delta \Phi = g^{ij} \left(\Phi,_{ij} - \Gamma^k_{ij} \Phi,_k \right) = \Phi,_i|^i .$$

For the cylindrical coordinates it takes in view of (2.30) and (2.90) the following form

$$\Delta \Phi = r^{-2} \Phi,_{11} + \Phi,_{22} + \Phi,_{33} + r^{-1} \Phi,_3 = \frac{1}{r^2} \frac{\partial^2 \Phi}{\partial \varphi^2} + \frac{\partial^2 \Phi}{\partial z^2} + \frac{\partial^2 \Phi}{\partial r^2} + \frac{1}{r} \frac{\partial \Phi}{\partial r} .$$

For the spherical coordinates we use (9.5)–(9.8). Thus,

$$\begin{aligned}
\Delta \Phi &= \frac{1}{r^2 \sin^2 \phi} \Phi,_{11} + r^{-2} \Phi,_{22} + \Phi,_{33} + \frac{\cos \phi}{r^2 \sin \phi} \Phi,_2 + 2r^{-1} \Phi,_3 \\
&= \frac{1}{r^2 \sin^2 \phi} \frac{\partial^2 \Phi}{\partial \varphi^2} + r^{-2} \frac{\partial^2 \Phi}{\partial \phi^2} + \frac{\partial^2 \Phi}{\partial r^2} + r^{-2} \cot \phi \frac{\partial \Phi}{\partial \phi} + 2r^{-1} \frac{\partial \Phi}{\partial r} .
\end{aligned}$$

2.16 According to the solution of Exercise 2.14

$$\Delta \boldsymbol{t} = g^{ij} \left(\boldsymbol{t},_{ij} - \Gamma^m_{ij} \boldsymbol{t},_m \right), \tag{9.11}$$

where in view of $(2.72)_1$

$$\boldsymbol{t},_i = t^k|_i \, \boldsymbol{g}_k, \qquad \boldsymbol{t},_{ij} = t^k|_{ij} \, \boldsymbol{g}_k .$$

By virtue of $(2.93)_1$ we further write $t^k|_i = t^k,_i + \Gamma^k_{li} t^l$ and consequently

$$t^k|_{ij} = t^k|_i,_j + \Gamma^k_{mj} t^m|_i = t^k,_{ij} + \Gamma^k_{li},_j t^l + \Gamma^k_{li} t^l,_j + \Gamma^k_{mj} t^m,_i + \Gamma^k_{mj} \Gamma^m_{li} t^l .$$

Substituting these results into the expression of the Laplacian (9.11) finally yields

$$\Delta \boldsymbol{t} = g^{ij} \left(t^k,_{ij} + 2\Gamma^k_{li} t^l,_j - \Gamma^m_{ij} t^k,_m + \Gamma^k_{li},_j t^l + \Gamma^k_{mj} \Gamma^m_{li} t^l - \Gamma^m_{ij} \Gamma^k_{lm} t^l \right) \boldsymbol{g}_k .$$

Taking (9.5)–(9.8) into account we thus obtain for the spherical coordinates (2.157)

$$\Delta t = \left(\frac{t^1{}_{,\varphi\varphi}}{r^2 \sin^2 \phi} + \frac{t^1{}_{,\phi\phi}}{r^2} + t^1{}_{,rr} \right.$$

$$\left. + \frac{3 \cot \phi}{r^2} t^1{}_{,\phi} + \frac{2 \cos \phi}{r^2 \sin^3 \phi} t^2{}_{,\varphi} + \frac{4 t^1{}_{,r}}{r} + \frac{2 t^3{}_{,\varphi}}{r^3 \sin^2 \phi} \right) \boldsymbol{g}_1$$

$$+ \left(\frac{t^2{}_{,\varphi\varphi}}{r^2 \sin^2 \phi} + \frac{t^2{}_{,\phi\phi}}{r^2} + t^2{}_{,rr} \right.$$

$$\left. - \frac{2 \cot \phi}{r^2} t^1{}_{,\varphi} + \frac{\cot \phi}{r^2} t^2{}_{,\phi} + \frac{4 t^2{}_{,r}}{r} + \frac{2 t^3{}_{,\phi}}{r^3} + \frac{1 - \cot^2 \phi}{r^2} t^2 \right) \boldsymbol{g}_2$$

$$+ \left(\frac{t^3{}_{,\varphi\varphi}}{r^2 \sin^2 \phi} + \frac{t^3{}_{,\phi\phi}}{r^2} + t^3{}_{,rr} \right.$$

$$\left. - \frac{2 t^1{}_{,\varphi}}{r} - \frac{2 t^2{}_{,\phi}}{r} + \frac{\cot \phi}{r^2} t^3{}_{,\phi} + \frac{2 t^3{}_{,r}}{r} - \frac{2 \cot \phi}{r} t^2 - \frac{2 t^3}{r^2} \right) \boldsymbol{g}_3 .$$

9.3 Exercises of Chap. 3

3.1 (C.4) and (3.18): $\boldsymbol{a}_1 = \mathrm{d}\boldsymbol{r}/\mathrm{d}s = const$. Hence, $\boldsymbol{r}(s) = \boldsymbol{b} + s\boldsymbol{a}_1$.

3.2 Using the fact that $\mathrm{d}/\mathrm{d}(-s) = -\mathrm{d}/\mathrm{d}s$ we can write by means of (3.15), (3.18), (3.20), (3.21) and (3.27): $\boldsymbol{a}_1'(s) = -\boldsymbol{a}_1(s)$, $\boldsymbol{a}_2'(s) = \boldsymbol{a}_2(s)$, $\boldsymbol{a}_3'(s) = -\boldsymbol{a}_3(s)$, $\varkappa'(s) = \varkappa(s)$ and $\tau'(s) = \tau(s)$.

3.3 Let us show that the curve $\boldsymbol{r}(s)$ with the zero torsion $\tau(s) \equiv 0$ belongs to the plane $\boldsymbol{p}(t^1, t^2) = \boldsymbol{r}(s_0) + t^1 \boldsymbol{a}_1(s_0) + t^2 \boldsymbol{a}_2(s_0)$, where $\boldsymbol{a}_1(s_0)$ and $\boldsymbol{a}_2(s_0)$ are, respectively, the unit tangent vector and the principal normal vector at a point s_0. For any arbitrary point we can write using (3.15)

$$\boldsymbol{r}(s) = \boldsymbol{r}(s_0) + \int_{\boldsymbol{r}(s_0)}^{\boldsymbol{r}(s)} \mathrm{d}\boldsymbol{r} = \boldsymbol{r}(s_0) + \int_{s_0}^{s} \boldsymbol{a}_1(s)\,\mathrm{d}s. \tag{9.12}$$

The vector $\boldsymbol{a}_1(s)$ can further be represented with respect to the trihedron at s_0 as $\boldsymbol{a}_1(s) = \alpha^i(s)\,\boldsymbol{a}_i(s_0)$. Taking (3.26) into account we observe that $\boldsymbol{a}_3{}_{,s} = \boldsymbol{0}$ and consequently $\boldsymbol{a}_3(s) \equiv \boldsymbol{a}_3(s_0)$. In view of $(3.23)_2$ it yields $\boldsymbol{a}_1(s) \cdot \boldsymbol{a}_3(s_0) = 0$, so that $\boldsymbol{a}_1(s) = \alpha^1(s)\,\boldsymbol{a}_1(s_0) + \alpha^2(s)\,\boldsymbol{a}_2(s_0)$. Inserting this result into (9.12) we have

$$\boldsymbol{r}(s) = \boldsymbol{r}(s_0) + \boldsymbol{a}_1(s_0) \int\limits_{s_0}^{s} \alpha^1(s)\,\mathrm{d}s + \boldsymbol{a}_2(s_0) \int\limits_{s_0}^{s} \alpha^2(s)\,\mathrm{d}s$$

$$= \boldsymbol{r}(s_0) + t^1 \boldsymbol{a}_1(s_0) + t^2 \boldsymbol{a}_2(s_0),$$

where we set $t^i = \int_{s_0}^{s} \alpha^i(s)\,\mathrm{d}s$ $(i = 1, 2)$.

3.4 Setting in (2.30) $r = R$ yields

$$[g_{\alpha\beta}] = \begin{bmatrix} R^2 & 0 \\ 0 & 1 \end{bmatrix}.$$

By means of (2.90), (3.74), (3.79), (3.90) and (3.93) we further obtain

$$[b_{\alpha\beta}] = \begin{bmatrix} -R & 0 \\ 0 & 0 \end{bmatrix}, \quad [b_\alpha^\beta] = \begin{bmatrix} -R^{-1} & 0 \\ 0 & 0 \end{bmatrix}, \quad \Gamma^1_{\alpha\beta} = \Gamma^2_{\alpha\beta} = 0, \quad \alpha, \beta = 1, 2,$$

$$K = \left|b_\alpha^\beta\right| = 0, \quad H = \frac{1}{2} b_\alpha^\alpha = -\frac{1}{2} R^{-1}. \tag{9.13}$$

3.5 Keeping in mind the results of Exercise 2.1 and using (9.6)–(9.8), (3.58), (3.62), (3.67), (3.74), (3.79), (3.90) and (3.93) we write

$$\boldsymbol{g}_1 = R\cos t^1 \sin t^2 \boldsymbol{e}_1 - R\sin t^1 \sin t^2 \boldsymbol{e}_3,$$

$$\boldsymbol{g}_2 = R\sin t^1 \cos t^2 \boldsymbol{e}_1 - R\sin t^2 \boldsymbol{e}_2 + R\cos t^1 \cos t^2 \boldsymbol{e}_3,$$

$$\boldsymbol{g}_3 = -\sin t^1 \sin t^2 \boldsymbol{e}_1 - \cos t^2 \boldsymbol{e}_2 - \cos t^1 \sin t^2 \boldsymbol{e}_3,$$

$$[g_{\alpha\beta}] = \begin{bmatrix} R^2 \sin^2 t^2 & 0 \\ 0 & R^2 \end{bmatrix}, \quad [b_{\alpha\beta}] = \begin{bmatrix} R\sin^2 t^2 & 0 \\ 0 & R \end{bmatrix}, \quad [b_\alpha^\beta] = \begin{bmatrix} R^{-1} & 0 \\ 0 & R^{-1} \end{bmatrix},$$

$$\left[\Gamma^1_{\alpha\beta}\right] = \begin{bmatrix} 0 & \cot t^2 \\ \cot t^2 & 0 \end{bmatrix}, \quad \left[\Gamma^2_{\alpha\beta}\right] = \begin{bmatrix} -\sin t^2 \cos t^2 & 0 \\ 0 & 0 \end{bmatrix},$$

$$K = \left|b_\alpha^\beta\right| = R^{-2}, \quad H = \frac{1}{2} b_\alpha^\alpha = R^{-1}. \tag{9.14}$$

3.6 Equations (3.62), (3.67) and (3.143):

$$\boldsymbol{g}_1 = \frac{\partial \boldsymbol{r}}{\partial t^1} = \boldsymbol{e}_1 + \bar{t}^2 \boldsymbol{e}_3, \quad \boldsymbol{g}_2 = \frac{\partial \boldsymbol{r}}{\partial t^2} = \boldsymbol{e}_2 + \bar{t}^1 \boldsymbol{e}_3,$$

$$\boldsymbol{g}_3 = \frac{\boldsymbol{g}_1 \times \boldsymbol{g}_2}{\|\boldsymbol{g}_1 \times \boldsymbol{g}_2\|} = \frac{1}{\sqrt{1 + (\bar{t}^1)^2 + (\bar{t}^2)^2}} \left(-\bar{t}^2 \boldsymbol{e}_1 - \bar{t}^1 \boldsymbol{e}_2 + \boldsymbol{e}_3\right),$$

where $\bar{t}^i = \dfrac{t^i}{c}$ $(i = 1, 2)$. Thus, the coefficients of the first fundamental form are

$$g_{11} = \boldsymbol{g}_1 \cdot \boldsymbol{g}_1 = 1 + \left(\bar{t}^2\right)^2, \; g_{12} = g_{21} = \boldsymbol{g}_1 \cdot \boldsymbol{g}_2 = \bar{t}^1 \bar{t}^2, \; g_{22} = \boldsymbol{g}_2 \cdot \boldsymbol{g}_2 = 1 + \left(\bar{t}^1\right)^2.$$

For the coefficients of the inversed matrix $\left[g^{\alpha\beta}\right]$ we have

$$\left[g^{\alpha\beta}\right] = \frac{1}{1 + (\bar{t}^1)^2 + (\bar{t}^2)^2} \begin{bmatrix} 1 + \left(\bar{t}^1\right)^2 & -\bar{t}^1\bar{t}^2 \\ -\bar{t}^1\bar{t}^2 & 1 + \left(\bar{t}^2\right)^2 \end{bmatrix}.$$

The derivatives of the tangent vectors result in

$$\boldsymbol{g}_{1,1} = \boldsymbol{0}, \quad \boldsymbol{g}_{1,2} = \boldsymbol{g}_{2,1} = \frac{1}{c}\boldsymbol{e}_3, \quad \boldsymbol{g}_{2,2} = \boldsymbol{0}.$$

By (3.74), (3.79), (3.90) and (3.93) we further obtain

$$b_{11} = \boldsymbol{g}_{1,1} \cdot \boldsymbol{g}_3 = 0, \quad b_{12} = b_{21} = \boldsymbol{g}_{1,2} \cdot \boldsymbol{g}_3 = \frac{1}{c\sqrt{1 + (\bar{t}^1)^2 + (\bar{t}^2)^2}},$$

$$b_{22} = \boldsymbol{g}_{2,2} \cdot \boldsymbol{g}_3 = 0,$$

$$\left[b_{\alpha \cdot}^{\cdot \beta}\right] = \frac{1}{c\left[1 + (\bar{t}^1)^2 + (\bar{t}^2)^2\right]^{3/2}} \begin{bmatrix} -\bar{t}^1\bar{t}^2 & 1 + \left(\bar{t}^2\right)^2 \\ 1 + \left(\bar{t}^1\right)^2 & -\bar{t}^1\bar{t}^2 \end{bmatrix},$$

$$K = \left|b_\alpha^\beta\right| = -\frac{1}{c^2\left[1 + (\bar{t}^1)^2 + (\bar{t}^2)^2\right]^2} = -\left[c^2 + \left(t^1\right)^2 + \left(t^2\right)^2\right]^{-2},$$

$$H = \frac{1}{2} b_\alpha^\alpha = -\frac{\bar{t}^1\bar{t}^2}{c\left[1 + (\bar{t}^1)^2 + (\bar{t}^2)^2\right]^{3/2}}.$$

3.7 Equations (3.62), (3.67) and (3.144):

$$\boldsymbol{g}_1 = \frac{\partial \boldsymbol{r}}{\partial t^1} = -ct^2 \sin t^1 \boldsymbol{e}_1 + ct^2 \cos t^1 \boldsymbol{e}_2,$$

$$\boldsymbol{g}_2 = \frac{\partial \boldsymbol{r}}{\partial t^2} = c \cos t^1 \boldsymbol{e}_1 + c \sin t^1 \boldsymbol{e}_2 + \boldsymbol{e}_3,$$

$$\boldsymbol{g}_3 = \frac{\boldsymbol{g}_1 \times \boldsymbol{g}_2}{\|\boldsymbol{g}_1 \times \boldsymbol{g}_2\|} = \frac{1}{\sqrt{1 + c^2}} \left(\cos t^1 \boldsymbol{e}_1 + \sin t^1 \boldsymbol{e}_2 - c\boldsymbol{e}_3\right).$$

Thus, the coefficients of the first fundamental form are calculated as

$$g_{11} = \boldsymbol{g}_1 \cdot \boldsymbol{g}_1 = c^2 \left(t^2\right)^2, \; g_{12} = g_{21} = \boldsymbol{g}_1 \cdot \boldsymbol{g}_2 = 0, \; g_{22} = \boldsymbol{g}_2 \cdot \boldsymbol{g}_2 = 1 + c^2,$$

so that

$$\left[g^{\alpha\beta}\right] = \begin{bmatrix} \left(ct^2\right)^{-2} & 0 \\ 0 & \left(1+c^2\right)^{-1} \end{bmatrix}.$$

The derivatives of the tangent vectors take the form

$$\boldsymbol{g}_{1,1} = -ct^2 \cos t^1 \boldsymbol{e}_1 - ct^2 \sin t^1 \boldsymbol{e}_2, \quad \boldsymbol{g}_{1,2} = \boldsymbol{g}_{2,1} = -c \sin t^1 \boldsymbol{e}_1 + c \cos t^1 \boldsymbol{e}_2,$$

$$\boldsymbol{g}_{2,2} = \boldsymbol{0}.$$

By means of (3.74), (3.79), (3.90) and (3.93) this leads to

$$b_{11} = \boldsymbol{g}_{1,1} \cdot \boldsymbol{g}_3 = -\frac{ct^2}{\sqrt{1+c^2}}, \; b_{12} = b_{21} = \boldsymbol{g}_{1,2} \cdot \boldsymbol{g}_3 = 0, \; b_{22} = \boldsymbol{g}_{2,2} \cdot \boldsymbol{g}_3 = 0,$$

$$\left[b_\alpha^\beta\right] = \begin{bmatrix} -\dfrac{1}{ct^2\sqrt{1+c^2}} & 0 \\ 0 & 0 \end{bmatrix}, \quad K = \left|b_\alpha^\beta\right| = 0, \quad H = \frac{1}{2} b_\alpha^\alpha = -\frac{1}{2ct^2\sqrt{1+c^2}}.$$

3.8 Taking (3.105) into account we can write

$$\left[g_{ij}^*\right] = \begin{bmatrix} g_{11}^* & g_{12}^* & 0 \\ g_{21}^* & g_{22}^* & 0 \\ 0 & 0 & 1 \end{bmatrix}, \quad \left[g^{*ij}\right] = \left[g_{ij}^*\right]^{-1} = \frac{1}{\left|g_{ij}^*\right|} \begin{bmatrix} g_{22}^* & -g_{21}^* & 0 \\ -g_{12}^* & g_{11}^* & 0 \\ 0 & 0 & 1 \end{bmatrix},$$

which immediately implies (3.111).

3.9 For a cylindrical shell equilibrium equations (3.140) and (3.141) take by means of $(3.77)_1$ and (9.13) the form

$$f^{11}{}_{,1} + f^{12}{}_{,2} + p^1 = 0, \quad f^{12}{}_{,1} + f^{22}{}_{,2} + p^2 = 0, \quad -Rf^{11} + p^3 = 0.$$

For a spherical shell we further obtain by virtue of (9.14)

$$f^{11}{}_{,1} + f^{12}{}_{,2} + 3 \cot t^2 f^{12} + p^1 = 0,$$

$$f^{12}{}_{,1} + f^{22}{}_{,2} - \sin t^2 \cos t^2 f^{11} + \cot t^2 f^{22} + p^2 = 0,$$

$$R \sin^2 t^2 f^{11} + Rf^{22} + p^3 = 0.$$

9.4 Exercises of Chap. 4

4.1 In the case of simple shear the right Cauchy-Green tensor $\mathbf{C}$ has the form:

$$\mathbf{C} = \mathrm{C}^{i}_{\cdot j}\boldsymbol{e}_i \otimes \boldsymbol{e}^j, \quad \left[\mathrm{C}^{i}_{\cdot j}\right] = \left[\mathrm{C}_{ij}\right] = \begin{bmatrix} 1 & \gamma & 0 \\ \gamma & 1+\gamma^2 & 0 \\ 0 & 0 & 1 \end{bmatrix}.$$

The characteristic equation can then be written by

$$\begin{vmatrix} 1-\Lambda & \gamma & 0 \\ \gamma & 1+\gamma^2-\Lambda & 0 \\ 0 & 0 & 1-\Lambda \end{vmatrix} = 0 \quad \Rightarrow \quad (1-\Lambda)\left\{\Lambda^2 - \Lambda\left(2+\gamma^2\right)+1\right\} = 0.$$

Solving the latter equation with respect to Λ we obtain the eigenvalues of $\mathbf{C}$ as

$$\Lambda_{1/2} = 1 + \frac{\gamma^2 \pm \sqrt{4\gamma^2+\gamma^4}}{2} = \left(\frac{\sqrt{4+\gamma^2}\pm\gamma}{2}\right)^2, \quad \Lambda_3 = 1. \tag{9.15}$$

The eigenvectors $\boldsymbol{a} = a^i \boldsymbol{e}_i$ corresponding to the first two eigenvalues result from the equation system $(4.16)_1$

$$\begin{cases} \dfrac{-\gamma^2 \mp \sqrt{4\gamma^2+\gamma^4}}{2} a^1 + \gamma a^2 = 0, \\ \gamma a^1 + \dfrac{\gamma^2 \mp \sqrt{4\gamma^2+\gamma^4}}{2} a^2 = 0, \\ \dfrac{-\gamma^2 \mp \sqrt{4\gamma^2+\gamma^4}}{2} a^3 = 0. \end{cases}$$

Since the first and second equation are equivalent we only obtain

$$a^2 = \frac{\gamma \pm \sqrt{4+\gamma^2}}{2} a^1, \quad a^3 = 0,$$

so that $a^2 = \sqrt{\Lambda_1} a^1$ or $a^2 = -\sqrt{\Lambda_2} a^1$. In order to ensure the unit length of the eigenvectors we also require that

$$\left(a^1\right)^2 + \left(a^2\right)^2 + \left(a^3\right)^2 = 1.$$

This yields

$$\boldsymbol{a}_1 = \frac{1}{\sqrt{1+\Lambda_1}}\boldsymbol{e}_1 + \sqrt{\frac{\Lambda_1}{1+\Lambda_1}}\boldsymbol{e}_2, \quad \boldsymbol{a}_2 = \frac{1}{\sqrt{1+\Lambda_2}}\boldsymbol{e}_1 - \sqrt{\frac{\Lambda_2}{1+\Lambda_2}}\boldsymbol{e}_2. \tag{9.16}$$

Applying the above procedure for the third eigenvector corresponding to the eigenvalue $\Lambda_3 = 1$ we easily obtain: $\boldsymbol{a}_3 = \boldsymbol{e}_3$.

4.2 Since the vectors $\boldsymbol{g}_i$ $(i = 1, 2, 3)$ are linearly independent, they form a basis in $\mathbb{E}^3$. Thus, by means of the representation $\mathbf{A} = \mathrm{A}^{i}_{\cdot j} \boldsymbol{g}_i \otimes \boldsymbol{g}^j$ we obtain using (1.39)

$$[\mathbf{A}\boldsymbol{g}_1 \, \mathbf{A}\boldsymbol{g}_2 \, \mathbf{A}\boldsymbol{g}_3] = \left[\mathrm{A}^{i}_{\cdot 1}\boldsymbol{g}_i \, \mathrm{A}^{j}_{\cdot 2}\boldsymbol{g}_j \, \mathrm{A}^{k}_{\cdot 3}\boldsymbol{g}_k\right] = \mathrm{A}^{i}_{\cdot 1}\mathrm{A}^{j}_{\cdot 2}\mathrm{A}^{k}_{\cdot 3}\left[\boldsymbol{g}_i \, \boldsymbol{g}_j \, \boldsymbol{g}_k\right]$$

$$= \mathrm{A}^{i}_{\cdot 1}\mathrm{A}^{j}_{\cdot 2}\mathrm{A}^{k}_{\cdot 3} e_{ijk} \, g = \left|\mathrm{A}^{i}_{\cdot j}\right| g = \det\mathbf{A} \, g.$$

4.3 Using $(4.26)_{1-3}$ we write

$$\mathrm{I}_{\mathbf{A}} = \mathrm{tr}\mathbf{A},$$

$$\mathrm{II}_{\mathbf{A}} = \frac{1}{2}\left[(\mathrm{tr}\mathbf{A})^2 - \mathrm{tr}\mathbf{A}^2\right],$$

$$\mathrm{III}_{\mathbf{A}} = \frac{1}{3}\left[\mathrm{II}_{\mathbf{A}}\mathrm{tr}\mathbf{A} - \mathrm{I}_{\mathbf{A}}\mathrm{tr}\mathbf{A}^2 + \mathrm{tr}\mathbf{A}^3\right].$$

Inserting the first and second expression into the third one we obtain

$$\mathrm{III}_{\mathbf{A}} = \frac{1}{3}\left\{\frac{1}{2}\left[(\mathrm{tr}\mathbf{A})^2 - \mathrm{tr}\mathbf{A}^2\right]\mathrm{tr}\mathbf{A} - \mathrm{tr}\mathbf{A}\mathrm{tr}\mathbf{A}^2 + \mathrm{tr}\mathbf{A}^3\right\}$$

$$= \frac{1}{3}\left[\mathrm{tr}\mathbf{A}^3 - \frac{3}{2}\mathrm{tr}\mathbf{A}^2\mathrm{tr}\mathbf{A} + \frac{1}{2}(\mathrm{tr}\mathbf{A})^3\right].$$

4.4 Since $r_i = t_i$ for every eigenvalue λ_i we have exactly $n = \sum_{i=1}^{s} r_i$ eigenvectors, say $\boldsymbol{a}_i^{(k)}$ $(i = 1, 2, \ldots, s; k = 1, 2, \ldots, r_i)$. Let us assume, on the contrary, that they are linearly dependent so that

$$\sum_{i=1}^{s}\sum_{k=1}^{r_i} \alpha_i^{(k)} \boldsymbol{a}_i^{(k)} = \mathbf{0},$$

where not all $\alpha_i^{(k)}$ are zero. Linear combinations $\boldsymbol{a}_i = \sum_{k=1}^{r_i} \alpha_i^{(k)} \boldsymbol{a}_i^{(k)}$ of the eigenvectors $\boldsymbol{a}_i^{(k)}$ associated with the same eigenvalue λ_i are again eigenvectors or zero vectors. Thus, we arrive at

$$\sum_{i=1}^{s} \varepsilon_i \boldsymbol{a}_i = \mathbf{0},$$

where ε_i are either one or zero (but not all). This relation establishes the linear dependence between eigenvectors corresponding to distinct eigenvalues, which contradicts the statement of Theorem 4.2. Applying then Theorem 1.3 for the space $\mathbb{C}^n$ instead of $\mathbb{V}$ we infer that the eigenvectors $\boldsymbol{a}_i^{(k)}$ form a basis of $\mathbb{C}^n$.

4.5 Let $\bar{\boldsymbol{a}} = a^i \boldsymbol{g}_i$. Then, $\boldsymbol{a} \cdot \bar{\boldsymbol{a}} = a^i (\boldsymbol{a} \cdot \boldsymbol{g}_i)$. Thus, if $\boldsymbol{a} \cdot \boldsymbol{g}_i = 0$ $(i = 1, 2, \ldots, n)$, then $\boldsymbol{a} \cdot \bar{\boldsymbol{a}} = \boldsymbol{0}$ and according to (4.9) $\boldsymbol{a} = \boldsymbol{0}$ (sufficiency). Conversely, if $\boldsymbol{a} = \boldsymbol{0}$, then according to the results of Exercise 1.5 (also valid for complex vectors in $\mathbb{C}^n$) $\boldsymbol{a} \cdot \boldsymbol{g}_i = 0$ $(i = 1, 2, \ldots, n)$ (necessity).

4.6 Equations (4.40) and (4.42):

$$\mathbf{P}_i \mathbf{P}_j = \left(\sum_{k=1}^{r_i} \boldsymbol{a}_i^{(k)} \otimes \boldsymbol{b}_i^{(k)} \right) \left(\sum_{l=1}^{r_j} \boldsymbol{a}_j^{(l)} \otimes \boldsymbol{b}_j^{(l)} \right) = \sum_{k=1}^{r_i} \sum_{l=1}^{r_j} \delta_{ij} \delta^{kl} \boldsymbol{a}_i^{(k)} \otimes \boldsymbol{b}_j^{(l)}$$

$$= \delta_{ij} \sum_{k=1}^{r_i} \boldsymbol{a}_i^{(k)} \otimes \boldsymbol{b}_j^{(k)} = \begin{cases} \mathbf{P}_i & \text{if } i = j, \\ \mathbf{0} & \text{if } i \neq j. \end{cases}$$

4.7 By means of (4.40) and (4.42) we infer that $\mathbf{P}_i \boldsymbol{a}_j^{(l)} = \delta_{ij} \boldsymbol{a}_j^{(l)}$. Every vector $\boldsymbol{x}$ in $\mathbb{C}^n$ can be represented with respect the basis of this space $\boldsymbol{a}_i^{(k)}$ $(i = 1, 2, \ldots, s; k = 1, 2, \ldots, r_i)$ by $\boldsymbol{x} = \sum_{j=1}^{s} \sum_{k=1}^{r_j} x_j^{(k)} \boldsymbol{a}_j^{(k)}$. Hence,

$$\left(\sum_{i=1}^{s} \mathbf{P}_i \right) \boldsymbol{x} = \sum_{i=1}^{s} \sum_{j=1}^{s} \sum_{k=1}^{r_j} x_j^{(k)} \mathbf{P}_i \boldsymbol{a}_j^{(k)}$$

$$= \sum_{i=1}^{s} \sum_{j=1}^{s} \sum_{k=1}^{r_j} x_j^{(k)} \delta_{ij} \boldsymbol{a}_j^{(k)} = \sum_{j=1}^{s} \sum_{k=1}^{r_j} x_j^{(k)} \boldsymbol{a}_j^{(k)} = \boldsymbol{x}, \quad \forall \boldsymbol{x} \in \mathbb{C}^n,$$

which immediately implies (4.46).

4.8 Let $\lambda_1, \lambda_2, \ldots, \lambda_n$ be eigenvalues of $\mathbf{A} \in \mathbf{Lin}^n$. By the spectral mapping theorem (Theorem 4.1) we infer that $\exp(\lambda_i)$ $(i = 1, 2, \ldots, n)$ are eigenvalues of $\exp \mathbf{A}$. On use of (4.24) and (4.26) we can thus write: $\det[\exp(\mathbf{A})] = \prod_{i=1}^{n} \exp \lambda_i = \exp\left(\sum_{i=1}^{n} \lambda_i \right) = \exp(\mathrm{tr}\mathbf{A})$.

4.9 By Theorem 1.8 it suffices to prove that all eigenvalues of a second-order tensor $\mathbf{A}$ are non-zero (statement A) if and only if $\mathbf{A}\boldsymbol{x} = \boldsymbol{0}$ implies that $\boldsymbol{x} = \boldsymbol{0}$ (statement B). Indeed, if $\mathbf{A}\boldsymbol{x} = \boldsymbol{0}$ then either $\boldsymbol{x} = \boldsymbol{0}$ or $\boldsymbol{x}$ is an eigenvector corresponding to a zero eigenvalue. Thus, if A is true then B is also true. Conversely, if A is not true, the eigenvector corresponding to a zero eigenvalues does not satisfies the statement B.

4.10 Let $\boldsymbol{a}_i$ be a (right) eigenvector corresponding to an eigenvalue λ_i. Then, $\mathbf{A}\boldsymbol{a}_i = \lambda_i \boldsymbol{a}_i$. According to (1.129) $\mathbf{A}^{-1}(\lambda_i \boldsymbol{a}_i) = \boldsymbol{a}_i$, which implies that λ_i^{-1} is the eigenvalue of $\mathbf{A}^{-1}$.

4.11 Let for example $\mathbf{M}$ be positive-definite. Setting in (4.66) $\alpha = 1/2$ and $\alpha = -1/2$ we can define $\mathbf{M}^{1/2}$ and its inverse $\mathbf{M}^{-1/2}$, respectively. Now, consider a symmetric tensor $\mathbf{S} = \mathbf{M}^{1/2}\mathbf{N}\mathbf{M}^{1/2} = \mathbf{S}^{\mathrm{T}}$ with the spectral representation $\mathbf{S} = \sum_{i=1}^{n} \lambda_i \boldsymbol{d}_i \otimes \boldsymbol{d}_i$, where $\boldsymbol{d}_i \cdot \boldsymbol{d}_j = \delta_{ij}$ $(i, j = 1, 2, \ldots, n)$. Then, $\boldsymbol{a}_i = \mathbf{M}^{1/2}\boldsymbol{d}_i$ is the right eigenvector of $\mathbf{MN}$ associated with its eigenvalue λ_i $(i = 1, 2, \ldots, n)$. Indeed,

$$\mathbf{MN}\boldsymbol{a}_i = \mathbf{MN}\left(\mathbf{M}^{1/2}\boldsymbol{d}_i\right) = \mathbf{M}^{1/2}\mathbf{S}\boldsymbol{d}_i = \lambda_i \mathbf{M}^{1/2}\boldsymbol{d}_i = \lambda_i \boldsymbol{a}_i.$$

In the same manner, one verifies that $\boldsymbol{b}_i = \mathbf{M}^{-1/2}\boldsymbol{d}_i$ $(i = 1, 2, \ldots, n)$ is the corresponding left eigenvector of $\mathbf{MN}$, such that $\boldsymbol{a}_i \cdot \boldsymbol{b}_j = \delta_{ij}$ $(i, j = 1, 2, \ldots, n)$. The eigenvectors $\boldsymbol{d}_i$ $(i = 1, 2, \ldots, n)$ of $\mathbf{S} \in \mathbf{Sym}^n$ form a basis of $\mathbb{E}^n$. This is also the case both for the vectors $\boldsymbol{a}_i$ and $\boldsymbol{b}_i$ $(i = 1, 2, \ldots, n)$ since the tensor $\mathbf{M}^{1/2}$ is invertible (see proof of Theorem 1.8). This implies the spectral decomposition of $\mathbf{MN}$ by (4.39) as

$$\mathbf{MN} = \sum_{i=1}^{n} \lambda_i \boldsymbol{a}_i \otimes \boldsymbol{b}_i.$$

4.12 Let us consider the right hand side of (4.55) for example for $i = 1$. In this case we have

$$\prod_{\substack{j=1 \\ j \neq 1}}^{3} \frac{\mathbf{A} - \lambda_j \mathbf{I}}{\lambda_1 - \lambda_j} = \frac{\mathbf{A} - \lambda_2 \mathbf{I}}{\lambda_1 - \lambda_2} \frac{\mathbf{A} - \lambda_3 \mathbf{I}}{\lambda_1 - \lambda_3}.$$

On use of (4.43), (4.44) and (4.46) we further obtain

$$\begin{aligned}
\frac{\mathbf{A} - \lambda_2 \mathbf{I}}{\lambda_1 - \lambda_2} \frac{\mathbf{A} - \lambda_3 \mathbf{I}}{\lambda_1 - \lambda_3} &= \frac{\sum\limits_{i=1}^{3} (\lambda_i - \lambda_2) \mathbf{P}_i}{\lambda_1 - \lambda_2} \frac{\sum\limits_{j=1}^{3} (\lambda_j - \lambda_3) \mathbf{P}_j}{\lambda_1 - \lambda_3} \\
&= \frac{\sum\limits_{i,j=1}^{3} (\lambda_i - \lambda_2)(\lambda_j - \lambda_3)\delta_{ij}\mathbf{P}_i}{(\lambda_1 - \lambda_2)(\lambda_1 - \lambda_3)} = \frac{\sum\limits_{i=1}^{3} (\lambda_i - \lambda_2)(\lambda_i - \lambda_3)\mathbf{P}_i}{(\lambda_1 - \lambda_2)(\lambda_1 - \lambda_3)} \\
&= \frac{(\lambda_1 - \lambda_2)(\lambda_1 - \lambda_3)\mathbf{P}_1}{(\lambda_1 - \lambda_2)(\lambda_1 - \lambda_3)} = \mathbf{P}_1.
\end{aligned}$$

In a similar way, one verifies the Sylvester formula also for $i = 2$ and $i = 3$.

4.13 By (4.42), (9.15) and (9.16) we first obtain

$$\begin{aligned}
\mathbf{P}_1 &= \boldsymbol{a}_1 \otimes \boldsymbol{a}_1 \\
&= \left(\frac{1}{\sqrt{1 + \Lambda_1}} \boldsymbol{e}_1 + \sqrt{\frac{\Lambda_1}{1 + \Lambda_1}} \boldsymbol{e}_2\right) \otimes \left(\frac{1}{\sqrt{1 + \Lambda_1}} \boldsymbol{e}_1 + \sqrt{\frac{\Lambda_1}{1 + \Lambda_1}} \boldsymbol{e}_2\right)
\end{aligned}$$

$$= \frac{1}{1+\Lambda_1} e_1 \otimes e_1 + \frac{\Lambda_1}{1+\Lambda_1} e_2 \otimes e_2 + \frac{\sqrt{\Lambda_1}}{1+\Lambda_1} (e_1 \otimes e_2 + e_2 \otimes e_1)$$

$$= \begin{bmatrix} \frac{2}{\gamma^2+4+\gamma\sqrt{\gamma^2+4}} & \frac{1}{\sqrt{\gamma^2+4}} & 0 \\ \frac{1}{\sqrt{\gamma^2+4}} & \frac{2}{\gamma^2+4-\gamma\sqrt{\gamma^2+4}} & 0 \\ 0 & 0 & 0 \end{bmatrix} e_i \otimes e^j,$$

$$\mathbf{P}_2 = a_2 \otimes a_2$$

$$= \left(\frac{1}{\sqrt{1+\Lambda_2}} e_1 - \sqrt{\frac{\Lambda_2}{1+\Lambda_2}} e_2 \right) \otimes \left(\frac{1}{\sqrt{1+\Lambda_2}} e_1 - \sqrt{\frac{\Lambda_2}{1+\Lambda_2}} e_2 \right)$$

$$= \frac{1}{1+\Lambda_2} e_1 \otimes e_1 + \frac{\Lambda_2}{1+\Lambda_2} e_2 \otimes e_2 - \frac{\sqrt{\Lambda_2}}{1+\Lambda_2} (e_1 \otimes e_2 + e_2 \otimes e_1)$$

$$= \begin{bmatrix} \frac{2}{\gamma^2+4-\gamma\sqrt{\gamma^2+4}} & -\frac{1}{\sqrt{\gamma^2+4}} & 0 \\ -\frac{1}{\sqrt{\gamma^2+4}} & \frac{2}{\gamma^2+4+\gamma\sqrt{\gamma^2+4}} & 0 \\ 0 & 0 & 0 \end{bmatrix} e_i \otimes e^j,$$

$$\mathbf{P}_3 = a_3 \otimes a_3 = e_3 \otimes e_3 = \begin{bmatrix} 0 & 0 & 0 \\ 0 & 0 & 0 \\ 0 & 0 & 1 \end{bmatrix} e_i \otimes e^j.$$

The same expressions result also from the Sylvester formula (4.55) as follows

$$\mathbf{P}_1 = \frac{(\mathbf{C} - \Lambda_2 \mathbf{I})(\mathbf{C} - \Lambda_3 \mathbf{I})}{(\Lambda_1 - \Lambda_2)(\Lambda_1 - \Lambda_3)}$$

$$= \begin{bmatrix} \frac{2}{\gamma^2+4+\gamma\sqrt{\gamma^2+4}} & \frac{1}{\sqrt{\gamma^2+4}} & 0 \\ \frac{1}{\sqrt{\gamma^2+4}} & \frac{2}{\gamma^2+4-\gamma\sqrt{\gamma^2+4}} & 0 \\ 0 & 0 & 0 \end{bmatrix} e_i \otimes e^j,$$

$$\mathbf{P}_2 = \frac{(\mathbf{C} - \Lambda_3\mathbf{I})(\mathbf{C} - \Lambda_1\mathbf{I})}{(\Lambda_2 - \Lambda_3)(\Lambda_2 - \Lambda_1)}$$

$$= \begin{bmatrix} \dfrac{2}{\gamma^2 + 4 - \gamma\sqrt{\gamma^2+4}} & -\dfrac{1}{\sqrt{\gamma^2+4}} & 0 \\ -\dfrac{1}{\sqrt{\gamma^2+4}} & \dfrac{2}{\gamma^2 + 4 + \gamma\sqrt{\gamma^2+4}} & 0 \\ 0 & 0 & 0 \end{bmatrix} \boldsymbol{e}_i \otimes \boldsymbol{e}^j,$$

$$\mathbf{P}_3 = \frac{(\mathbf{C} - \Lambda_1\mathbf{I})(\mathbf{C} - \Lambda_2\mathbf{I})}{(\Lambda_3 - \Lambda_1)(\Lambda_3 - \Lambda_2)} = \begin{bmatrix} 0\,0\,0 \\ 0\,0\,0 \\ 0\,0\,1 \end{bmatrix} \boldsymbol{e}_i \otimes \boldsymbol{e}^j. \tag{9.17}$$

4.14 The characteristic equation of the tensor **A** takes the form:

$$\begin{vmatrix} -2-\lambda & 2 & 2 \\ 2 & 1-\lambda & 4 \\ 2 & 4 & 1-\lambda \end{vmatrix} = 0.$$

Writing out this determinant we get after some algebraic manipulations

$$\lambda^3 - 27\lambda - 54 = 0.$$

Comparing this equation with (4.28) we see that

$$\mathrm{I}_{\mathbf{A}} = 0, \quad \mathrm{II}_{\mathbf{A}} = -27, \quad \mathrm{III}_{\mathbf{A}} = 54. \tag{9.18}$$

Inserting this result into the Cardano formula (4.31) and (4.32) we obtain

$$\vartheta = \arccos\left[\frac{2\mathrm{I}_{\mathbf{A}}^3 - 9\mathrm{I}_{\mathbf{A}}\mathrm{II}_{\mathbf{A}} + 27\mathrm{III}_{\mathbf{A}}}{2\left(\mathrm{I}_{\mathbf{A}}^2 - 3\mathrm{II}_{\mathbf{A}}\right)^{3/2}}\right]$$

$$= \arccos\left[\frac{27 \cdot 54}{2\,(3 \cdot 27)^{3/2}}\right] = \arccos(1) = 0,$$

$$\lambda_k = \frac{1}{3}\left\{\mathrm{I}_{\mathbf{A}} + 2\sqrt{\mathrm{I}_{\mathbf{A}}^2 - 3\mathrm{II}_{\mathbf{A}}}\cos\frac{1}{3}\left[\vartheta + 2\pi(k-1)\right]\right\}$$

$$= \frac{2}{3}\sqrt{3 \cdot 27}\cos\left(\frac{2}{3}\pi(k-1)\right) = 6\cos\left(\frac{2}{3}\pi(k-1)\right), \quad k = 1, 2, 3.$$

Thus, we obtain two pairwise distinct eigenvalues ($s = 2$):

$$\lambda_1 = 6, \quad \lambda_2 = \lambda_3 = -3. \tag{9.19}$$

The Sylvester formula (4.55) further yields

$$\mathbf{P}_1 = \prod_{\substack{j=1\\j\neq 1}}^{2} \frac{\mathbf{A} - \lambda_j \mathbf{I}}{\lambda_i - \lambda_j} = \frac{\mathbf{A} - \lambda_2 \mathbf{I}}{\lambda_1 - \lambda_2} = \frac{\mathbf{A} + 3\mathbf{I}}{9} = \frac{1}{9}\begin{bmatrix} 1 & 2 & 2 \\ 2 & 4 & 4 \\ 2 & 4 & 4 \end{bmatrix} \boldsymbol{e}_i \otimes \boldsymbol{e}^j,$$

$$\mathbf{P}_2 = \prod_{\substack{j=1\\j\neq 2}}^{2} \frac{\mathbf{A} - \lambda_j \mathbf{I}}{\lambda_i - \lambda_j} = \frac{\mathbf{A} - \lambda_1 \mathbf{I}}{\lambda_2 - \lambda_1} = \frac{\mathbf{A} - 6\mathbf{I}}{-9} = \frac{1}{9}\begin{bmatrix} 8 & -2 & -2 \\ -2 & 5 & -4 \\ -2 & -4 & 5 \end{bmatrix} \boldsymbol{e}_i \otimes \boldsymbol{e}^j.$$

4.15 The spectral representation of **A** takes the form

$$\mathbf{A} = \sum_{i=1}^{s} \lambda_i \mathbf{P}_i = \lambda_1 \mathbf{P}_1 + \lambda_2 \mathbf{P}_2 = 6\mathbf{P}_1 - 3\mathbf{P}_2.$$

Thus,

$$\begin{aligned} \exp \mathbf{A} &= \sum_{i=1}^{s} \exp(\lambda_i)\, \mathbf{P}_i \\ &= \exp(\lambda_1)\, \mathbf{P}_1 + \exp(\lambda_2)\, \mathbf{P}_2 = \exp(6)\, \mathbf{P}_1 + \exp(-3)\, \mathbf{P}_2 \\ &= \frac{\mathrm{e}^6}{9}\begin{bmatrix} 1 & 2 & 2 \\ 2 & 4 & 4 \\ 2 & 4 & 4 \end{bmatrix} \boldsymbol{e}_i \otimes \boldsymbol{e}_j + \frac{\mathrm{e}^{-3}}{9}\begin{bmatrix} 8 & -2 & -2 \\ -2 & 5 & -4 \\ -2 & -4 & 5 \end{bmatrix} \boldsymbol{e}_i \otimes \boldsymbol{e}_j \\ &= \frac{1}{9}\begin{bmatrix} \mathrm{e}^6 + 8\mathrm{e}^{-3} & 2\mathrm{e}^6 - 2\mathrm{e}^{-3} & 2\mathrm{e}^6 - 2\mathrm{e}^{-3} \\ 2\mathrm{e}^6 - 2\mathrm{e}^{-3} & 4\mathrm{e}^6 + 5\mathrm{e}^{-3} & 4\mathrm{e}^6 - 4\mathrm{e}^{-3} \\ 2\mathrm{e}^6 - 2\mathrm{e}^{-3} & 4\mathrm{e}^6 - 4\mathrm{e}^{-3} & 4\mathrm{e}^6 + 5\mathrm{e}^{-3} \end{bmatrix} \boldsymbol{e}_i \otimes \boldsymbol{e}_j. \end{aligned}$$

4.16 Components of the eigenvectors $\boldsymbol{a} = a^i \boldsymbol{e}_i$ result from the equation system (4.16)

$$\left(\mathrm{A}^i_{\cdot j} - \delta^i_j \lambda\right) a^j = 0, \quad i = 1, 2, 3. \tag{9.20}$$

Setting $\lambda = 6$ we obtain only two independent equations

$$\begin{cases} -8a^1 + 2a^2 + 2a^3 = 0, \\ 2a^1 - 5a^2 + 4a^3 = 0. \end{cases}$$

Multiplying the first equation by two and subtracting from the second one we get $a^2 = 2a^1$ and consequently $a^3 = 2a^1$. Additionally we require that the eigenvectors have unit length so that

$$\left(a^1\right)^2 + \left(a^2\right)^2 + \left(a^3\right)^2 = 1, \tag{9.21}$$

which leads to

$$\boldsymbol{a}_1 = \frac{1}{3}\boldsymbol{e}_1 + \frac{2}{3}\boldsymbol{e}_2 + \frac{2}{3}\boldsymbol{e}_3.$$

Further, setting in the equation system (9.20) $\lambda = -3$ we obtain only one independent linear equation

$$a^1 + 2a^2 + 2a^3 = 0 \tag{9.22}$$

with respect to the components of the eigenvectors corresponding to this double eigenvalue. One of these eigenvectors can be obtained by setting for example $a^1 = 0$. In this case, $a^2 = -a^3$ and in view of (9.21)

$$\boldsymbol{a}_2^{(1)} = \frac{1}{\sqrt{2}}\boldsymbol{e}_2 - \frac{1}{\sqrt{2}}\boldsymbol{e}_3.$$

Requiring that the eigenvectors $\boldsymbol{a}_2^{(1)}$ and $\boldsymbol{a}_2^{(2)}$ corresponding to the double eigenvalue $\lambda = -3$ are orthogonal we get an additional condition $a^2 = a^3$ for the components of $\boldsymbol{a}_2^{(2)}$. Taking into account (9.21) and (9.22) this yields

$$\boldsymbol{a}_2^{(2)} = -\frac{4}{3\sqrt{2}}\boldsymbol{e}_1 + \frac{1}{3\sqrt{2}}\boldsymbol{e}_2 + \frac{1}{3\sqrt{2}}\boldsymbol{e}_3.$$

With the aid of the eigenvectors we can construct eigenprojections without the Sylvester formula by (4.42):

$$\begin{aligned}
\mathbf{P}_1 &= \boldsymbol{a}_1 \otimes \boldsymbol{a}_1 \\
&= \left(\frac{1}{3}\boldsymbol{e}_1 + \frac{2}{3}\boldsymbol{e}_2 + \frac{2}{3}\boldsymbol{e}_3\right) \otimes \left(\frac{1}{3}\boldsymbol{e}_1 + \frac{2}{3}\boldsymbol{e}_2 + \frac{2}{3}\boldsymbol{e}_3\right) = \frac{1}{9}\begin{bmatrix} 1 & 2 & 2 \\ 2 & 4 & 4 \\ 2 & 4 & 4 \end{bmatrix} \boldsymbol{e}_i \otimes \boldsymbol{e}_j,
\end{aligned}$$

$$\begin{aligned}
\mathbf{P}_2 = \boldsymbol{a}_2^{(1)} \otimes \boldsymbol{a}_2^{(1)} + \boldsymbol{a}_2^{(2)} \otimes \boldsymbol{a}_2^{(2)} &= \left(\frac{1}{\sqrt{2}}\boldsymbol{e}_2 - \frac{1}{\sqrt{2}}\boldsymbol{e}_3\right) \otimes \left(\frac{1}{\sqrt{2}}\boldsymbol{e}_2 - \frac{1}{\sqrt{2}}\boldsymbol{e}_3\right) \\
&+ \left(-\frac{4}{3\sqrt{2}}\boldsymbol{e}_1 + \frac{1}{3\sqrt{2}}\boldsymbol{e}_2 + \frac{1}{3\sqrt{2}}\boldsymbol{e}_3\right) \otimes \left(-\frac{4}{3\sqrt{2}}\boldsymbol{e}_1 + \frac{1}{3\sqrt{2}}\boldsymbol{e}_2 + \frac{1}{3\sqrt{2}}\boldsymbol{e}_3\right)
\end{aligned}$$

$$= \frac{1}{9}\begin{bmatrix} 8 & -2 & -2 \\ -2 & 5 & -4 \\ -2 & -4 & 5 \end{bmatrix} \boldsymbol{e}_i \otimes \boldsymbol{e}_j .$$

4.17 Since linearly independent vectors are non-zero it follows from (4.9) that $\boldsymbol{c}_i \cdot \overline{\boldsymbol{c}}_i \neq 0$ $(i = 1, 2, \ldots, m)$. Thus, the first vector can be given by

$$\boldsymbol{a}_1 = \frac{\boldsymbol{c}_1}{\sqrt{\boldsymbol{c}_1 \cdot \overline{\boldsymbol{c}}_1}},$$

such that $\boldsymbol{a}_1 \cdot \overline{\boldsymbol{a}}_1 = 1$. Next, we set

$$\boldsymbol{a}_2' = \boldsymbol{c}_2 - (\boldsymbol{c}_2 \cdot \overline{\boldsymbol{a}}_1)\, \boldsymbol{a}_1 ,$$

so that $\boldsymbol{a}_2' \cdot \overline{\boldsymbol{a}}_1 = 0$. Further, $\boldsymbol{a}_2' \neq \boldsymbol{0}$ because otherwise $\boldsymbol{c}_2 = (\boldsymbol{c}_2 \cdot \overline{\boldsymbol{a}}_1)\, \boldsymbol{a}_1 = (\boldsymbol{c}_2 \cdot \overline{\boldsymbol{a}}_1)\, (\boldsymbol{c}_1 \cdot \overline{\boldsymbol{c}}_1)^{-1/2}\, \boldsymbol{c}_1$ which implies a linear dependence between $\boldsymbol{c}_1$ and $\boldsymbol{c}_2$. Thus, we can set $\boldsymbol{a}_2 = \boldsymbol{a}_2' / \sqrt{\boldsymbol{a}_2' \cdot \overline{\boldsymbol{a}}_2'}$. The third vector can be given by

$$\boldsymbol{a}_3 = \frac{\boldsymbol{a}_3'}{\sqrt{\boldsymbol{a}_3' \cdot \overline{\boldsymbol{a}}_3'}}, \quad \text{where} \quad \boldsymbol{a}_3' = \boldsymbol{c}_3 - (\boldsymbol{c}_3 \cdot \overline{\boldsymbol{a}}_2)\, \boldsymbol{a}_2 - (\boldsymbol{c}_3 \cdot \overline{\boldsymbol{a}}_1)\, \boldsymbol{a}_1 ,$$

so that $\boldsymbol{a}_3 \cdot \overline{\boldsymbol{a}}_1 = \boldsymbol{a}_3 \cdot \overline{\boldsymbol{a}}_2 = 0$. Repeating this procedure we finally obtain the set of vectors $\boldsymbol{a}_i$ satisfying the condition $\boldsymbol{a}_i \cdot \overline{\boldsymbol{a}}_j = \delta_{ij}, (i, j = 1, 2, \ldots, m)$. One can easily show that these vectors are linearly independent. Indeed, otherwise $\sum_{i=1}^{m} \alpha_i \boldsymbol{a}_i = \boldsymbol{0}$, where not all α_i are zero. Multiplying this vector equation scalarly by $\overline{\boldsymbol{a}}_j$ $(j = 1, 2, \ldots, m)$ yields, however, $\alpha_j = 0$ $(j = 1, 2, \ldots, m)$.

4.18 Comparing $(4.67)_1$ with $(4.72)_1$ we infer that the right eigenvectors $\boldsymbol{a}_i^{(k)}$ $(k = 1, 2, \ldots, t_i)$ associated with a complex eigenvalue λ_i are simultaneously the left eigenvalues associated with $\overline{\lambda}_i$. Since $\overline{\lambda}_i \neq \lambda_i$ Theorem 4.3 implies that $\boldsymbol{a}_i^{(k)} \cdot \boldsymbol{a}_i^{(l)} = 0$ $(k, l = 1, 2, \ldots, t_i)$.

4.19 Taking into account the identities $\mathrm{tr}\mathbf{W}^k = 0$, where $k = 1, 3, 5, \ldots$ (see Exercise 1.49) we obtain from (4.29)

$$\mathrm{I}_{\mathbf{W}} = \mathrm{tr}\mathbf{W} = 0,$$

$$\begin{aligned} \mathrm{II}_{\mathbf{W}} &= \frac{1}{2}\left[(\mathrm{tr}\mathbf{W})^2 - \mathrm{tr}\mathbf{W}^2\right] \\ &= -\frac{1}{2}\mathrm{tr}\mathbf{W}^2 = -\frac{1}{2}\left(\mathbf{W} : \mathbf{W}^{\mathrm{T}}\right) = \frac{1}{2}\left(\mathbf{W} : \mathbf{W}\right) = \frac{1}{2}\left\|\mathbf{W}\right\|^2 , \\ \mathrm{III}_{\mathbf{W}} &= \frac{1}{3}\left[\mathrm{tr}\mathbf{W}^3 - \frac{3}{2}\mathrm{tr}\mathbf{W}^2\mathrm{tr}\mathbf{W} + \frac{1}{2}(\mathrm{tr}\mathbf{W})^3\right] = 0, \end{aligned}$$

or in another way

$$\mathrm{III}_{\mathbf{W}} = \det\mathbf{W} = \det\mathbf{W}^{\mathrm{T}} = \det(-\mathbf{W}) = (-1)^3\det\mathbf{W} = -\mathrm{III}_{\mathbf{W}} = 0.$$

4.20 Eigenvalues of the rotation tensor (Exercise 1.24)

$$\mathbf{R} = \mathrm{R}^{i}_{\cdot j}\boldsymbol{e}_i \otimes \boldsymbol{e}^j, \quad \text{where} \quad \left[\mathrm{R}^{i}_{\cdot j}\right] = \left[\mathrm{R}^{ij}\right] = \begin{bmatrix} \cos\alpha & -\sin\alpha & 0 \\ \sin\alpha & \cos\alpha & 0 \\ 0 & 0 & 1 \end{bmatrix}$$

result from the characteristic equation

$$\begin{vmatrix} \cos\alpha - \lambda & -\sin\alpha & 0 \\ \sin\alpha & \cos\alpha - \lambda & 0 \\ 0 & 0 & 1-\lambda \end{vmatrix} = 0.$$

Writing out this determinant we have

$$(1-\lambda)\left(\lambda^2 - 2\lambda\cos\alpha + 1\right) = 0$$

and consequently

$$\lambda_1 = 1, \quad \lambda_{2/3} = \cos\alpha \pm \mathrm{i}\sin\alpha.$$

Components of the right eigenvectors $\boldsymbol{a} = a^i\boldsymbol{e}_i$ result from the equation system (4.16)

$$\left(\mathrm{R}^{i}_{\cdot j} - \delta^{i}_{j}\lambda\right)a^j = 0, \quad i = 1,2,3. \tag{9.23}$$

Setting first $\lambda = 1$ we obtain a homogeneous equation system

$$a^1(\cos\alpha - 1) - a^2\sin\alpha = 0,$$
$$a^1\sin\alpha + a^2(\cos\alpha - 1) = 0,$$

leading to $a^1 = a^2 = 0$. Thus, $\boldsymbol{a}_1 = a^3\boldsymbol{e}_3$, where a^3 can be an arbitrary real number. The unit length condition requires further that

$$\boldsymbol{a}_1 = \boldsymbol{e}_3.$$

Next, inserting $\lambda = \cos\alpha \pm \mathrm{i}\sin\alpha$ into (9.23) yields

$$a^2 = \mp\mathrm{i}a^1, \quad a^3 = 0.$$

Thus, the right eigenvectors associated with the complex conjugate eigenvalues $\lambda_{2/3}$ are of the form $\boldsymbol{a}_{2/3} = a^1(\boldsymbol{e}_1 \mp \mathrm{i}\boldsymbol{e}_2)$. Bearing in mind that any rotation tensor is orthogonal we infer that $\overline{\boldsymbol{a}}_{2/3} = \boldsymbol{a}_{3/2} = a^1(\boldsymbol{e}_1 \pm \mathrm{i}\boldsymbol{e}_2)$ are the left eigenvectors associated with $\lambda_{2/3}$. Imposing the additional condition $\boldsymbol{a}_2 \cdot \overline{\boldsymbol{a}}_2 = \boldsymbol{a}_2 \cdot \boldsymbol{a}_3 = 1$ (4.38) we finally obtain

$$\boldsymbol{a}_2 = \frac{\sqrt{2}}{2}\left(\boldsymbol{e}_1 - \mathrm{i}\boldsymbol{e}_2\right), \quad \boldsymbol{a}_3 = \frac{\sqrt{2}}{2}\left(\boldsymbol{e}_1 + \mathrm{i}\boldsymbol{e}_2\right).$$

The eigenprojections can further be expressed by (4.42) as

$$\mathbf{P}_1 = \boldsymbol{a}_1 \otimes \boldsymbol{a}_1 = \boldsymbol{e}_3 \otimes \boldsymbol{e}_3,$$

$$\mathbf{P}_2 = \boldsymbol{a}_2 \otimes \overline{\boldsymbol{a}}_2 = \frac{\sqrt{2}}{2}\left(\boldsymbol{e}_1 - \mathrm{i}\boldsymbol{e}_2\right) \otimes \frac{\sqrt{2}}{2}\left(\boldsymbol{e}_1 + \mathrm{i}\boldsymbol{e}_2\right)$$

$$= \frac{1}{2}\left(\boldsymbol{e}_1 \otimes \boldsymbol{e}_1 + \boldsymbol{e}_2 \otimes \boldsymbol{e}_2\right) + \frac{1}{2}\mathrm{i}\left(\boldsymbol{e}_1 \otimes \boldsymbol{e}_2 - \boldsymbol{e}_2 \otimes \boldsymbol{e}_1\right),$$

$$\mathbf{P}_3 = \boldsymbol{a}_3 \otimes \overline{\boldsymbol{a}}_3 = \frac{\sqrt{2}}{2}\left(\boldsymbol{e}_1 + \mathrm{i}\boldsymbol{e}_2\right) \otimes \frac{\sqrt{2}}{2}\left(\boldsymbol{e}_1 - \mathrm{i}\boldsymbol{e}_2\right)$$

$$= \frac{1}{2}\left(\boldsymbol{e}_1 \otimes \boldsymbol{e}_1 + \boldsymbol{e}_2 \otimes \boldsymbol{e}_2\right) - \frac{1}{2}\mathrm{i}\left(\boldsymbol{e}_1 \otimes \boldsymbol{e}_2 - \boldsymbol{e}_2 \otimes \boldsymbol{e}_1\right).$$

4.21 First, we write

$$\left[\left(\mathbf{A}^2\right)^i_{\ j}\right] = \begin{bmatrix} -2 & 2 & 2 \\ 2 & 1 & 4 \\ 2 & 4 & 1 \end{bmatrix}\begin{bmatrix} -2 & 2 & 2 \\ 2 & 1 & 4 \\ 2 & 4 & 1 \end{bmatrix} = \begin{bmatrix} 12 & 6 & 6 \\ 6 & 21 & 12 \\ 6 & 12 & 21 \end{bmatrix},$$

$$\left[\left(\mathbf{A}^3\right)^i_{\ j}\right] = \begin{bmatrix} 12 & 6 & 6 \\ 6 & 21 & 12 \\ 6 & 12 & 21 \end{bmatrix}\begin{bmatrix} -2 & 2 & 2 \\ 2 & 1 & 4 \\ 2 & 4 & 1 \end{bmatrix} = \begin{bmatrix} 0 & 54 & 54 \\ 54 & 81 & 108 \\ 54 & 108 & 81 \end{bmatrix}.$$

Then,

$$p_{\mathbf{A}}\left(\mathbf{A}\right) = \mathbf{A}^3 - 27\mathbf{A} - 54\mathbf{I} = \begin{bmatrix} 0 & 54 & 54 \\ 54 & 81 & 108 \\ 54 & 108 & 81 \end{bmatrix} \boldsymbol{e}_i \otimes \boldsymbol{e}_j$$

$$-27\begin{bmatrix} -2 & 2 & 2 \\ 2 & 1 & 4 \\ 2 & 4 & 1 \end{bmatrix} \boldsymbol{e}_i \otimes \boldsymbol{e}_j - 54\begin{bmatrix} 1 & 0 & 0 \\ 0 & 1 & 0 \\ 0 & 0 & 1 \end{bmatrix} \boldsymbol{e}_i \otimes \boldsymbol{e}_j = \begin{bmatrix} 0 & 0 & 0 \\ 0 & 0 & 0 \\ 0 & 0 & 0 \end{bmatrix} \boldsymbol{e}_i \otimes \boldsymbol{e}_j.$$

4.22 The characteristic polynomial of $\mathbf{F}$ (2.69) can be represented by $p_{\mathbf{A}}\left(\lambda\right) = \left(1 - \lambda\right)^3$. Hence,

$$p_{\mathbf{F}}\left(\mathbf{F}\right) = \left(\mathbf{I} - \mathbf{F}\right)^3 = \begin{bmatrix} 0 & -\gamma & 0 \\ 0 & 0 & 0 \\ 0 & 0 & 0 \end{bmatrix}^3 \boldsymbol{e}_i \otimes \boldsymbol{e}^j = \begin{bmatrix} 0 & 0 & 0 \\ 0 & 0 & 0 \\ 0 & 0 & 0 \end{bmatrix} \boldsymbol{e}_i \otimes \boldsymbol{e}^j = \mathbf{0}.$$

9.5 Exercises of Chap. 5

5.1 By using $(1.106)_1$, (D.2) and (5.17) one can verify for example $(5.20)_1$ and $(5.21)_1$ within the following steps

$$\mathbf{A} \otimes (\mathbf{B} + \mathbf{C}) : \mathbf{X} = \mathbf{AX}(\mathbf{B} + \mathbf{C}) = \mathbf{AXB} + \mathbf{AXC} = (\mathbf{A} \otimes \mathbf{B} + \mathbf{A} \otimes \mathbf{C}) : \mathbf{X},$$

$$\begin{aligned} \mathbf{A} \odot (\mathbf{B} + \mathbf{C}) : \mathbf{X} &= \mathbf{A}\left[(\mathbf{B} + \mathbf{C}) : \mathbf{X}\right] = \mathbf{A}\left(\mathbf{B} : \mathbf{X} + \mathbf{C} : \mathbf{X}\right) \\ &= \mathbf{A}\left(\mathbf{B} : \mathbf{X}\right) + \mathbf{A}\left(\mathbf{C} : \mathbf{X}\right) \\ &= (\mathbf{A} \odot \mathbf{B} + \mathbf{A} \odot \mathbf{C}) : \mathbf{X}, \quad \forall \mathbf{X} \in \mathbf{Lin}^n. \end{aligned}$$

The proof of $(5.20)_2$ and $(5.21)_2$ is similar.

5.2 With the aid of (5.16), (5.17) and (1.145) we can write

$$(\mathbf{Y} : \mathbf{A} \otimes \mathbf{B}) : \mathbf{X} = \mathbf{Y} : (\mathbf{A} \otimes \mathbf{B} : \mathbf{X}) = \mathbf{Y} : \mathbf{AXB} = \mathbf{A}^{\mathrm{T}}\mathbf{Y}\mathbf{B}^{\mathrm{T}} : \mathbf{X},$$

$$\begin{aligned} (\mathbf{Y} : \mathbf{A} \odot \mathbf{B}) : \mathbf{X} &= \mathbf{Y} : (\mathbf{A} \odot \mathbf{B} : \mathbf{X}) = \mathbf{Y} : [\mathbf{A}(\mathbf{B} : \mathbf{X})] \\ &= (\mathbf{Y} : \mathbf{A})(\mathbf{B} : \mathbf{X}) = [(\mathbf{Y} : \mathbf{A})\mathbf{B}] : \mathbf{X}, \quad \forall \mathbf{X}, \mathbf{Y} \in \mathbf{Lin}^n. \end{aligned}$$

5.3 Using the definition of the simple composition (5.40) and taking (5.17) into account we obtain

$$\begin{aligned} \mathbf{A}(\mathbf{B} \otimes \mathbf{C})\mathbf{D} : \mathbf{X} &= \mathbf{A}(\mathbf{B} \otimes \mathbf{C} : \mathbf{X})\mathbf{D} = \mathbf{A}(\mathbf{BXC})\mathbf{D} \\ &= (\mathbf{AB})\mathbf{X}(\mathbf{CD}) = (\mathbf{AB}) \otimes (\mathbf{CD}) : \mathbf{X}, \\ \mathbf{A}(\mathbf{B} \odot \mathbf{C})\mathbf{D} : \mathbf{X} &= \mathbf{A}(\mathbf{B} \odot \mathbf{C} : \mathbf{X})\mathbf{D} = \mathbf{A}[\mathbf{B}(\mathbf{C} : \mathbf{X})]\mathbf{D} \\ &= \mathbf{ABD}(\mathbf{C} : \mathbf{X}) = (\mathbf{ABD}) \odot \mathbf{C} : \mathbf{X}, \quad \forall \mathbf{X} \in \mathbf{Lin}^n. \end{aligned}$$

5.4 By means of (1.147), (5.17), (5.22) and (5.45) we can write

$$(\mathbf{A} \otimes \mathbf{B})^{\mathrm{T}} : \mathbf{X} = \mathbf{X} : (\mathbf{A} \otimes \mathbf{B}) = \mathbf{A}^{\mathrm{T}}\mathbf{X}\mathbf{B}^{\mathrm{T}} = \left(\mathbf{A}^{\mathrm{T}} \otimes \mathbf{B}^{\mathrm{T}}\right) : \mathbf{X},$$

$$(\mathbf{A} \odot \mathbf{B})^{\mathrm{T}} : \mathbf{X} = \mathbf{X} : (\mathbf{A} \odot \mathbf{B}) = (\mathbf{X} : \mathbf{A})\mathbf{B} = (\mathbf{B} \odot \mathbf{A}) : \mathbf{X},$$

$$\begin{aligned} (\mathbf{A} \odot \mathbf{B})^{\mathrm{t}} : \mathbf{X} &= (\mathbf{A} \odot \mathbf{B}) : \mathbf{X}^{\mathrm{T}} = \mathbf{A}\left(\mathbf{B} : \mathbf{X}^{\mathrm{T}}\right) \\ &= \mathbf{A}\left(\mathbf{B}^{\mathrm{T}} : \mathbf{X}\right) = \left(\mathbf{A} \odot \mathbf{B}^{\mathrm{T}}\right) : \mathbf{X}, \quad \forall \mathbf{X} \in \mathbf{Lin}^n. \end{aligned}$$

Identities (5.51) and (5.52) follow immediately from (1.121) (5.23), (5.24) $(5.49)_1$ and (5.50) by setting $\mathbf{A} = \boldsymbol{a} \otimes \boldsymbol{b}$, $\mathbf{B} = \boldsymbol{c} \otimes \boldsymbol{d}$ or $\mathbf{A} = \boldsymbol{a} \otimes \boldsymbol{d}$, $\mathbf{B} = \boldsymbol{b} \otimes \boldsymbol{c}$, respectively.

5.5 Using (5.51) and (5.52) we obtain for the left and right hand sides different results:

$$(\boldsymbol{a} \otimes \boldsymbol{b} \otimes \boldsymbol{c} \otimes \boldsymbol{d})^{\mathrm{tT}} = (\boldsymbol{a} \otimes \boldsymbol{c} \otimes \boldsymbol{b} \otimes \boldsymbol{d})^{\mathrm{T}} = \boldsymbol{c} \otimes \boldsymbol{a} \otimes \boldsymbol{d} \otimes \boldsymbol{b},$$

$$(\boldsymbol{a} \otimes \boldsymbol{b} \otimes \boldsymbol{c} \otimes \boldsymbol{d})^{\mathrm{tT}} = (\boldsymbol{b} \otimes \boldsymbol{a} \otimes \boldsymbol{d} \otimes \boldsymbol{c})^{\mathrm{t}} = \boldsymbol{b} \otimes \boldsymbol{d} \otimes \boldsymbol{a} \otimes \boldsymbol{c}.$$

5.6 Equations (5.31), (5.32) and (5.45):

$$\begin{aligned}(\mathcal{A} : \mathcal{B})^{\mathrm{T}} : \mathbf{X} &= \mathbf{X} : (\mathcal{A} : \mathcal{B}) = (\mathbf{X} : \mathcal{A}) : \mathcal{B} \\ &= \mathcal{B}^{\mathrm{T}} : (\mathbf{X} : \mathcal{A}) = \mathcal{B}^{\mathrm{T}} : \left(\mathcal{A}^{\mathrm{T}} : \mathbf{X}\right) = \left(\mathcal{B}^{\mathrm{T}} : \mathcal{A}^{\mathrm{T}}\right) : \mathbf{X}, \\ (\mathcal{A} : \mathcal{B})^{\mathrm{t}} : \mathbf{X} &= (\mathcal{A} : \mathcal{B}) : \mathbf{X}^{\mathrm{T}} = \mathcal{A} : \left(\mathcal{B} : \mathbf{X}^{\mathrm{T}}\right) \\ &= \mathcal{A} : \left(\mathcal{B}^{\mathrm{t}} : \mathbf{X}\right) = \left(\mathcal{A} : \mathcal{B}^{\mathrm{t}}\right) : \mathbf{X}, \quad \forall \mathbf{X} \in \mathbf{Lin}^n.\end{aligned}$$

5.7 In view of (1.120), (5.17) and (5.45) we write for an arbitrary tensor $\mathbf{X} \in \mathbf{Lin}^n$

$$(\mathbf{A} \otimes \mathbf{B})^{\mathrm{t}} : (\mathbf{C} \otimes \mathbf{D}) : \mathbf{X} = (\mathbf{A} \otimes \mathbf{B})^{\mathrm{t}} : (\mathbf{CXD}) = (\mathbf{A} \otimes \mathbf{B}) : (\mathbf{CXD})^{\mathrm{T}}$$

$$= (\mathbf{A} \otimes \mathbf{B}) : \left(\mathbf{D}^{\mathrm{T}}\mathbf{X}^{\mathrm{T}}\mathbf{C}^{\mathrm{T}}\right) = \mathbf{AD}^{\mathrm{T}}\mathbf{X}^{\mathrm{T}}\mathbf{C}^{\mathrm{T}}\mathbf{B}$$

$$= \left[\left(\mathbf{AD}^{\mathrm{T}}\right) \otimes \left(\mathbf{C}^{\mathrm{T}}\mathbf{B}\right)\right] : \mathbf{X}^{\mathrm{T}} = \left[\left(\mathbf{AD}^{\mathrm{T}}\right) \otimes \left(\mathbf{C}^{\mathrm{T}}\mathbf{B}\right)\right]^{\mathrm{t}} : \mathbf{X},$$

$$(\mathbf{A} \otimes \mathbf{B})^{\mathrm{t}} : (\mathbf{C} \odot \mathbf{D}) : \mathbf{X} = (\mathbf{A} \otimes \mathbf{B})^{\mathrm{t}} : [(\mathbf{D} : \mathbf{X})\,\mathbf{C}]$$

$$= (\mathbf{A} \otimes \mathbf{B}) : \left[(\mathbf{D} : \mathbf{X})\,\mathbf{C}^{\mathrm{T}}\right] = (\mathbf{D} : \mathbf{X})\,\mathbf{AC}^{\mathrm{T}}\mathbf{B} = \left(\mathbf{AC}^{\mathrm{T}}\mathbf{B}\right) \odot \mathbf{D} : \mathbf{X}.$$

5.8 By virtue of (5.51) and (5.52) we can write

$$\begin{aligned}\mathcal{C}^{\mathrm{T}} &= \left(\mathcal{C}^{ijkl}\boldsymbol{g}_i \otimes \boldsymbol{g}_j \otimes \boldsymbol{g}_k \otimes \boldsymbol{g}_l\right)^{\mathrm{T}} = \mathcal{C}^{ijkl}\boldsymbol{g}_j \otimes \boldsymbol{g}_i \otimes \boldsymbol{g}_l \otimes \boldsymbol{g}_k \\ &= \mathcal{C}^{jilk}\boldsymbol{g}_i \otimes \boldsymbol{g}_j \otimes \boldsymbol{g}_k \otimes \boldsymbol{g}_l, \\ \mathcal{C}^{\mathrm{t}} &= \left(\mathcal{C}^{ijkl}\boldsymbol{g}_i \otimes \boldsymbol{g}_j \otimes \boldsymbol{g}_k \otimes \boldsymbol{g}_l\right)^{\mathrm{t}} = \mathcal{C}^{ijkl}\boldsymbol{g}_i \otimes \boldsymbol{g}_k \otimes \boldsymbol{g}_j \otimes \boldsymbol{g}_l \\ &= \mathcal{C}^{ikjl}\boldsymbol{g}_i \otimes \boldsymbol{g}_j \otimes \boldsymbol{g}_k \otimes \boldsymbol{g}_l.\end{aligned}$$

According to (5.60) and (5.61) $\mathcal{C}^{\mathrm{T}} = \mathcal{C}^{\mathrm{t}} = \mathcal{C}$. Taking also into account that the tensors $\boldsymbol{g}_i \otimes \boldsymbol{g}_j \otimes \boldsymbol{g}_k \otimes \boldsymbol{g}_l$ $(i, j, k, l = 1, 2, \ldots, n)$ are linearly independent we thus write

$$\mathcal{C}^{ijkl} = \mathcal{C}^{jilk} = \mathcal{C}^{ikjl}.$$

The remaining relations (5.70) are obtained in the same manner by applying the identities $\mathcal{C} = \mathcal{C}^{\mathrm{TtT}}$ and $\mathcal{C} = \mathcal{C}^{\mathrm{TtT}}$.

5.9 With the aid of (1.147), (5.16) and (5.81) we get

$$(\mathbf{Y} : \mathcal{T}) : \mathbf{X} = \mathbf{Y} : (\mathcal{T} : \mathbf{X}) = \mathbf{Y} : \mathbf{X}^{\mathrm{T}} = \mathbf{Y}^{\mathrm{T}} : \mathbf{X}, \quad \forall \mathbf{X}, \mathbf{Y} \in \mathrm{Lin}^n.$$

5.10 On use of (5.31), $(5.45)_2$ and (5.81) we obtain

$$(\mathcal{A} : \mathcal{T}) : \mathbf{X} = \mathcal{A} : (\mathcal{T} : \mathbf{X}) = \mathcal{A} : \mathbf{X}^{\mathrm{T}} = \mathcal{A}^{\mathrm{t}} : \mathbf{X}, \quad \forall \mathbf{X} \in \mathrm{Lin}^n.$$

The second identity (5.85) can be derived by means of (5.54), (5.80) and (5.83) as follows

$$\mathcal{A}^{\mathrm{TtT}} = (\mathcal{I} : \mathcal{A})^{\mathrm{TtT}} = \left(\mathcal{A}^{\mathrm{T}} : \mathcal{I}\right)^{\mathrm{tT}} = \left(\mathcal{A}^{\mathrm{T}} : \mathcal{I}^{\mathrm{t}}\right)^{\mathrm{T}} = \left(\mathcal{A}^{\mathrm{T}} : \mathcal{T}\right)^{\mathrm{T}} = \mathcal{T} : \mathcal{A}.$$

The last identity (5.85) can finally be proved by

$$(\mathcal{T} : \mathcal{T}) : \mathbf{X} = \mathcal{T} : (\mathcal{T} : \mathbf{X}) = \mathcal{T} : \mathbf{X}^{\mathrm{T}} = \mathbf{X} = \mathcal{I} : \mathbf{X}, \quad \forall \mathbf{X} \in \mathrm{Lin}^n.$$

5.11 $\mathcal{C}$ possesses the minor symmetry (5.61) by the very definition. In order to prove the major symmetry (5.60) we show that $\mathcal{C} : \mathbf{X} = \mathbf{X} : \mathcal{C}$, $\forall \mathbf{X} \in \mathrm{Lin}^n$. Indeed, in view of $(5.17)_1$, $(5.22)_1$ and (5.48)

$$\begin{aligned}\mathcal{C} : \mathbf{X} &= (\mathbf{M}_1 \otimes \mathbf{M}_2 + \mathbf{M}_2 \otimes \mathbf{M}_1)^{\mathrm{s}} : \mathbf{X} = (\mathbf{M}_1 \otimes \mathbf{M}_2 + \mathbf{M}_2 \otimes \mathbf{M}_1) : \mathrm{sym}\mathbf{X} \\ &= \mathbf{M}_1 (\mathrm{sym}\mathbf{X}) \mathbf{M}_2 + \mathbf{M}_2 (\mathrm{sym}\mathbf{X}) \mathbf{M}_1,\end{aligned}$$

$$\begin{aligned}\mathbf{X} : \mathcal{C} &= \mathbf{X} : (\mathbf{M}_1 \otimes \mathbf{M}_2 + \mathbf{M}_2 \otimes \mathbf{M}_1)^{\mathrm{s}} \\ &= \mathrm{sym}\left[\mathbf{X} : (\mathbf{M}_1 \otimes \mathbf{M}_2 + \mathbf{M}_2 \otimes \mathbf{M}_1)\right] \\ &= \mathrm{sym}\left(\mathbf{M}_1 \mathbf{X} \mathbf{M}_2 + \mathbf{M}_2 \mathbf{X} \mathbf{M}_1\right) = \mathbf{M}_1 (\mathrm{sym}\mathbf{X}) \mathbf{M}_2 + \mathbf{M}_2 (\mathrm{sym}\mathbf{X}) \mathbf{M}_1.\end{aligned}$$

5.12

(a) Let $\boldsymbol{e}_i$ $(i = 1, 2, 3)$ be an orthonormal basis in $\mathbb{E}^3$. By virtue of (5.77), (5.84) and (5.86) we can write

$$\mathcal{I}^{\mathrm{s}} = \sum_{i,j=1}^{3} \frac{1}{2} \boldsymbol{e}_i \otimes \left(\boldsymbol{e}_i \otimes \boldsymbol{e}_j + \boldsymbol{e}_j \otimes \boldsymbol{e}_i\right) \otimes \boldsymbol{e}_j.$$

Using the notation

$$\mathbf{M}_i = \boldsymbol{e}_i \otimes \boldsymbol{e}_i, \quad i = 1,2,3, \quad \mathbf{M}_4 = \frac{\boldsymbol{e}_1 \otimes \boldsymbol{e}_2 + \boldsymbol{e}_2 \otimes \boldsymbol{e}_1}{\sqrt{2}},$$

$$\mathbf{M}_5 = \frac{\boldsymbol{e}_1 \otimes \boldsymbol{e}_3 + \boldsymbol{e}_3 \otimes \boldsymbol{e}_1}{\sqrt{2}}, \qquad \mathbf{M}_6 = \frac{\boldsymbol{e}_3 \otimes \boldsymbol{e}_2 + \boldsymbol{e}_2 \otimes \boldsymbol{e}_3}{\sqrt{2}} \tag{9.24}$$

and taking (5.23) into account one thus obtains the spectral decomposition of $\mathfrak{I}^{\mathrm{s}}$ as

$$\mathfrak{I}^{\mathrm{s}} = \sum_{p=1}^{6} \mathbf{M}_p \odot \mathbf{M}_p.$$

The only eigenvalue 1 is of multiplicity 6. Note that the corresponding eigentensors $\mathbf{M}_p$ $(p = 1,2,\ldots,6)$ form an orthonormal basis of $\mathbf{Lin}^3$.

(b) Using the orthonormal basis (9.24) we can write keeping (1.92) and $(5.89)_1$ in mind

$$\mathcal{P}_{\mathrm{sph}} = \frac{1}{3}\left(\mathbf{M}_1 + \mathbf{M}_2 + \mathbf{M}_3\right) \odot \left(\mathbf{M}_1 + \mathbf{M}_2 + \mathbf{M}_3\right)$$

$$= \sum_{p,q=1}^{6} \mathcal{P}_{\mathrm{sph}}^{pq} \mathbf{M}_p \odot \mathbf{M}_q, \quad \text{where} \quad \left[\mathcal{P}_{\mathrm{sph}}^{pq}\right] = \frac{1}{3}\begin{bmatrix} 1&1&1&0&0&0 \\ 1&1&1&0&0&0 \\ 1&1&1&0&0&0 \\ 0&0&0&0&0&0 \\ 0&0&0&0&0&0 \\ 0&0&0&0&0&0 \end{bmatrix}.$$

Due to the structure of this matrix the eigenvalue problem can be solved separately for the upper left and lower right 3×3 submatrices. For the latter (zero) matrix all three eigenvalues are zero and every set of three linearly independent vectors forms eigenvectors (eigentensors). The characteristic equation of the upper left submatrix is written by

$$\begin{vmatrix} \frac{1}{3} - \Lambda & \frac{1}{3} & \frac{1}{3} \\ \frac{1}{3} & \frac{1}{3} - \Lambda & \frac{1}{3} \\ \frac{1}{3} & \frac{1}{3} & \frac{1}{3} - \Lambda \end{vmatrix} = 0 \quad \Rightarrow \quad -\Lambda^3 + \Lambda^2 = 0$$

and yields the following eigenvalues

$$\Lambda_1 = 1, \quad \Lambda_{2/3} = 0.$$

The eigenvector (eigentensor) $\widehat{\mathbf{M}}_1 = A^i \mathbf{M}_i$ corresponding to the first eigenvalue results from the equation system $(4.16)_1$

$$\begin{cases} -\frac{2}{3}A^1 + \frac{1}{3}A^2 + \frac{1}{3}A^3 = 0, \\ \frac{1}{3}A^1 - \frac{2}{3}A^2 + \frac{1}{3}A^3 = 0, \\ \frac{1}{3}A^1 + \frac{1}{3}A^2 - \frac{2}{3}A^3 = 0, \end{cases}$$

which leads to $A^1 = A^2 = A^3$. Thus, the unit eigenvector (eigentensor) corresponding to the eigenvalue $\Lambda_1 = 1$ is $\widehat{\mathbf{M}}_1 = \frac{1}{\sqrt{3}}(\mathbf{M}_1 + \mathbf{M}_2 + \mathbf{M}_3)$. Components of the eigenvectors (eigentensors) corresponding to the double eigenvalue $\Lambda_{2/3} = 0$ satisfy the single equation

$$\frac{1}{3}A^1 + \frac{1}{3}A^2 + \frac{1}{3}A^3 = 0. \tag{9.25}$$

One of the eigenvectors (eigentensors) can be obtained by setting for example $A^3 = 0$. Thus, $A^1 = -A^2$ which results in a unit eigenvector (eigentensor) $\widehat{\mathbf{M}}_2 = -\frac{\sqrt{2}}{2}\mathbf{M}_1 + \frac{\sqrt{2}}{2}\mathbf{M}_2$. For the spectral representation the third eigenvector (eigentensor) should be orthogonal to the second one so that

$$-\frac{\sqrt{2}}{2}A^1 + \frac{\sqrt{2}}{2}A^2 = 0 \quad \Rightarrow \quad A^1 = A^2.$$

Solving this equation together with (9.25) we obtain the third unit eigenvector (eigentensor) of the form $\widehat{\mathbf{M}}_3 = -\frac{\sqrt{6}}{6}\mathbf{M}_1 - \frac{\sqrt{6}}{6}\mathbf{M}_2 + \frac{\sqrt{6}}{3}\mathbf{M}_3$. Summarizing these results the solution of the eigenvalue problem for the tensor $\mathcal{P}_{\text{sph}}$ can be represented in the following form.

$$\Lambda_1 = 1, \quad \widehat{\mathbf{M}}_1 = \frac{1}{\sqrt{3}}(\mathbf{M}_1 + \mathbf{M}_2 + \mathbf{M}_3),$$

$$\Lambda_2 = \Lambda_3 = \Lambda_4 = \Lambda_5 = \Lambda_6 = 0, \quad \widehat{\mathbf{M}}_2 = -\frac{\sqrt{2}}{2}\mathbf{M}_1 + \frac{\sqrt{2}}{2}\mathbf{M}_2,$$

$$\widehat{\mathbf{M}}_3 = -\frac{\sqrt{6}}{6}\mathbf{M}_1 - \frac{\sqrt{6}}{6}\mathbf{M}_2 + \frac{\sqrt{6}}{3}\mathbf{M}_3, \quad \widehat{\mathbf{M}}_p = \mathbf{M}_p, \quad p = 4, 5, 6, \tag{9.26}$$

where the tensors $\mathbf{M}_q$, $(q = 1, 2, \ldots, 6)$ are defined by (9.24).

(c) For the super-symmetric counterpart of the deviatoric projection tensor $(5.89)_2$ $(n = 3)$ we can write

$$\mathcal{P}^{\mathrm{s}}_{\mathrm{dev}} = \sum_{p,q=1}^{6} \mathcal{P}^{pq}_{\mathrm{dev}} \mathbf{M}_p \odot \mathbf{M}_q, \quad \text{where} \quad [\mathcal{P}^{pq}_{\mathrm{dev}}] = \begin{bmatrix} \frac{2}{3} & -\frac{1}{3} & -\frac{1}{3} & 0 & 0 & 0 \\ -\frac{1}{3} & \frac{2}{3} & -\frac{1}{3} & 0 & 0 & 0 \\ -\frac{1}{3} & -\frac{1}{3} & \frac{2}{3} & 0 & 0 & 0 \\ 0 & 0 & 0 & 1 & 0 & 0 \\ 0 & 0 & 0 & 0 & 1 & 0 \\ 0 & 0 & 0 & 0 & 0 & 1 \end{bmatrix}.$$

The eigenvalues of $\mathcal{P}^{\mathrm{s}}_{\mathrm{dev}}$ can be obtained as linear combinations of those ones of $\mathcal{P}_{\mathrm{sph}}$ and $\mathcal{I}^{\mathrm{s}}$ as $\Lambda_1 = 0$, $\Lambda_q = 1$ $(q = 2, 3, \ldots, 6)$. The corresponding eigentensors are again given by (9.26).

(d) With respect to the orthonormal basis (9.24) the elasticity tensor (5.93) can be represented by

$$\mathcal{C} = \sum_{p,q=1}^{6} \mathcal{C}^{pq} \mathbf{M}_p \odot \mathbf{M}_q,$$

where

$$[\mathcal{C}^{pq}] = \begin{bmatrix} 2G+\lambda & \lambda & \lambda & 0 & 0 & 0 \\ \lambda & 2G+\lambda & \lambda & 0 & 0 & 0 \\ \lambda & \lambda & 2G+\lambda & 0 & 0 & 0 \\ 0 & 0 & 0 & 2G & 0 & 0 \\ 0 & 0 & 0 & 0 & 2G & 0 \\ 0 & 0 & 0 & 0 & 0 & 2G \end{bmatrix}.$$

The eigentensors of $\mathcal{C}$ are the same as of $\mathcal{P}_{\mathrm{sph}}$ and $\mathcal{P}^{\mathrm{s}}_{\mathrm{dev}}$ and are given by (9.26). The eigenvalues are as follows: $\Lambda_1 = 2G + 3\lambda$, $\Lambda_q = 2G$ $(q = 2, 3, \ldots, 6)$. They can be obtained as linear combinations of those ones of $\mathcal{P}_{\mathrm{sph}}$ and $\mathcal{P}^{\mathrm{s}}_{\mathrm{dev}}$.

9.6 Exercises of Chap. 6

6.1

(a) $f\left(\mathbf{Q}\mathbf{A}\mathbf{Q}^{\mathrm{T}}\right) = \boldsymbol{a}\mathbf{Q}\mathbf{A}\mathbf{Q}^{\mathrm{T}}\boldsymbol{b} \neq \boldsymbol{a}\mathbf{A}\boldsymbol{b}$.
(b) Since the components of $\mathbf{A}$ are related to an orthonormal basis we can write

$$f(\mathbf{A}) = \mathrm{A}^{11} + \mathrm{A}^{22} + \mathrm{A}^{33} = \mathrm{A}^{1}_{\cdot 1} + \mathrm{A}^{2}_{\cdot 2} + \mathrm{A}^{3}_{\cdot 3} = \mathrm{A}^{i}_{\cdot i} = \mathrm{tr}\mathbf{A}.$$

Trace of a tensor represents its isotropic function.

(c) For an isotropic tensor function the condition (6.1) $f(\mathbf{QAQ}^{\mathrm{T}}) = f(\mathbf{A})$ must hold on the whole definition domain of the arguments $\mathbf{A}$ and $\forall \mathbf{Q} \in \mathbf{Orth}^3$. Let us consider a special case where

$$\mathbf{A} = \begin{bmatrix} 1 & 0 & 0 \\ 0 & 0 & 0 \\ 0 & 0 & 0 \end{bmatrix} \boldsymbol{e}_i \otimes \boldsymbol{e}_j, \quad \mathbf{Q} = \begin{bmatrix} 0 & 1 & 0 \\ -1 & 0 & 0 \\ 0 & 0 & 1 \end{bmatrix} \boldsymbol{e}_i \otimes \boldsymbol{e}_j.$$

Thus,

$$\mathbf{A}' = \mathbf{QAQ}^{\mathrm{T}} = \begin{bmatrix} 0 & 0 & 0 \\ 0 & 1 & 0 \\ 0 & 0 & 0 \end{bmatrix} \boldsymbol{e}_i \otimes \boldsymbol{e}_j$$

and consequently

$$f(\mathbf{A}) = \mathrm{A}^{11} + \mathrm{A}^{12} + \mathrm{A}^{13} = 1 \neq 0 = \mathrm{A}'^{11} + \mathrm{A}'^{12} + \mathrm{A}'^{13} = f\left(\mathbf{QAQ}^{\mathrm{T}}\right),$$

which means that the function $f(\mathbf{A})$ is not isotropic.

(d) $\det\mathbf{A}$ is the last principal invariant of $\mathbf{A}$ and represents thus its isotropic tensor function. Isotropy of the determinant can, however, be shown directly using the relation $\det(\mathbf{BC}) = \det\mathbf{B}\det\mathbf{C}$. Indeed,

$$\begin{aligned} \det\left(\mathbf{QAQ}^{\mathrm{T}}\right) &= \det\mathbf{Q}\det\mathbf{A}\det\mathbf{Q}^{\mathrm{T}} = \det\mathbf{Q}\det\mathbf{Q}^{\mathrm{T}}\det\mathbf{A} \\ &= \det\left(\mathbf{QQ}^{\mathrm{T}}\right)\det\mathbf{A} = \det\mathbf{I}\det\mathbf{A} = \det\mathbf{A}, \quad \forall \mathbf{Q} \in \mathbf{Orth}^n. \end{aligned}$$

(e) Eigenvalues of a second-order tensor are uniquely defined by its principal invariants and represent thus its isotropic functions. This can also be shown in a direct way considering the eigenvalue problem for the tensor $\mathbf{QAQ}^{\mathrm{T}}$ as

$$\left(\mathbf{QAQ}^{\mathrm{T}}\right)\boldsymbol{a} = \lambda\boldsymbol{a}.$$

Mapping both sides of this vector equation by $\mathbf{Q}^{\mathrm{T}}$ yields

$$\left(\mathbf{Q}^{\mathrm{T}}\mathbf{QAQ}^{\mathrm{T}}\right)\boldsymbol{a} = \lambda\mathbf{Q}^{\mathrm{T}}\boldsymbol{a}.$$

Using the abbreviation $\boldsymbol{a}' = \mathbf{Q}^{\mathrm{T}}\boldsymbol{a}$ we finally obtain

$$\mathbf{A}\boldsymbol{a}' = \lambda\boldsymbol{a}'.$$

Thus, every eigenvalue of $\mathbf{QAQ}^{\mathrm{T}}$ is the eigenvalue of $\mathbf{A}$ and vice versa. In other words, the eigenvalues of these tensors are pairwise equal which immediately implies that they are characterized by the same value of $\lambda_{\max}$. The tensors obtained by the operation $\mathbf{QAQ}^\mathrm{T}$ from the original one $\mathbf{A}$ are called similar tensors.

6.2 Inserting

$$\mathbf{M} = \frac{1}{2}\left(\mathbf{A} + \mathbf{A}^\mathrm{T}\right), \quad \mathbf{W} = \frac{1}{2}\left(\mathbf{A} - \mathbf{A}^\mathrm{T}\right)$$

into (6.17) we obtain

$$\mathrm{tr}\mathbf{M} = \frac{1}{2}\left(\mathrm{tr}\mathbf{A} + \mathrm{tr}\mathbf{A}^\mathrm{T}\right) = \mathrm{tr}\mathbf{A},$$

$$\begin{aligned}
\mathrm{tr}\mathbf{M}^2 &= \frac{1}{4}\mathrm{tr}\left(\mathbf{A} + \mathbf{A}^\mathrm{T}\right)^2 \\
&= \frac{1}{4}\left[\mathrm{tr}\mathbf{A}^2 + \mathrm{tr}\left(\mathbf{A}\mathbf{A}^\mathrm{T}\right) + \mathrm{tr}\left(\mathbf{A}^\mathrm{T}\mathbf{A}\right) + \mathrm{tr}\left(\mathbf{A}^\mathrm{T}\right)^2\right] \\
&= \frac{1}{2}\left[\mathrm{tr}\mathbf{A}^2 + \mathrm{tr}\left(\mathbf{A}\mathbf{A}^\mathrm{T}\right)\right],
\end{aligned}$$

$$\begin{aligned}
\mathrm{tr}\mathbf{M}^3 &= \frac{1}{8}\mathrm{tr}\left(\mathbf{A} + \mathbf{A}^\mathrm{T}\right)^3 \\
&= \frac{1}{8}\left\{\mathrm{tr}\mathbf{A}^3 + \mathrm{tr}\left(\mathbf{A}^2\mathbf{A}^\mathrm{T}\right) + \mathrm{tr}\left(\mathbf{A}\mathbf{A}^\mathrm{T}\mathbf{A}\right) + \mathrm{tr}\left[\mathbf{A}\left(\mathbf{A}^\mathrm{T}\right)^2\right]\right. \\
&\quad \left. + \,\mathrm{tr}\left(\mathbf{A}^\mathrm{T}\mathbf{A}^2\right) + \mathrm{tr}\left(\mathbf{A}^\mathrm{T}\mathbf{A}\mathbf{A}^\mathrm{T}\right) + \mathrm{tr}\left[\left(\mathbf{A}^\mathrm{T}\right)^2\mathbf{A}\right] + \mathrm{tr}\left(\mathbf{A}^\mathrm{T}\right)^3\right\} \\
&= \frac{1}{4}\left[\mathrm{tr}\mathbf{A}^3 + 3\mathrm{tr}\left(\mathbf{A}^2\mathbf{A}^\mathrm{T}\right)\right],
\end{aligned}$$

$$\begin{aligned}
\mathrm{tr}\mathbf{W}^2 &= \frac{1}{4}\mathrm{tr}\left(\mathbf{A} - \mathbf{A}^\mathrm{T}\right)^2 \\
&= \frac{1}{4}\left[\mathrm{tr}\mathbf{A}^2 - \mathrm{tr}\left(\mathbf{A}\mathbf{A}^\mathrm{T}\right) - \mathrm{tr}\left(\mathbf{A}^\mathrm{T}\mathbf{A}\right) + \mathrm{tr}\left(\mathbf{A}^\mathrm{T}\right)^2\right] \\
&= \frac{1}{2}\left[\mathrm{tr}\mathbf{A}^2 - \mathrm{tr}\left(\mathbf{A}\mathbf{A}^\mathrm{T}\right)\right],
\end{aligned}$$

$$\begin{aligned}
\mathrm{tr}\left(\mathbf{M}\mathbf{W}^2\right) &= \frac{1}{8}\mathrm{tr}\left[\left(\mathbf{A} + \mathbf{A}^\mathrm{T}\right)\left(\mathbf{A} - \mathbf{A}^\mathrm{T}\right)^2\right] \\
&= \frac{1}{8}\left\{\mathrm{tr}\mathbf{A}^3 - \mathrm{tr}\left(\mathbf{A}^2\mathbf{A}^\mathrm{T}\right) - \mathrm{tr}\left(\mathbf{A}\mathbf{A}^\mathrm{T}\mathbf{A}\right) + \mathrm{tr}\left[\mathbf{A}\left(\mathbf{A}^\mathrm{T}\right)^2\right]\right. \\
&\quad \left. + \,\mathrm{tr}\left(\mathbf{A}^\mathrm{T}\mathbf{A}^2\right) - \mathrm{tr}\left(\mathbf{A}^\mathrm{T}\mathbf{A}\mathbf{A}^\mathrm{T}\right) - \mathrm{tr}\left[\left(\mathbf{A}^\mathrm{T}\right)^2\mathbf{A}\right] + \mathrm{tr}\left(\mathbf{A}^\mathrm{T}\right)^3\right\} \\
&= \frac{1}{4}\left[\mathrm{tr}\mathbf{A}^3 - \mathrm{tr}\left(\mathbf{A}^2\mathbf{A}^\mathrm{T}\right)\right],
\end{aligned}$$

$$\mathrm{tr}\left(\mathbf{M}^2\mathbf{W}^2\right) = \frac{1}{16}\mathrm{tr}\left[\left(\mathbf{A}+\mathbf{A}^{\mathrm{T}}\right)^2\left(\mathbf{A}-\mathbf{A}^{\mathrm{T}}\right)^2\right]$$

$$= \frac{1}{16}\mathrm{tr}\left\{\left[\mathbf{A}^2+\mathbf{A}\mathbf{A}^{\mathrm{T}}+\mathbf{A}^{\mathrm{T}}\mathbf{A}+\left(\mathbf{A}^{\mathrm{T}}\right)^2\right]\left[\mathbf{A}^2-\mathbf{A}\mathbf{A}^{\mathrm{T}}-\mathbf{A}^{\mathrm{T}}\mathbf{A}+\left(\mathbf{A}^{\mathrm{T}}\right)^2\right]\right\}$$

$$= \frac{1}{16}\mathrm{tr}\left[\mathbf{A}^4-\mathbf{A}^3\mathbf{A}^{\mathrm{T}}-\mathbf{A}^2\mathbf{A}^{\mathrm{T}}\mathbf{A}+\mathbf{A}^2\left(\mathbf{A}^{\mathrm{T}}\right)^2+\mathbf{A}\mathbf{A}^{\mathrm{T}}\mathbf{A}^2-\mathbf{A}\mathbf{A}^{\mathrm{T}}\mathbf{A}\mathbf{A}^{\mathrm{T}}\right.$$

$$-\mathbf{A}\left(\mathbf{A}^{\mathrm{T}}\right)^2\mathbf{A}+\mathbf{A}\left(\mathbf{A}^{\mathrm{T}}\right)^3+\mathbf{A}^{\mathrm{T}}\mathbf{A}^3-\mathbf{A}^{\mathrm{T}}\mathbf{A}^2\mathbf{A}^{\mathrm{T}}-\mathbf{A}^{\mathrm{T}}\mathbf{A}\mathbf{A}^{\mathrm{T}}\mathbf{A}$$

$$\left.+\mathbf{A}^{\mathrm{T}}\mathbf{A}\left(\mathbf{A}^{\mathrm{T}}\right)^2+\left(\mathbf{A}^{\mathrm{T}}\right)^2\mathbf{A}^2-\left(\mathbf{A}^{\mathrm{T}}\right)^2\mathbf{A}\mathbf{A}^{\mathrm{T}}-\left(\mathbf{A}^{\mathrm{T}}\right)^3\mathbf{A}+\left(\mathbf{A}^{\mathrm{T}}\right)^4\right]$$

$$= \frac{1}{8}\left[\mathrm{tr}\mathbf{A}^4-\mathrm{tr}\left(\mathbf{A}\mathbf{A}^{\mathrm{T}}\right)^2\right],$$

$$\mathrm{tr}\left(\mathbf{M}^2\mathbf{W}^2\mathbf{M}\mathbf{W}\right)$$

$$= \frac{1}{64}\mathrm{tr}\left[\mathbf{A}^4-\mathbf{A}^3\mathbf{A}^{\mathrm{T}}-\mathbf{A}^2\mathbf{A}^{\mathrm{T}}\mathbf{A}+\mathbf{A}^2\left(\mathbf{A}^{\mathrm{T}}\right)^2+\mathbf{A}\mathbf{A}^{\mathrm{T}}\mathbf{A}^2-\mathbf{A}\mathbf{A}^{\mathrm{T}}\mathbf{A}\mathbf{A}^{\mathrm{T}}\right.$$

$$-\mathbf{A}\left(\mathbf{A}^{\mathrm{T}}\right)^2\mathbf{A}+\mathbf{A}\left(\mathbf{A}^{\mathrm{T}}\right)^3+\mathbf{A}^{\mathrm{T}}\mathbf{A}^3-\mathbf{A}^{\mathrm{T}}\mathbf{A}^2\mathbf{A}^{\mathrm{T}}-\mathbf{A}^{\mathrm{T}}\mathbf{A}\mathbf{A}^{\mathrm{T}}\mathbf{A}$$

$$\left.+\mathbf{A}^{\mathrm{T}}\mathbf{A}\left(\mathbf{A}^{\mathrm{T}}\right)^2+\left(\mathbf{A}^{\mathrm{T}}\right)^2\mathbf{A}^2-\left(\mathbf{A}^{\mathrm{T}}\right)^2\mathbf{A}\mathbf{A}^{\mathrm{T}}-\left(\mathbf{A}^{\mathrm{T}}\right)^3\mathbf{A}+\left(\mathbf{A}^{\mathrm{T}}\right)^4\right]$$

$$\left.\left[\mathbf{A}^2-\mathbf{A}\mathbf{A}^{\mathrm{T}}+\mathbf{A}^{\mathrm{T}}\mathbf{A}-\left(\mathbf{A}^{\mathrm{T}}\right)^2\right]\right\}$$

$$= \frac{1}{16}\mathrm{tr}\left[\left(\mathbf{A}^{\mathrm{T}}\right)^2\mathbf{A}^2\mathbf{A}^{\mathrm{T}}\mathbf{A}-\mathbf{A}^2\left(\mathbf{A}^{\mathrm{T}}\right)^2\mathbf{A}\mathbf{A}^{\mathrm{T}}\right].$$

Finally, $\mathrm{tr}\mathbf{A}^4$ should be expressed in terms of the principal traces $\mathrm{tr}\mathbf{A}^i$ $(i = 1, 2, 3)$ presented in the functional basis (6.18). To this end, we apply the Cayley-Hamilton equation (4.95). Its composition with $\mathbf{A}$ yields

$$\mathbf{A}^4 - \mathrm{I}_{\mathbf{A}}\mathbf{A}^3 + \mathrm{II}_{\mathbf{A}}\mathbf{A}^2 - \mathrm{III}_{\mathbf{A}}\mathbf{A} = \mathbf{0},$$

so that

$$\mathrm{tr}\mathbf{A}^4 = \mathrm{I}_{\mathbf{A}}\mathrm{tr}\mathbf{A}^3 - \mathrm{II}_{\mathbf{A}}\mathrm{tr}\mathbf{A}^2 + \mathrm{III}_{\mathbf{A}}\mathrm{tr}\mathbf{A},$$

where $\mathrm{I}_{\mathbf{A}}$, $\mathrm{II}_{\mathbf{A}}$ and $\mathrm{III}_{\mathbf{A}}$ are given by (4.29). Thus, all the invariants of the functional basis (6.17) are expressed in a unique form in terms of (6.18).

6.3 Equation (6.44):

$$\frac{\mathrm{d}}{\mathrm{d}t} f(\mathbf{A}+t\mathbf{X})\, g(\mathbf{A}+t\mathbf{X})\Big|_{t=0}$$

$$= \frac{\mathrm{d}}{\mathrm{d}t} f(\mathbf{A}+t\mathbf{X})\Big|_{t=0} g(\mathbf{A}) + f(\mathbf{A}) \frac{\mathrm{d}}{\mathrm{d}t} g(\mathbf{A}+t\mathbf{X})\Big|_{t=0}$$

$$= [f(\mathbf{A})_{,\mathbf{A}} : \mathbf{X}]\, g(\mathbf{A}) + f(\mathbf{A})\, [g(\mathbf{A})_{,\mathbf{A}} : \mathbf{X}]$$

$$= [g(\mathbf{A})\, f(\mathbf{A})_{,\mathbf{A}} + f(\mathbf{A})\, g(\mathbf{A})_{,\mathbf{A}}] : \mathbf{X}.$$

6.4 By Theorem 6.1 ψ is an isotropic function of $\mathbf{C}$ and $\mathbf{L}_i$ $(i = 1, 2, 3)$. Applying further (6.15) and taking into account the identities

$$\mathbf{L}_i \mathbf{L}_j = \mathbf{0}, \quad \mathbf{L}_i^k = \mathbf{L}_i, \quad \mathrm{tr}\mathbf{L}_i^k = 1,\ i \neq j,\ i, j = 1, 2, 3;\ k = 1, 2, \ldots \tag{9.27}$$

we obtain the following orthotropic invariants

$$\begin{gathered}
\mathrm{tr}\mathbf{C}, \quad \mathrm{tr}\mathbf{C}^2, \quad \mathrm{tr}\mathbf{C}^3, \\
\mathrm{tr}(\mathbf{C}\mathbf{L}_1) = \mathrm{tr}\left(\mathbf{C}\mathbf{L}_1^2\right), \quad \mathrm{tr}(\mathbf{C}\mathbf{L}_2) = \mathrm{tr}\left(\mathbf{C}\mathbf{L}_2^2\right), \quad \mathrm{tr}(\mathbf{C}\mathbf{L}_3) = \mathrm{tr}\left(\mathbf{C}\mathbf{L}_3^2\right), \\
\mathrm{tr}\left(\mathbf{C}^2\mathbf{L}_1\right) = \mathrm{tr}\left(\mathbf{C}^2\mathbf{L}_1^2\right), \ \mathrm{tr}\left(\mathbf{C}^2\mathbf{L}_2\right) = \mathrm{tr}\left(\mathbf{C}^2\mathbf{L}_2^2\right), \ \mathrm{tr}\left(\mathbf{C}^2\mathbf{L}_3\right) = \mathrm{tr}\left(\mathbf{C}^2\mathbf{L}_3^2\right), \\
\mathrm{tr}\left(\mathbf{L}_i\mathbf{C}\mathbf{L}_j\right) = \mathrm{tr}\left(\mathbf{C}\mathbf{L}_j\mathbf{L}_i\right) = \mathrm{tr}\left(\mathbf{L}_j\mathbf{L}_i\mathbf{C}\right) = 0, \quad i \neq j = 1, 2, 3.
\end{gathered} \tag{9.28}$$

Using the relation

$$\sum_{i=1}^{3} \mathbf{L}_i = \mathbf{I} \tag{9.29}$$

one can further write

$$\begin{aligned}
\mathrm{tr}(\mathbf{C}) &= \mathbf{C} : \mathbf{I} = \mathbf{C} : (\mathbf{L}_1 + \mathbf{L}_2 + \mathbf{L}_3) \\
&= \mathbf{C} : \mathbf{L}_1 + \mathbf{C} : \mathbf{L}_2 + \mathbf{C} : \mathbf{L}_3 = \mathrm{tr}(\mathbf{C}\mathbf{L}_1) + \mathrm{tr}(\mathbf{C}\mathbf{L}_2) + \mathrm{tr}(\mathbf{C}\mathbf{L}_3).
\end{aligned}$$

In the same manner we also obtain

$$\mathrm{tr}\left(\mathbf{C}^2\right) = \mathrm{tr}\left(\mathbf{C}^2\mathbf{L}_1\right) + \mathrm{tr}\left(\mathbf{C}^2\mathbf{L}_2\right) + \mathrm{tr}\left(\mathbf{C}^2\mathbf{L}_3\right).$$

Thus, the invariants $\mathrm{tr}\mathbf{C}$ and $\mathrm{tr}\mathbf{C}^2$ are redundant and can be excluded from the functional basis (9.28). Finally, the orthotropic strain energy function can be represented by

$$\begin{aligned}
\psi = \tilde{\psi}\, [&\mathrm{tr}(\mathbf{C}\mathbf{L}_1), \mathrm{tr}(\mathbf{C}\mathbf{L}_2), \mathrm{tr}(\mathbf{C}\mathbf{L}_3), \\
&\mathrm{tr}\left(\mathbf{C}^2\mathbf{L}_1\right), \mathrm{tr}\left(\mathbf{C}^2\mathbf{L}_2\right), \mathrm{tr}\left(\mathbf{C}^2\mathbf{L}_3\right), \mathrm{tr}\mathbf{C}^3\right].
\end{aligned} \tag{9.30}$$

Alternatively, a functional basis for the orthotropic material symmetry can be obtained in the component form. To this end, we represent the right Cauchy-Green tensor by $\mathbf{C} = \mathrm{C}^{ij} \boldsymbol{l}_i \otimes \boldsymbol{l}_j$. Then,

$$\operatorname{tr}(\mathbf{C}\mathbf{L}_i) = \left(\mathrm{C}^{kl} \boldsymbol{l}_k \otimes \boldsymbol{l}_l\right) : \boldsymbol{l}_i \otimes \boldsymbol{l}_i = \mathrm{C}^{ii}, \quad i = 1, 2, 3,$$

$$\operatorname{tr}\left(\mathbf{C}^2\mathbf{L}_i\right) = \left(\mathrm{C}^{i1}\right)^2 + \left(\mathrm{C}^{i2}\right)^2 + \left(\mathrm{C}^{i3}\right)^2, \quad i = 1, 2, 3,$$

$$\operatorname{tr}\left(\mathbf{C}^3\right) = \left(\mathrm{C}^{11}\right)^3 + \left(\mathrm{C}^{22}\right)^3 + \left(\mathrm{C}^{33}\right)^3 + 3\left(\mathrm{C}^{12}\right)^2\left(\mathrm{C}^{11} + \mathrm{C}^{22}\right)$$
$$+ 3\left(\mathrm{C}^{13}\right)^2\left(\mathrm{C}^{11} + \mathrm{C}^{33}\right) + 3\left(\mathrm{C}^{23}\right)^2\left(\mathrm{C}^{22} + \mathrm{C}^{33}\right) + 6\mathrm{C}^{12}\mathrm{C}^{13}\mathrm{C}^{23}.$$

Thus, the orthotropic strain energy function (9.30) can be given in another form as

$$\psi = \hat{\psi}\left[\mathrm{C}^{11}, \mathrm{C}^{22}, \mathrm{C}^{33}, \left(\mathrm{C}^{12}\right)^2, \left(\mathrm{C}^{13}\right)^2, \left(\mathrm{C}^{23}\right)^2, \mathrm{C}^{12}\mathrm{C}^{13}\mathrm{C}^{23}\right].$$

6.5 Equations (6.52), (6.54), (6.80), (6.136), (6.140), (6.144) (6.149) and (9.30):

$$\mathbf{S} = 6\frac{\partial\tilde{\psi}}{\partial \operatorname{tr}\mathbf{C}^3}\mathbf{C}^2 + 2\sum_{i=1}^{3}\frac{\partial\tilde{\psi}}{\partial \operatorname{tr}(\mathbf{C}\mathbf{L}_i)}\mathbf{L}_i + 2\sum_{i=1}^{3}\frac{\partial\tilde{\psi}}{\partial \operatorname{tr}(\mathbf{C}^2\mathbf{L}_i)}\left(\mathbf{C}\mathbf{L}_i + \mathbf{L}_i\mathbf{C}\right),$$

$$\mathcal{C} = 36\frac{\partial^2\tilde{\psi}}{\partial \operatorname{tr}\mathbf{C}^3\,\partial \operatorname{tr}\mathbf{C}^3}\mathbf{C}^2 \odot \mathbf{C}^2 + 4\sum_{i,j=1}^{3}\frac{\partial^2\tilde{\psi}}{\partial \operatorname{tr}(\mathbf{C}\mathbf{L}_i)\,\partial \operatorname{tr}(\mathbf{C}\mathbf{L}_j)}\mathbf{L}_i \odot \mathbf{L}_j$$

$$+ 4\sum_{i,j=1}^{3}\frac{\partial^2\tilde{\psi}}{\partial \operatorname{tr}(\mathbf{C}^2\mathbf{L}_i)\,\partial \operatorname{tr}(\mathbf{C}^2\mathbf{L}_j)}\left(\mathbf{C}\mathbf{L}_i + \mathbf{L}_i\mathbf{C}\right) \odot \left(\mathbf{C}\mathbf{L}_j + \mathbf{L}_j\mathbf{C}\right)$$

$$+ 12\sum_{i=1}^{3}\frac{\partial^2\tilde{\psi}}{\partial \operatorname{tr}(\mathbf{C}\mathbf{L}_i)\,\partial \operatorname{tr}\mathbf{C}^3}\left(\mathbf{L}_i \odot \mathbf{C}^2 + \mathbf{C}^2 \odot \mathbf{L}_i\right)$$

$$+ 12\sum_{i=1}^{3}\frac{\partial^2\tilde{\psi}}{\partial \operatorname{tr}(\mathbf{C}^2\mathbf{L}_i)\,\partial \operatorname{tr}\mathbf{C}^3}\left[\mathbf{C}^2 \odot \left(\mathbf{C}\mathbf{L}_i + \mathbf{L}_i\mathbf{C}\right) + \left(\mathbf{C}\mathbf{L}_i + \mathbf{L}_i\mathbf{C}\right) \odot \mathbf{C}^2\right]$$

$$+ 4\sum_{i,j=1}^{3}\frac{\partial^2\tilde{\psi}}{\partial \operatorname{tr}(\mathbf{C}\mathbf{L}_i)\,\partial \operatorname{tr}(\mathbf{C}^2\mathbf{L}_j)}\left[\mathbf{L}_i \odot \left(\mathbf{C}\mathbf{L}_j + \mathbf{L}_j\mathbf{C}\right) + \left(\mathbf{C}\mathbf{L}_j + \mathbf{L}_j\mathbf{C}\right) \odot \mathbf{L}_i\right]$$

$$+ 12\frac{\partial\tilde{\psi}}{\partial \operatorname{tr}\mathbf{C}^3}\left(\mathbf{C} \otimes \mathbf{I} + \mathbf{I} \otimes \mathbf{C}\right)^{\mathrm{s}} + 4\sum_{i=1}^{3}\frac{\partial\tilde{\psi}}{\partial \operatorname{tr}(\mathbf{C}^2\mathbf{L}_i)}\left(\mathbf{L}_i \otimes \mathbf{I} + \mathbf{I} \otimes \mathbf{L}_i\right)^{\mathrm{s}}.$$

6.6 Any orthotropic function $\mathbf{S}(\mathbf{C})$ is an isotropic function of $\mathbf{C}$ and $\mathbf{L}_i$ $(i = 1, 2, 3)$. The latter function can be represented by (6.117). Taking (9.27) and (9.29) into account we thus obtain

$$\mathbf{S} = \sum_{i=1}^{3} \left[\alpha_i \mathbf{L}_i + \beta_i (\mathbf{C}\mathbf{L}_i + \mathbf{L}_i \mathbf{C}) + \gamma_i \left(\mathbf{C}^2 \mathbf{L}_i + \mathbf{L}_i \mathbf{C}^2 \right) \right],$$

where α_i, β_i and γ_i $(i = 1, 2, 3)$ are some scalar-valued orthotropic functions of $\mathbf{C}$ (isotropic functions of $\mathbf{C}$ and $\mathbf{L}_i$ $(i = 1, 2, 3)$).

6.7 Applying (6.15) and taking into account the identities $\mathbf{L}_i^m = \mathbf{L}_i$, $\mathrm{tr}\mathbf{L}_i^m = 1$ $(i = 1, 2;\ m = 1, 2, \ldots)$ we obtain similarly to (9.28)

$$\mathrm{tr}\mathbf{C}, \quad \mathrm{tr}\mathbf{C}^2, \quad \mathrm{tr}\mathbf{C}^3,$$

$$\mathrm{tr}(\mathbf{C}\mathbf{L}_1) = \mathrm{tr}\left(\mathbf{C}\mathbf{L}_1^2\right), \quad \mathrm{tr}(\mathbf{C}\mathbf{L}_2) = \mathrm{tr}\left(\mathbf{C}\mathbf{L}_2^2\right),$$

$$\mathrm{tr}(\mathbf{L}_1\mathbf{L}_2) = \mathrm{tr}\left(\mathbf{L}_1\mathbf{L}_2^2\right) = \mathrm{tr}\left(\mathbf{L}_1^2\mathbf{L}_2\right)$$

$$= (\boldsymbol{l}_1 \otimes \boldsymbol{l}_1) : (\boldsymbol{l}_2 \otimes \boldsymbol{l}_2) = (\boldsymbol{l}_1 \cdot \boldsymbol{l}_2)^2 = \cos^2\phi,$$

$$\mathrm{tr}\left(\mathbf{C}^2\mathbf{L}_1\right) = \mathrm{tr}\left(\mathbf{C}^2\mathbf{L}_1^2\right), \ \mathrm{tr}\left(\mathbf{C}^2\mathbf{L}_2\right) = \mathrm{tr}\left(\mathbf{C}^2\mathbf{L}_2^2\right), \ \mathrm{tr}(\mathbf{L}_1\mathbf{C}\mathbf{L}_2),$$

where ϕ denotes the angle between the fiber directions $\boldsymbol{l}_1$ and $\boldsymbol{l}_2$. Thus, we can write

$$\psi = \tilde{\psi}\left[\mathrm{tr}\mathbf{C}, \mathrm{tr}\mathbf{C}^2, \mathrm{tr}\mathbf{C}^3, \mathrm{tr}(\mathbf{C}\mathbf{L}_1), \mathrm{tr}(\mathbf{C}\mathbf{L}_2),\right.$$

$$\left.\mathrm{tr}\left(\mathbf{C}^2\mathbf{L}_1\right), \mathrm{tr}\left(\mathbf{C}^2\mathbf{L}_2\right), \mathrm{tr}(\mathbf{L}_1\mathbf{L}_2), \mathrm{tr}(\mathbf{L}_1\mathbf{C}\mathbf{L}_2)\right].$$

6.8 Using (6.59), (6.137), (6.139) and (6.144) we obtain

$$\mathbf{S} = 2\frac{\partial\psi}{\partial\mathbf{C}} + p\mathbf{C}^{-1} = 2c_1\mathrm{I}_{\mathbf{C}},_{\mathbf{C}} + 2c_2\mathrm{II}_{\mathbf{C}},_{\mathbf{C}} + p\mathbf{C}^{-1}$$

$$= 2c_1\mathbf{I} + 2c_2(\mathrm{I}_{\mathbf{C}}\mathbf{I} - \mathbf{C}) + p\mathbf{C}^{-1} = 2(c_1 + c_2\mathrm{I}_{\mathbf{C}})\mathbf{I} - 2c_2\mathbf{C} + p\mathbf{C}^{-1},$$

$$\mathcal{C} = 2\frac{\partial\mathbf{S}}{\partial\mathbf{C}} = 4c_2(\mathbf{I} \odot \mathbf{I} - \mathfrak{J}^{\mathrm{s}}) - 2p\left(\mathbf{C}^{-1} \otimes \mathbf{C}^{-1}\right)^{\mathrm{s}}.$$

6.9 Using the abbreviation $\Lambda_i = \lambda_i^2$ $(i = 1, 2, 3)$ for the eigenvalues of $\mathbf{C}$ we can write

$$\psi(\mathbf{C}) = \sum_{r=1}^{m} \frac{\mu_r}{\alpha_r}\left(\Lambda_1^{\alpha_r/2} + \Lambda_2^{\alpha_r/2} + \Lambda_3^{\alpha_r/2} - 3\right).$$

Assuming further that $\Lambda_1 \neq \Lambda_2 \neq \Lambda_3 \neq \Lambda_1$ and applying (6.74) we obtain

$$\mathbf{S} = 2\frac{\partial\psi}{\partial\mathbf{C}} = \sum_{r=1}^{m} \mu_r \left(\Lambda_1^{\alpha_r/2-1} \Lambda_{1,\mathbf{C}} + \Lambda_2^{\alpha_r/2-1} \Lambda_{2,\mathbf{C}} + \Lambda_3^{\alpha_r/2-1} \Lambda_{3,\mathbf{C}} \right)$$

$$= \sum_{r=1}^{m} \mu_r \left(\Lambda_1^{\alpha_r/2-1} \mathbf{P}_1 + \Lambda_2^{\alpha_r/2-1} \mathbf{P}_2 + \Lambda_3^{\alpha_r/2-1} \mathbf{P}_3 \right) = \sum_{r=1}^{m} \mu_r \mathbf{C}^{\alpha_r/2-1}. \quad (9.31)$$

Note that the latter expression is obtained by means of (7.2).

6.10 Using the identities

$$\mathbf{Q}^{\mathrm{T}}\mathbf{L}_i\mathbf{Q} = \mathbf{Q}\mathbf{L}_i\mathbf{Q}^{\mathrm{T}} = \mathbf{L}_i, \quad \forall \mathbf{Q} \in \mathfrak{g}_o$$

and taking (1.151) into account we can write

$$\mathrm{tr}\left(\mathbf{Q}\mathbf{C}\mathbf{Q}^{\mathrm{T}}\mathbf{L}_i\mathbf{Q}\mathbf{C}\mathbf{Q}^{\mathrm{T}}\mathbf{L}_j\right) = \mathrm{tr}\left(\mathbf{C}\mathbf{Q}^{\mathrm{T}}\mathbf{L}_i\mathbf{Q}\mathbf{C}\mathbf{Q}^{\mathrm{T}}\mathbf{L}_j\mathbf{Q}\right)$$

$$= \mathrm{tr}\left(\mathbf{C}\mathbf{L}_i\mathbf{C}\mathbf{L}_j\right), \quad \forall \mathbf{Q} \in \mathfrak{g}_o.$$

Further, one can show that

$$\mathrm{tr}\left(\mathbf{C}\mathbf{L}_i\mathbf{C}\mathbf{L}_i\right) = \mathrm{tr}^2\left(\mathbf{C}\mathbf{L}_i\right), \quad i = 1, 2, 3, \quad (9.32)$$

where we use the abbreviation $\mathrm{tr}^2(\bullet) = [\mathrm{tr}(\bullet)]^2$. Indeed, in view of the relation $\mathrm{tr}(\mathbf{C}\mathbf{L}_i) = \mathbf{C} : (\boldsymbol{l}_i \otimes \boldsymbol{l}_i) = \boldsymbol{l}_i\mathbf{C}\boldsymbol{l}_i$ we have

$$\mathrm{tr}\left(\mathbf{C}\mathbf{L}_i\mathbf{C}\mathbf{L}_i\right) = \mathrm{tr}\left(\mathbf{C}\boldsymbol{l}_i \otimes \boldsymbol{l}_i\mathbf{C}\boldsymbol{l}_i \otimes \boldsymbol{l}_i\right) = \boldsymbol{l}_i\mathbf{C}\boldsymbol{l}_i\,\mathrm{tr}\left(\mathbf{C}\boldsymbol{l}_i \otimes \boldsymbol{l}_i\right)$$

$$= \boldsymbol{l}_i\mathbf{C}\boldsymbol{l}_i\,\mathrm{tr}\left(\mathbf{C}\mathbf{L}_i\right) = \mathrm{tr}^2\left(\mathbf{C}\mathbf{L}_i\right), \quad i = 1, 2, 3.$$

Next, we obtain

$$\mathrm{tr}\left(\mathbf{C}^2\mathbf{L}_i\right) = \mathrm{tr}\left(\mathbf{C}\mathbf{I}\mathbf{C}\mathbf{L}_i\right) = \mathrm{tr}\left[\mathbf{C}\left(\mathbf{L}_1 + \mathbf{L}_2 + \mathbf{L}_3\right)\mathbf{C}\mathbf{L}_i\right]$$

$$= \sum_{j=1}^{3} \mathrm{tr}\left(\mathbf{C}\mathbf{L}_j\mathbf{C}\mathbf{L}_i\right), \quad i = 1, 2, 3$$

and consequently

$$\mathrm{tr}\left(\mathbf{C}\mathbf{L}_2\mathbf{C}\mathbf{L}_1\right) + \mathrm{tr}\left(\mathbf{C}\mathbf{L}_3\mathbf{C}\mathbf{L}_1\right) = \mathrm{tr}\left(\mathbf{C}^2\mathbf{L}_1\right) - \mathrm{tr}^2\left(\mathbf{C}\mathbf{L}_1\right),$$

$$\mathrm{tr}\left(\mathbf{C}\mathbf{L}_3\mathbf{C}\mathbf{L}_2\right) + \mathrm{tr}\left(\mathbf{C}\mathbf{L}_1\mathbf{C}\mathbf{L}_2\right) = \mathrm{tr}\left(\mathbf{C}^2\mathbf{L}_2\right) - \mathrm{tr}^2\left(\mathbf{C}\mathbf{L}_2\right),$$

$$\mathrm{tr}\left(\mathbf{C}\mathbf{L}_1\mathbf{C}\mathbf{L}_3\right) + \mathrm{tr}\left(\mathbf{C}\mathbf{L}_2\mathbf{C}\mathbf{L}_3\right) = \mathrm{tr}\left(\mathbf{C}^2\mathbf{L}_3\right) - \mathrm{tr}^2\left(\mathbf{C}\mathbf{L}_3\right).$$

The latter relations can be given briefly by

$$\mathrm{tr}\left(\mathbf{C}\mathbf{L}_j\mathbf{C}\mathbf{L}_i\right) + \mathrm{tr}\left(\mathbf{C}\mathbf{L}_k\mathbf{C}\mathbf{L}_i\right)$$
$$= \mathrm{tr}\left(\mathbf{C}^2\mathbf{L}_i\right) - \mathrm{tr}^2\left(\mathbf{C}\mathbf{L}_i\right), \quad i \neq j \neq k \neq i;\ i, j, k = 1, 2, 3.$$

Their linear combinations finally yield:

$$\mathrm{tr}\left(\mathbf{C}\mathbf{L}_i\mathbf{C}\mathbf{L}_j\right) = \frac{1}{2}\left[\mathrm{tr}\left(\mathbf{C}^2\mathbf{L}_i\right) + \mathrm{tr}\left(\mathbf{C}^2\mathbf{L}_j\right) - \mathrm{tr}\left(\mathbf{C}^2\mathbf{L}_k\right)\right]$$
$$- \frac{1}{2}\left[\mathrm{tr}^2\left(\mathbf{C}\mathbf{L}_i\right) + \mathrm{tr}^2\left(\mathbf{C}\mathbf{L}_j\right) - \mathrm{tr}^2\left(\mathbf{C}\mathbf{L}_k\right)\right],$$

where $i \neq j \neq k \neq i;\ i, j, k = 1, 2, 3$.

6.11 We begin with the directional derivative of $\mathrm{tr}\left(\tilde{\mathbf{E}}\mathbf{L}_i\tilde{\mathbf{E}}\mathbf{L}_j\right)$:

$$\frac{\mathrm{d}}{\mathrm{d}t}\mathrm{tr}\left[\left(\tilde{\mathbf{E}} + t\mathbf{X}\right)\mathbf{L}_i\left(\tilde{\mathbf{E}} + t\mathbf{X}\right)\mathbf{L}_j\right]\Big|_{t=0}$$
$$= \frac{\mathrm{d}}{\mathrm{d}t}\left[\tilde{\mathbf{E}}\mathbf{L}_i\tilde{\mathbf{E}}\mathbf{L}_j + t\left(\mathbf{X}\mathbf{L}_i\tilde{\mathbf{E}}\mathbf{L}_j + \tilde{\mathbf{E}}\mathbf{L}_i\mathbf{X}\mathbf{L}_j\right) + t^2\mathbf{X}\mathbf{L}_i\mathbf{X}\mathbf{L}_j\right]\Big|_{t=0} : \mathbf{I}$$
$$= \left(\mathbf{X}\mathbf{L}_i\tilde{\mathbf{E}}\mathbf{L}_j + \tilde{\mathbf{E}}\mathbf{L}_i\mathbf{X}\mathbf{L}_j\right) : \mathbf{I} = \left(\mathbf{X}\mathbf{L}_i\tilde{\mathbf{E}}\mathbf{L}_j + \mathbf{L}_j\tilde{\mathbf{E}}\mathbf{L}_i\mathbf{X}\right) : \mathbf{I}$$
$$= \left(\mathbf{L}_i\tilde{\mathbf{E}}\mathbf{L}_j + \mathbf{L}_j\tilde{\mathbf{E}}\mathbf{L}_i\right) : \mathbf{X}^{\mathrm{T}} = \left(\mathbf{L}_i\tilde{\mathbf{E}}\mathbf{L}_j + \mathbf{L}_j\tilde{\mathbf{E}}\mathbf{L}_i\right)^{\mathrm{T}} : \mathbf{X}.$$

Hence,

$$\mathrm{tr}\left(\tilde{\mathbf{E}}\mathbf{L}_i\tilde{\mathbf{E}}\mathbf{L}_j\right)_{,\tilde{\mathbf{E}}} = \mathbf{L}_i\tilde{\mathbf{E}}\mathbf{L}_j + \mathbf{L}_j\tilde{\mathbf{E}}\mathbf{L}_i.$$

For the second Piola-Kirchhoff stress tensor $\mathbf{S}$ we thus obtain

$$\mathbf{S} = \frac{\partial\psi}{\partial\tilde{\mathbf{E}}} = \frac{1}{2}\sum_{i,j=1}^{3} a_{ij}\mathbf{L}_i\mathrm{tr}\left(\tilde{\mathbf{E}}\mathbf{L}_j\right) + \frac{1}{2}\sum_{i,j=1}^{3} a_{ij}\mathrm{tr}\left(\tilde{\mathbf{E}}\mathbf{L}_i\right)\mathbf{L}_j$$
$$+ \sum_{\substack{i,j=1\\ j\neq i}}^{3} G_{ij}\left(\mathbf{L}_i\tilde{\mathbf{E}}\mathbf{L}_j + \mathbf{L}_j\tilde{\mathbf{E}}\mathbf{L}_i\right)$$
$$= \sum_{i,j=1}^{3} a_{ij}\mathbf{L}_i\mathrm{tr}\left(\tilde{\mathbf{E}}\mathbf{L}_j\right) + 2\sum_{\substack{i,j=1\\ j\neq i}}^{3} G_{ij}\mathbf{L}_i\tilde{\mathbf{E}}\mathbf{L}_j.$$

By virtue of (5.42), (6.137), (6.140) and (6.144) the tangent moduli finally take the form

$$\mathcal{C} = \frac{\partial \mathbf{S}}{\partial \tilde{\mathbf{E}}} = \sum_{i,j=1}^{3} a_{ij} \mathbf{L}_i \odot \mathbf{L}_j + 2 \sum_{\substack{i,j=1 \\ j \neq i}}^{3} G_{ij} \left(\mathbf{L}_i \otimes \mathbf{L}_j \right)^{\mathrm{s}}.$$

6.12 Setting (6.167) in (6.166) yields

$$\begin{aligned}
\psi\left(\tilde{\mathbf{E}}\right) = & \frac{1}{2} a_{11} \mathrm{tr}^2 \left(\tilde{\mathbf{E}}\mathbf{L}_1\right) + \frac{1}{2} a_{22} \left[\mathrm{tr}^2 \left(\tilde{\mathbf{E}}\mathbf{L}_2\right) + \mathrm{tr}^2 \left(\tilde{\mathbf{E}}\mathbf{L}_3\right)\right] \\
& + a_{12} \left[\mathrm{tr}\left(\tilde{\mathbf{E}}\mathbf{L}_1\right) \mathrm{tr}\left(\tilde{\mathbf{E}}\mathbf{L}_2\right) + \mathrm{tr}\left(\tilde{\mathbf{E}}\mathbf{L}_1\right) \mathrm{tr}\left(\tilde{\mathbf{E}}\mathbf{L}_3\right)\right] \\
& + a_{23} \mathrm{tr}\left(\tilde{\mathbf{E}}\mathbf{L}_2\right) \mathrm{tr}\left(\tilde{\mathbf{E}}\mathbf{L}_3\right) + \left(a_{22} - a_{23}\right) \mathrm{tr}\left(\tilde{\mathbf{E}}\mathbf{L}_2\tilde{\mathbf{E}}\mathbf{L}_3\right) \\
& + 2G_{12} \left[\mathrm{tr}\left(\tilde{\mathbf{E}}\mathbf{L}_1\tilde{\mathbf{E}}\mathbf{L}_2\right) + \mathrm{tr}\left(\tilde{\mathbf{E}}\mathbf{L}_1\tilde{\mathbf{E}}\mathbf{L}_3\right)\right].
\end{aligned}$$

Thus, we can write keeping in mind (9.32)

$$\begin{aligned}
\psi\left(\tilde{\mathbf{E}}\right) = & \frac{1}{2} a_{11} \mathrm{tr}^2 \left(\tilde{\mathbf{E}}\mathbf{L}_1\right) + \frac{1}{2} a_{23} \left[\mathrm{tr}\left(\tilde{\mathbf{E}}\mathbf{L}_2\right) + \mathrm{tr}\left(\tilde{\mathbf{E}}\mathbf{L}_3\right)\right]^2 \\
& + a_{12} \mathrm{tr}\left(\tilde{\mathbf{E}}\mathbf{L}_1\right) \left[\mathrm{tr}\left(\tilde{\mathbf{E}}\mathbf{L}_2\right) + \mathrm{tr}\left(\tilde{\mathbf{E}}\mathbf{L}_3\right)\right] \\
& + \frac{1}{2} \left(a_{22} - a_{23}\right) \mathrm{tr}\left(\tilde{\mathbf{E}}\mathbf{L}_2 + \tilde{\mathbf{E}}\mathbf{L}_3\right)^2 + 2G_{12} \mathrm{tr}\left[\tilde{\mathbf{E}}\mathbf{L}_1\tilde{\mathbf{E}}\left(\mathbf{L}_2 + \mathbf{L}_3\right)\right].
\end{aligned}$$

Using the abbreviation $\mathbf{L} = \mathbf{L}_1$ and taking (9.29) into account one thus obtains

$$\begin{aligned}
\psi\left(\tilde{\mathbf{E}}\right) = & \frac{1}{2} a_{11} \mathrm{tr}^2 \left(\tilde{\mathbf{E}}\mathbf{L}\right) + \frac{1}{2} a_{23} \left[\mathrm{tr}\tilde{\mathbf{E}} - \mathrm{tr}\left(\tilde{\mathbf{E}}\mathbf{L}\right)\right]^2 \\
& + a_{12} \mathrm{tr}\left(\tilde{\mathbf{E}}\mathbf{L}\right) \left[\mathrm{tr}\tilde{\mathbf{E}} - \mathrm{tr}\left(\tilde{\mathbf{E}}\mathbf{L}\right)\right] + 2G_{12} \left[\mathrm{tr}\left(\tilde{\mathbf{E}}^2\mathbf{L}\right) - \mathrm{tr}^2\left(\tilde{\mathbf{E}}\mathbf{L}\right)\right] \\
& + \frac{1}{2} \left(a_{22} - a_{23}\right) \left[\mathrm{tr}\tilde{\mathbf{E}}^2 - 2\mathrm{tr}\left(\tilde{\mathbf{E}}^2\mathbf{L}\right) + \mathrm{tr}^2\left(\tilde{\mathbf{E}}\mathbf{L}\right)\right].
\end{aligned}$$

Collecting the terms with the transversely isotropic invariants delivers

$$\begin{aligned}
\psi\left(\tilde{\mathbf{E}}\right) = & \frac{1}{2} a_{23} \mathrm{tr}^2 \tilde{\mathbf{E}} + \frac{1}{2} \left(a_{22} - a_{23}\right) \mathrm{tr}\tilde{\mathbf{E}}^2 + \left(a_{23} - a_{22} + 2G_{12}\right) \mathrm{tr}\left(\tilde{\mathbf{E}}^2\mathbf{L}\right) \\
& + \left(\frac{1}{2} a_{11} + \frac{1}{2} a_{22} - a_{12} - 2G_{12}\right) \mathrm{tr}^2\left(\tilde{\mathbf{E}}\mathbf{L}\right) + \left(a_{12} - a_{23}\right) \mathrm{tr}\tilde{\mathbf{E}} \mathrm{tr}\left(\tilde{\mathbf{E}}\mathbf{L}\right).
\end{aligned}$$

It is seen that the function $\psi(\tilde{\mathbf{E}})$ is transversely isotropic in the sense of the representation (6.29). Finally, considering (6.168) in the latter relation we obtain the isotropic strain energy function of the form (6.116) as

$$\psi\left(\tilde{\mathbf{E}}\right) = \frac{1}{2}\lambda \mathrm{tr}^2\tilde{\mathbf{E}} + G\mathrm{tr}\tilde{\mathbf{E}}^2.$$

6.13 The tensor-valued function (6.120) can be shown to be isotropic. Indeed,

$$\hat{g}\left(\mathbf{Q}\mathbf{A}_i\mathbf{Q}^{\mathrm{T}}, \mathbf{Q}\mathbf{X}_j\mathbf{Q}^{\mathrm{T}}\right) = \mathbf{Q}''^{\mathrm{T}} g\left(\mathbf{Q}''\mathbf{Q}\mathbf{A}_i\mathbf{Q}^{\mathrm{T}}\mathbf{Q}''^{\mathrm{T}}\right)\mathbf{Q}'', \quad \forall \mathbf{Q} \in \mathbf{Orth}^n,$$

where $\mathbf{Q}''$ is defined by (6.39). Further, we can write taking (6.41) into account

$$\begin{aligned}
\mathbf{Q}''^{\mathrm{T}} g\left(\mathbf{Q}''\mathbf{Q}\mathbf{A}_i\mathbf{Q}^{\mathrm{T}}\mathbf{Q}''^{\mathrm{T}}\right)\mathbf{Q}'' &= \mathbf{Q}''^{\mathrm{T}} g\left(\mathbf{Q}^*\mathbf{Q}'\mathbf{A}_i\mathbf{Q}'^{\mathrm{T}}\mathbf{Q}^{*\mathrm{T}}\right)\mathbf{Q}'' \\
&= \mathbf{Q}''^{\mathrm{T}}\mathbf{Q}^* g\left(\mathbf{Q}'\mathbf{A}_i\mathbf{Q}'^{\mathrm{T}}\right)\mathbf{Q}^{*\mathrm{T}}\mathbf{Q}'' = \mathbf{Q}\mathbf{Q}'^{\mathrm{T}} g\left(\mathbf{Q}'\mathbf{A}_i\mathbf{Q}'^{\mathrm{T}}\right)\mathbf{Q}'\mathbf{Q}^{\mathrm{T}} \\
&= \mathbf{Q}\hat{g}\left(\mathbf{A}_i, \mathbf{X}_j\right)\mathbf{Q}^{\mathrm{T}},
\end{aligned}$$

which finally yields

$$\hat{g}\left(\mathbf{Q}\mathbf{A}_i\mathbf{Q}^{\mathrm{T}}, \mathbf{Q}\mathbf{X}_j\mathbf{Q}^{\mathrm{T}}\right) = \mathbf{Q}\hat{g}\left(\mathbf{A}_i, \mathbf{X}_j\right)\mathbf{Q}^{\mathrm{T}}, \quad \forall \mathbf{Q} \in \mathbf{Orth}^n.$$

Thus, the sufficiency is proved. The necessity is evident.

6.14 Consider the directional derivative of the identity $\mathbf{A}^{-k}\mathbf{A}^k = \mathbf{I}$. Taking into account (2.9) and using (6.133) we can write

$$\left.\frac{\mathrm{d}}{\mathrm{d}t}\left(\mathbf{A} + t\mathbf{X}\right)^{-k}\right|_{t=0}\mathbf{A}^k + \mathbf{A}^{-k}\left(\sum_{i=0}^{k-1}\mathbf{A}^i\mathbf{X}\mathbf{A}^{k-1-i}\right) = \mathbf{0}$$

and consequently

$$\begin{aligned}
\left.\frac{\mathrm{d}}{\mathrm{d}t}\left(\mathbf{A} + t\mathbf{X}\right)^{-k}\right|_{t=0} &= -\mathbf{A}^{-k}\left(\sum_{i=0}^{k-1}\mathbf{A}^i\mathbf{X}\mathbf{A}^{k-1-i}\right)\mathbf{A}^{-k} \\
&= -\sum_{i=0}^{k-1}\mathbf{A}^{i-k}\mathbf{X}\mathbf{A}^{-1-i}.
\end{aligned}$$

Hence, in view of $(5.17)_1$

$$\mathbf{A}^{-k},_{\mathbf{A}} = -\sum_{j=1}^{k}\mathbf{A}^{j-k-1} \otimes \mathbf{A}^{-j}. \tag{9.33}$$

6.15 Equations (2.4), (2.7), (5.16), $(5.17)_2$ and (6.128):

$$
\begin{aligned}
(f\mathbf{G})_{,\mathbf{A}} : \mathbf{X} &= \frac{\mathrm{d}}{\mathrm{d}t}\left[\hat{f}(\mathbf{A}+t\mathbf{X})\,g(\mathbf{A}+t\mathbf{X})\right]\Big|_{t=0} \\
&= \frac{\mathrm{d}}{\mathrm{d}t}\hat{f}(\mathbf{A}+t\mathbf{X})\Big|_{t=0}\mathbf{G} + f\frac{\mathrm{d}}{\mathrm{d}t}g(\mathbf{A}+t\mathbf{X})\Big|_{t=0} \\
&= (f_{,\mathbf{A}} : \mathbf{X})\,\mathbf{G} + f\,(\mathbf{G}_{,\mathbf{A}} : \mathbf{X}) \\
&= (\mathbf{G} \odot f_{,\mathbf{A}} + f\mathbf{G}_{,\mathbf{A}}) : \mathbf{X},
\end{aligned}
$$

$$
\begin{aligned}
(\mathbf{G} : \mathbf{H})_{,\mathbf{A}} : \mathbf{X} &= \frac{\mathrm{d}}{\mathrm{d}t}\left[g(\mathbf{A}+t\mathbf{X}) : h(\mathbf{A}+t\mathbf{X})\right]\Big|_{t=0} \\
&= \frac{\mathrm{d}}{\mathrm{d}t}g(\mathbf{A}+t\mathbf{X})\Big|_{t=0} : \mathbf{H} + \mathbf{G} : \frac{\mathrm{d}}{\mathrm{d}t}h(\mathbf{A}+t\mathbf{X})\Big|_{t=0} \\
&= (\mathbf{G}_{,\mathbf{A}} : \mathbf{X}) : \mathbf{H} + \mathbf{G} : (\mathbf{H}_{,\mathbf{A}} : \mathbf{X}) \\
&= (\mathbf{H} : \mathbf{G}_{,\mathbf{A}} + \mathbf{G} : \mathbf{H}_{,\mathbf{A}}) : \mathbf{X}, \quad \forall \mathbf{X} \in \mathbf{Lin}^n,
\end{aligned}
$$

where $f = \hat{f}(\mathbf{A})$, $\mathbf{G} = g(\mathbf{A})$ and $\mathbf{H} = h(\mathbf{A})$.

6.16 In the case $n = 2$ (6.159) takes the form

$$
\begin{aligned}
\mathbf{0} &= \sum_{k=1}^{2} \mathbf{A}^{2-k} \sum_{i=1}^{k} (-1)^{k-i}\,\mathrm{I}_{\mathbf{A}}^{(k-i)}\left[\mathrm{tr}\left(\mathbf{A}^{i-1}\mathbf{B}\right)\mathbf{I} - \mathbf{B}\mathbf{A}^{i-1}\right] \\
&= \mathbf{A}\left[\mathrm{tr}(\mathbf{B})\mathbf{I} - \mathbf{B}\right] - \mathrm{I}_{\mathbf{A}}^{(1)}\left[\mathrm{tr}(\mathbf{B})\mathbf{I} - \mathbf{B}\right] + \mathrm{tr}(\mathbf{A}\mathbf{B})\mathbf{I} - \mathbf{B}\mathbf{A}
\end{aligned}
$$

and finally

$$
\mathbf{A}\mathbf{B} + \mathbf{B}\mathbf{A} - \mathrm{tr}(\mathbf{B})\mathbf{A} - \mathrm{tr}(\mathbf{A})\mathbf{B} + \left[\mathrm{tr}(\mathbf{A})\,\mathrm{tr}(\mathbf{B}) - \mathrm{tr}(\mathbf{A}\mathbf{B})\right]\mathbf{I} = \mathbf{0}. \tag{9.34}
$$

9.7 Exercises of Chap. 7

7.1 By using (4.83) and (4.85) we can write

$$
\mathbf{R}(\omega) = \mathbf{P}_1 + \mathrm{e}^{\mathrm{i}\omega}\mathbf{P}_2 + \mathrm{e}^{-\mathrm{i}\omega}\mathbf{P}_3.
$$

Applying further (7.2) we get

$$\mathbf{R}^a(\omega) = 1^a\mathbf{P}_1 + \left(e^{i\omega}\right)^a \mathbf{P}_2 + \left(e^{-i\omega}\right)^a \mathbf{P}_3$$
$$= \mathbf{P}_1 + e^{ia\omega}\mathbf{P}_2 + e^{-ia\omega}\mathbf{P}_3 = \mathbf{R}(a\omega).$$

7.2 Equations $(7.5)_1$, (9.15) and (9.16):

$$\mathbf{U} = \sum_{i=1}^{s} \lambda_i \mathbf{P}_i = \sum_{i=1}^{s} \sqrt{\Lambda_i}\boldsymbol{a}_i \otimes \boldsymbol{a}_i = \boldsymbol{e}_3 \otimes \boldsymbol{e}_3$$
$$+\sqrt{\Lambda_1}\left(\frac{1}{\sqrt{1+\Lambda_1}}\boldsymbol{e}_1 + \sqrt{\frac{\Lambda_1}{1+\Lambda_1}}\boldsymbol{e}_2\right) \otimes \left(\frac{1}{\sqrt{1+\Lambda_1}}\boldsymbol{e}_1 + \sqrt{\frac{\Lambda_1}{1+\Lambda_1}}\boldsymbol{e}_2\right)$$
$$+\sqrt{\Lambda_2}\left(\frac{1}{\sqrt{1+\Lambda_2}}\boldsymbol{e}_1 - \sqrt{\frac{\Lambda_2}{1+\Lambda_2}}\boldsymbol{e}_2\right) \otimes \left(\frac{1}{\sqrt{1+\Lambda_2}}\boldsymbol{e}_1 - \sqrt{\frac{\Lambda_2}{1+\Lambda_2}}\boldsymbol{e}_2\right)$$
$$= \frac{2}{\sqrt{\gamma^2+4}}\boldsymbol{e}_1 \otimes \boldsymbol{e}_1 + \frac{\gamma}{\sqrt{\gamma^2+4}}\left(\boldsymbol{e}_1 \otimes \boldsymbol{e}_2 + \boldsymbol{e}_2 \otimes \boldsymbol{e}_1\right) + \frac{\gamma^2+2}{\sqrt{\gamma^2+4}}\boldsymbol{e}_2 \otimes \boldsymbol{e}_2$$
$$+\boldsymbol{e}_3 \otimes \boldsymbol{e}_3.$$

7.3 The proof of the first relation (7.21) directly results from the definition of the analytic tensor function (7.15) and is obvious. In order to prove $(7.21)_2$ we first write

$$f(\mathbf{A}) = \frac{1}{2\pi i}\oint_{\Gamma} f(\zeta)(\zeta\mathbf{I}-\mathbf{A})^{-1} d\zeta,\ h(\mathbf{A}) = \frac{1}{2\pi i}\oint_{\Gamma'} h(\zeta')\left(\zeta'\mathbf{I}-\mathbf{A}\right)^{-1} d\zeta',$$

where the closed curve Γ' of the second integral lies outside Γ which, in turn, includes all eigenvalues of $\mathbf{A}$. Using the identity

$$(\zeta\mathbf{I}-\mathbf{A})^{-1}\left(\zeta'\mathbf{I}-\mathbf{A}\right)^{-1} = \left(\zeta'-\zeta\right)^{-1}\left[(\zeta\mathbf{I}-\mathbf{A})^{-1} - \left(\zeta'\mathbf{I}-\mathbf{A}\right)^{-1}\right]$$

valid both on Γ and Γ' we thus obtain

$$f(\mathbf{A})h(\mathbf{A}) = \frac{1}{(2\pi i)^2}\oint_{\Gamma'}\oint_{\Gamma} f(\zeta)h\left(\zeta'\right)(\zeta\mathbf{I}-\mathbf{A})^{-1}\left(\zeta'\mathbf{I}-\mathbf{A}\right)^{-1} d\zeta d\zeta'$$
$$= \frac{1}{2\pi i}\oint_{\Gamma} f(\zeta)\frac{1}{2\pi i}\oint_{\Gamma'}\frac{h\left(\zeta'\right)}{\zeta'-\zeta}d\zeta'(\zeta\mathbf{I}-\mathbf{A})^{-1} d\zeta$$
$$+\frac{1}{2\pi i}\oint_{\Gamma'} h\left(\zeta'\right)\frac{1}{2\pi i}\oint_{\Gamma}\frac{f(\zeta)}{\zeta-\zeta'}d\zeta\left(\zeta'\mathbf{I}-\mathbf{A}\right)^{-1} d\zeta'.$$

Since the function $f(\zeta)(\zeta-\zeta')^{-1}$ is analytic in ζ inside Γ the Cauchy theorem (see, e.g. [5]) implies that

$$\frac{1}{2\pi \mathrm{i}}\oint_{\Gamma}\frac{f(\zeta)}{\zeta-\zeta'}\mathrm{d}\zeta=0.$$

Noticing further that

$$\frac{1}{2\pi \mathrm{i}}\oint_{\Gamma'}\frac{h(\zeta')}{\zeta'-\zeta}\mathrm{d}\zeta'=h(\zeta)$$

we obtain

$$\begin{aligned}f(\mathbf{A})h(\mathbf{A})&=\frac{1}{2\pi \mathrm{i}}\oint_{\Gamma}f(\zeta)\frac{1}{2\pi \mathrm{i}}\oint_{\Gamma'}\frac{h(\zeta')}{\zeta'-\zeta}\mathrm{d}\zeta'(\zeta\mathbf{I}-\mathbf{A})^{-1}\mathrm{d}\zeta\\&=\frac{1}{2\pi \mathrm{i}}\oint_{\Gamma}f(\zeta)h(\zeta)(\zeta\mathbf{I}-\mathbf{A})^{-1}\mathrm{d}\zeta\\&=\frac{1}{2\pi \mathrm{i}}\oint_{\Gamma}g(\zeta)(\zeta\mathbf{I}-\mathbf{A})^{-1}\mathrm{d}\zeta=g(\mathbf{A}).\end{aligned}$$

Finally, we focus on the third relation (7.21). It implies that the functions h and f are analytic on domains containing all the eigenvalues λ_i of $\mathbf{A}$ and $h(\lambda_i)$ $(i=1,2,\ldots,n)$ of $\mathbf{B}=h(\mathbf{A})$, respectively. Hence (cf. [25]),

$$f(h(\mathbf{A}))=f(\mathbf{B})=\frac{1}{2\pi \mathrm{i}}\oint_{\Gamma}f(\zeta)(\zeta\mathbf{I}-\mathbf{B})^{-1}\mathrm{d}\zeta, \tag{9.35}$$

where Γ encloses all the eigenvalues of $\mathbf{B}$. Further, we write

$$(\zeta\mathbf{I}-\mathbf{B})^{-1}=(\zeta\mathbf{I}-h(\mathbf{A}))^{-1}=\frac{1}{2\pi \mathrm{i}}\oint_{\Gamma'}\left(\zeta-h(\zeta')\right)^{-1}\left(\zeta'\mathbf{I}-\mathbf{A}\right)^{-1}\mathrm{d}\zeta', \tag{9.36}$$

where Γ' includes all the eigenvalues λ_i of $\mathbf{A}$ so that the image of Γ' under h lies within Γ. Thus, inserting (9.36) into (9.35) delivers

$$\begin{aligned}f(h(\mathbf{A}))&=\frac{1}{(2\pi \mathrm{i})^2}\oint_{\Gamma}\oint_{\Gamma'}f(\zeta)\left(\zeta-h(\zeta')\right)^{-1}\left(\zeta'\mathbf{I}-\mathbf{A}\right)^{-1}\mathrm{d}\zeta'\mathrm{d}\zeta\\&=\frac{1}{(2\pi \mathrm{i})^2}\oint_{\Gamma'}\oint_{\Gamma}f(\zeta)\left(\zeta-h(\zeta')\right)^{-1}\mathrm{d}\zeta\left(\zeta'\mathbf{I}-\mathbf{A}\right)^{-1}\mathrm{d}\zeta'\\&=\frac{1}{2\pi \mathrm{i}}\oint_{\Gamma'}f\left(h(\zeta')\right)\left(\zeta'\mathbf{I}-\mathbf{A}\right)^{-1}\mathrm{d}\zeta'\\&=\frac{1}{2\pi \mathrm{i}}\oint_{\Gamma'}g(\zeta')\left(\zeta'\mathbf{I}-\mathbf{A}\right)^{-1}\mathrm{d}\zeta'=g(\mathbf{A}).\end{aligned}$$

7.4 In view of (9.31) we can first write

$$\mathbf{S}(\mathbf{C}) = g(\mathbf{C}) = \sum_{r=1}^{m} \mu_r \mathbf{C}^{\alpha_r/2-1}.$$

Equations (6.149) and (7.49) further yield

$$\mathcal{C} = 2\mathbf{S},_{\mathbf{C}} = 2g(\mathbf{C}),_{\mathbf{C}} = 2\sum_{i,j=1}^{s} G_{ij}\mathbf{P}_i \otimes \mathbf{P}_j,$$

where $\mathbf{P}_i$, $(i = 1, 2, 3)$ are given by (9.17) while according to (7.50) and (9.15)

$$2G_{ii} = 2g'(\Lambda_i) = \sum_{r=1}^{m} \mu_r (\alpha_r - 2) \Lambda_i^{\alpha_r/2-2}$$

$$= \sum_{r=1}^{m} \mu_r (\alpha_r - 2) \left(\frac{\sqrt{4+\gamma^2} \pm \gamma}{2}\right)^{\alpha_r-4}, \quad i = 1, 2,$$

$$2G_{33} = 2g'(\Lambda_3) = \sum_{r=1}^{m} \mu_r (\alpha_r - 2) \Lambda_3^{\alpha_r/2-2} = \sum_{r=1}^{m} \mu_r (\alpha_r - 2),$$

$$2G_{12} = 2G_{21} = 2\frac{g(\Lambda_1) - g(\Lambda_2)}{\Lambda_1 - \Lambda_2} = \frac{2}{\Lambda_1 - \Lambda_2}\sum_{r=1}^{m} \mu_r \left(\Lambda_1^{\alpha_r/2-1} - \Lambda_2^{\alpha_r/2-1}\right)$$

$$= \frac{2}{\gamma\sqrt{\gamma^2+4}}\sum_{r=1}^{m} \mu_r \left[\left(\frac{\sqrt{4+\gamma^2}+\gamma}{2}\right)^{\alpha_r-2} - \left(\frac{\sqrt{4+\gamma^2}-\gamma}{2}\right)^{\alpha_r-2}\right],$$

$$2G_{i3} = 2G_{3i} = 2\frac{g(\Lambda_i) - g(\Lambda_3)}{\Lambda_i - \Lambda_3} = \frac{2}{\Lambda_i - \Lambda_3}\sum_{r=1}^{m} \mu_r \left(\Lambda_i^{\alpha_r/2-1} - \Lambda_3^{\alpha_r/2-1}\right)$$

$$= \frac{4}{\gamma^2 \pm \gamma\sqrt{\gamma^2+4}}\sum_{r=1}^{m} \mu_r \left[\left(\frac{\sqrt{4+\gamma^2} \pm \gamma}{2}\right)^{\alpha_r-2} - 1\right], \quad i = 1, 2.$$

7.5 Inserting into the right hand side of (7.54) the spectral decomposition in terms of eigenprojections (7.1) and taking (4.46) into account we can write similarly to (7.17)

$$\frac{1}{2\pi\mathrm{i}}\oint_{\Gamma_i}(\zeta\mathbf{I}-\mathbf{A})^{-1}\,\mathrm{d}\zeta=\frac{1}{2\pi\mathrm{i}}\oint_{\Gamma_i}\left(\zeta\mathbf{I}-\sum_{j=1}^{s}\lambda_j\mathbf{P}_j\right)^{-1}\mathrm{d}\zeta$$

$$=\frac{1}{2\pi\mathrm{i}}\oint_{\Gamma_i}\left[\sum_{j=1}^{s}(\zeta-\lambda_j)\,\mathbf{P}_j\right]^{-1}\mathrm{d}\zeta=\frac{1}{2\pi\mathrm{i}}\oint_{\Gamma_i}\sum_{j=1}^{s}(\zeta-\lambda_j)^{-1}\,\mathbf{P}_j\mathrm{d}\zeta$$

$$=\sum_{j=1}^{s}\left[\frac{1}{2\pi\mathrm{i}}\oint_{\Gamma_i}(\zeta-\lambda_j)^{-1}\,\mathrm{d}\zeta\right]\mathbf{P}_j.$$

In the case $i \neq j$ the closed curve Γ_i does not include any pole so that

$$\frac{1}{2\pi\mathrm{i}}\oint_{\Gamma_i}(\zeta-\lambda_j)^{-1}\,\mathrm{d}\zeta=\delta_{ij},\quad i,j=1,2,\ldots s.$$

This immediately leads to (7.54).

7.6 By means of (7.43) and (7.83) and using the result for the eigenvalues of $\mathbf{A}$ by (9.19), $\lambda_i = 6$, $\lambda = -3$ we write

$$\mathbf{P}_1=\sum_{p=0}^{2}\rho_{1p}\mathbf{A}^p=-\frac{\lambda}{(\lambda_i-\lambda)}\mathbf{I}+\frac{1}{(\lambda_i-\lambda)}\mathbf{A}=\frac{1}{3}\mathbf{I}+\frac{1}{9}\mathbf{A},$$

$$\mathbf{P}_2=\mathbf{I}-\mathbf{P}_1=\frac{2}{3}\mathbf{I}-\frac{1}{9}\mathbf{A}.$$

Taking symmetry of $\mathbf{A}$ into account we further obtain by virtue of (7.56) and (7.84)

$$\mathbf{P}_{1,\mathbf{A}}=\sum_{p,q=0}^{2}\upsilon_{1pq}\left(\mathbf{A}^p\otimes\mathbf{A}^q\right)^{\mathrm{s}}$$

$$=-\frac{2\lambda\lambda_i}{(\lambda_i-\lambda)^3}\mathfrak{J}^{\mathrm{s}}+\frac{\lambda_i+\lambda}{(\lambda_i-\lambda)^3}\left(\mathbf{I}\otimes\mathbf{A}+\mathbf{A}\otimes\mathbf{I}\right)^{\mathrm{s}}-\frac{2}{(\lambda_i-\lambda)^3}\left(\mathbf{A}\otimes\mathbf{A}\right)^{\mathrm{s}}$$

$$=\frac{4}{81}\mathfrak{J}^{\mathrm{s}}+\frac{1}{243}\left(\mathbf{I}\otimes\mathbf{A}+\mathbf{A}\otimes\mathbf{I}\right)^{\mathrm{s}}-\frac{2}{729}\left(\mathbf{A}\otimes\mathbf{A}\right)^{\mathrm{s}}.$$

The eigenprojection $\mathbf{P}_2$ corresponds to the double eigenvalue $\lambda = -3$ and for this reason is not differentiable.

7.7 Since $\mathbf{A}$ is a symmetric tensor and it is diagonalizable. Thus, taking double coalescence of eigenvalues (9.19) into account we can apply the representations (7.77) and (7.78). Setting there $\lambda_a = 6$, $\lambda = -3$ delivers

$$\exp(\mathbf{A}) = \frac{e^6 + 2e^{-3}}{3}\mathbf{I} + \frac{e^6 - e^{-3}}{9}\mathbf{A},$$

$$\exp(\mathbf{A})_{,\mathbf{A}} = \frac{13e^6 + 32e^{-3}}{81}\mathfrak{I}^s + \frac{10e^6 - 19e^{-3}}{243}(\mathbf{A} \otimes \mathbf{I} + \mathbf{I} \otimes \mathbf{A})^s$$

$$+\frac{7e^6 + 11e^{-3}}{729}(\mathbf{A} \otimes \mathbf{A})^s.$$

Inserting

$$\mathbf{A} = \begin{bmatrix} -2 & 2 & 2 \\ 2 & 1 & 4 \\ 2 & 4 & 1 \end{bmatrix} \boldsymbol{e}_i \otimes \boldsymbol{e}_j$$

into the expression for $\exp(\mathbf{A})$ we obtain

$$\exp(\mathbf{A}) = \frac{1}{9}\begin{bmatrix} e^6 + 8e^{-3} & 2e^6 - 2e^{-3} & 2e^6 - 2e^{-3} \\ 2e^6 - 2e^{-3} & 4e^6 + 5e^{-3} & 4e^6 - 4e^{-3} \\ 2e^6 - 2e^{-3} & 4e^6 - 4e^{-3} & 4e^6 + 5e^{-3} \end{bmatrix} \boldsymbol{e}_i \otimes \boldsymbol{e}_j,$$

which coincides with the result obtained in Exercise 4.15.

7.8 The computation of the coefficients series (7.89), (7.91) and (7.96), (7.97) with the precision parameter $\varepsilon = 1 \cdot 10^{-6}$ has required 23 iteration steps and has been carried out by using MAPLE-program. The results of the computation are summarized in Tables 9.1 and 9.2. On use of (7.90) and (7.92) we thus obtain

$$\exp(\mathbf{A}) = 44.96925\mathbf{I} + 29.89652\mathbf{A} + 4.974456\mathbf{A}^2,$$

$$\exp(\mathbf{A})_{,\mathbf{A}} = 16.20582\mathfrak{I}^s + 6.829754(\mathbf{I} \otimes \mathbf{A} + \mathbf{A} \otimes \mathbf{I})^s + 1.967368(\mathbf{A} \otimes \mathbf{A})^s$$
$$+1.039719\left(\mathbf{I} \otimes \mathbf{A}^2 + \mathbf{A}^2 \otimes \mathbf{I}\right)^s + 0.266328\left(\mathbf{A} \otimes \mathbf{A}^2 + \mathbf{A}^2 \otimes \mathbf{A}\right)^s$$
$$+0.034357\left(\mathbf{A}^2 \otimes \mathbf{A}^2\right)^s.$$

Taking into account double coalescence of eigenvalues of $\mathbf{A}$ we can further write

$$\mathbf{A}^2 = (\lambda_a + \lambda)\mathbf{A} - \lambda_a\lambda\mathbf{I} = 3\mathbf{A} + 18\mathbf{I}.$$

Inserting this relation into the above representations for $\exp(\mathbf{A})$ and $\exp(\mathbf{A})_{,\mathbf{A}}$ finally yields

$$\exp(\mathbf{A}) = 134.50946\mathbf{I} + 44.81989\mathbf{A},$$

$$\exp(\mathbf{A})_{,\mathbf{A}} = 64.76737\mathfrak{I}^s + 16.59809(\mathbf{I} \otimes \mathbf{A} + \mathbf{A} \otimes \mathbf{I})^s + 3.87638(\mathbf{A} \otimes \mathbf{A})^s.$$

Note that the relative error of this result in comparison to the closed-form solution used in Exercise 7.7 lies within 0.044%.

Table 9.1 Recurrent calculation of the coefficients $\omega_p^{(r)}$

r	$a_r\omega_0^{(r)}$	$a_r\omega_1^{(r)}$	$a_r\omega_2^{(r)}$
0	1	0	0
1	0	1	0
2	0	0	0.5
3	9.0	4.5	0
4	0	2.25	1.125
5	12.15	6.075	0.45
6	4.05	4.05	1.0125
...	...	...	...
23 $(\cdot 10^{-6})$	3.394287	2.262832	0.377134
φ_p	44.96925	29.89652	4.974456

Table 9.2 Recurrent calculation of the coefficients $\xi_{pq}^{(r)}$

r	$a_r\xi_{00}^{(r)}$	$a_r\xi_{01}^{(r)}$	$a_r\xi_{02}^{(r)}$	$a_r\xi_{11}^{(r)}$	$a_r\xi_{12}^{(r)}$	$a_r\xi_{22}^{(r)}$
1	1	0	0	0	0	0
2	0	0.5	0	0	0	0
3	0	0	0.166666	0.166666	0	0
4	4.5	1.125	0	0	0.041666	0
5	0	0.9	0.225	0.45	0	0.008333
6	4.05	1.0125	0.15	0.15	0.075	0
...	...	...	...			
23 $(\cdot 10^{-6})$	2.284387	1.229329	0.197840	0.623937	0.099319	0.015781
η_{pq}	16.20582	6.829754	1.039719	1.967368	0.266328	0.034357

9.8 Exercises of Chap. 8

8.1 By (8.2) we first calculate the right and left Cauchy-Green tensors as

$$\mathbf{C} = \mathbf{F}^{\mathrm{T}}\mathbf{F} = \begin{bmatrix} 5 & -2 & 0 \\ -2 & 8 & 0 \\ 0 & 0 & 1 \end{bmatrix} \boldsymbol{e}_i \otimes \boldsymbol{e}^j, \quad \mathbf{b} = \mathbf{F}\mathbf{F}^{\mathrm{T}} = \begin{bmatrix} 5 & 2 & 0 \\ 2 & 8 & 0 \\ 0 & 0 & 1 \end{bmatrix} \boldsymbol{e}_i \otimes \boldsymbol{e}^j,$$

with the following eigenvalues $\Lambda_1 = 1$, $\Lambda_2 = 4$, $\Lambda_3 = 9$. Thus, $\lambda_1 = \sqrt{\Lambda_1} = 1$, $\lambda_2 = \sqrt{\Lambda_2} = 2$, $\lambda_3 = \sqrt{\Lambda_3} = 3$. By means of (8.11) and (8.12) we further obtain $\varphi_0 = \frac{3}{5}$, $\varphi_1 = \frac{5}{12}$, $\varphi_2 = -\frac{1}{60}$ and

$$\mathbf{U} = \frac{3}{5}\mathbf{I} + \frac{5}{12}\mathbf{C} - \frac{1}{60}\mathbf{C}^2 = \frac{1}{5}\begin{bmatrix} 11 & -2 & 0 \\ -2 & 14 & 0 \\ 0 & 0 & 5 \end{bmatrix} \boldsymbol{e}_i \otimes \boldsymbol{e}^j,$$

$$\mathbf{v} = \frac{3}{5}\mathbf{I} + \frac{5}{12}\mathbf{b} - \frac{1}{60}\mathbf{b}^2 = \frac{1}{5}\begin{bmatrix} 11 & 2 & 0 \\ 2 & 14 & 0 \\ 0 & 0 & 5 \end{bmatrix} \boldsymbol{e}_i \otimes \boldsymbol{e}^j.$$

Equations (8.16) and (8.17) further yield $\varsigma_0 = \frac{37}{30}$, $\varsigma_1 = -\frac{1}{4}$, $\varsigma_2 = \frac{1}{60}$ and

$$\mathbf{R} = \mathbf{F}\left(\frac{37}{30}\mathbf{I} - \frac{1}{4}\mathbf{C} + \frac{1}{60}\mathbf{C}^2\right) = \frac{1}{5}\begin{bmatrix} 3 & 4 & 0 \\ -4 & 3 & 0 \\ 0 & 0 & 5 \end{bmatrix} \boldsymbol{e}_i \otimes \boldsymbol{e}^j.$$

8.2 Equations (4.44), (5.33), (5.47), (5.55) and $(5.85)_1$:

$$\begin{aligned}
\mathcal{P}_{ij} : \mathcal{P}_{kl} &= \left(\mathbf{P}_i \otimes \mathbf{P}_j + \mathbf{P}_j \otimes \mathbf{P}_i\right)^{\mathrm{s}} : \left(\mathbf{P}_k \otimes \mathbf{P}_l + \mathbf{P}_l \otimes \mathbf{P}_k\right)^{\mathrm{s}} \\
&= \left[\left(\mathbf{P}_i \otimes \mathbf{P}_j + \mathbf{P}_j \otimes \mathbf{P}_i\right)^{\mathrm{s}} : \left(\mathbf{P}_k \otimes \mathbf{P}_l + \mathbf{P}_l \otimes \mathbf{P}_k\right)\right]^{\mathrm{s}} \\
&= \frac{1}{2}\left\{\left[\mathbf{P}_i \otimes \mathbf{P}_j + \mathbf{P}_j \otimes \mathbf{P}_i + \left(\mathbf{P}_i \otimes \mathbf{P}_j\right)^{\mathrm{t}} + \left(\mathbf{P}_j \otimes \mathbf{P}_i\right)^{\mathrm{t}}\right]\right. \\
&\qquad \left. : \left(\mathbf{P}_k \otimes \mathbf{P}_l + \mathbf{P}_l \otimes \mathbf{P}_k\right)\right\}^{\mathrm{s}} \\
&= \left(\delta_{ik}\delta_{jl} + \delta_{il}\delta_{jk}\right)\left(\mathbf{P}_i \otimes \mathbf{P}_j + \mathbf{P}_j \otimes \mathbf{P}_i\right)^{\mathrm{s}}, \quad i \neq j,\ k \neq l.
\end{aligned}$$

In the case $i = j$ or $k = l$ the previous result should be divided by 2, whereas for $i = j$ and $k = l$ by 4, which immediately leads to (8.65).

8.3 Setting $f(\lambda) = \ln \lambda$ in (8.50) and $(8.56)_1$ one obtains

$$\dot{\mathbf{E}}^{(0)} = (\ln \mathbf{U})^{\cdot} = \sum_{i=1}^{s} \frac{1}{2\lambda_i^2}\mathbf{P}_i\dot{\mathbf{C}}\mathbf{P}_i + \sum_{\substack{i,j=1 \\ i \neq j}}^{s} \frac{\ln \lambda_i - \ln \lambda_j}{\lambda_i^2 - \lambda_j^2}\mathbf{P}_i\dot{\mathbf{C}}\mathbf{P}_j.$$

References

1. Başar Y, Krätzig WB (1985) Mechanik der Flächentragwerke. Vieweg Verlag, Braunschweig
2. Boehler JP (1977) Z Angew Math Mech 57:323–327
3. de Boer R (1982) Vektor- und Tensorrechnung für Ingenieure. Springer, Berlin/Heidelberg/ New York
4. Boulanger Ph, Hayes M (1993) Bivectors and waves in mechanics and optics. Chapman and Hall, London
5. Bronstein IN, Semendyayev KA, Musiol G, Muehlig H (2004) Handbook of mathematics. Springer, Berlin/Heidelberg/New York
6. Brousse P (1988) Optimization in mechanics: problems and methods. Elsevier, Amsterdam
7. Carlson DE, Hoger A (1986) J Elast 16:221–224
8. Chen Y, Wheeler L (1993) J Elast 32:175–182
9. Cheng H, Gupta KC (1989) J Appl Mech 56:139–145
10. Chrystal G (1980) Algebra. An elementary text-book. Part I. Chelsea, New York
11. Dui G, Chen Y-C (2004) J Elast 76:107–112
12. Friedberg SH, Insel AJ, Spence LE (2003) Linear algebra. Pearson, Upper Saddle River
13. Gantmacher FR (1959) The theory of matrices. Chelsea, New York
14. Guo ZH (1984) J Elast 14:263–267
15. Halmos PR (1958) Finite-dimensional vector spaces. Van Nostrand, New York
16. Hill R (1968) J Mech Phys Solids 16:229–242
17. Hill R (1978) Adv Appl Mech 18:1–75
18. Hoger A, Carlson DE (1984) J Elast 14:329–336
19. Itskov M (2002) Z Angew Math Mech 82:535–544
20. Itskov M (2003) Comput Meth Appl Mech Engrg 192:3985–3999
21. Itskov M (2003) Proc R Soc Lond A 459:1449–1457
22. Itskov M (2004) Mech Res Commun 31:507–517
23. Itskov M, Aksel N (2002) Int J Solids Struct 39:5963–5978
24. Kaplan W (2003) Advanced calculus. Addison Wesley, Boston
25. Kato T (1966) Perturbation theory for linear operators. Springer, New York
26. Kreyszig E (1991) Differential geometry. Dover, New York
27. Lax PD (1997) Linear algebra. Wiley, New York
28. Lew JS (1966) Z Angew Math Phys 17:650–653
29. Lubliner J (1990) Plasticity theory. Macmillan, New York
30. Ogden RW (1984) Non-Linear elastic deformations. Ellis Horwood, Chichester
31. Ortiz M, Radovitzky RA, Repetto EA (2001) Int J Numer Methods Eng 52:1431–1441
32. Papadopoulos P, Lu J (2001) Comput Methods Appl Mech Eng 190:4889–4910
33. Pennisi S, Trovato M (1987) Int J Eng Sci 25:1059–1065
34. Rinehart RF (1955) Am Math Mon 62:395–414

M. Itskov, *Tensor Algebra and Tensor Analysis for Engineers*, Mathematical Engineering,
DOI 10.1007/978-3-642-30879-6,

35. Rivlin RS (1955) J Ration Mech Anal 4:681–702
36. Rivlin RS, Ericksen JL (1955) J Ration Mech Anal 4:323–425
37. Rivlin RS, Smith GF (1975) Rendiconti di Matematica Serie VI 8:345–353
38. Rosati L (1999) J Elast 56:213–230
39. Sansour C, Kollmann FG (1998) Comput Mech 21:512–525
40. Seth BR (1964) Generalized strain measures with applications to physical problems. In: Reiner M, Abir D (eds) Second-order effects in elasticity, plasticity and fluid dynamics. Academic, Jerusalem
41. Smith GF (1971) Int J Eng Sci 9:899–916
42. Sokolnikoff IS (1964) Tensor analysis. Theory and applications to geometry and mechanics of continua. Wiley, New York
43. Spencer AJM (1984) Constitutive theory for strongly anisotropic solids. In: Spencer AJM (ed) Continuum theory of the mechanics of fibre-reinforced composites. Springer, Wien/New York
44. Steigmann DJ (2002) Math Mech Solids 7:393–404
45. Ting TCT (1985) J Elast 15:319–323
46. Truesdell C, Noll W (1965) The nonlinear field theories of mechanics. In: Flügge S (ed) Handbuch der Physik, vol III/3. Springer, Berlin
47. Wheeler L (1990) J Elast 24:129–133
48. Xiao H (1995) Int J Solids Struct 32:3327–3340
49. Xiao H, Bruhns OT, Meyers ATM (1998) J Elast 52:1–41
50. Zhang JM, Rychlewski J (1990) Arch Mech 42:267–277

Further Reading

51. Abraham R, Marsden JE, Ratiu T (1988) Manifolds, tensor analysis and applications. Springer, Berlin/Heidelberg/New York
52. Akivis MA, Goldberg VV (2003) Tensor calculus with applications. World Scientific, Singapore
53. Anton H, Rorres C (2000) Elementary linear algebra: application version. Wiley, New York
54. Başar Y, Weichert D (2000) Nonlinear continuum mechanics of solids. Fundamental mathematical and physical concepts. Springer, Berlin/Heidelberg/New York
55. Bertram A (2005) Elasticity and plasticity of large deformations. An introduction. Springer, Berlin/Heidelberg/New York
56. Betten J (1987) Tensorrechnung für Ingenieure. Teubner-Verlag, Stuttgart
57. Bishop RL, Goldberg SI (1968) Tensor analysis on manifolds. Macmillan, New York
58. Borisenko AI, Tarapov IE (1968) Vector and tensor analysis with applications. Prentice-Hall, Englewood Cliffs
59. Bowen RM, Wang C-C (1976) Introduction to vectors and tensors. Plenum, New York
60. Brillouin L (1964) Tensors in mechanics and elasticity. Academic, New York
61. Chadwick P (1976) Continuum mechanics. Concise theory and problems. George Allen and Unwin, London
62. Dimitrienko Yu I (2002) Tensor analysis and nonlinear tensor functions. Kluwer, Dordrecht
63. Flügge W (1972) Tensor analysis and continuum mechanics. Springer, Berlin/Heidelberg/New York
64. Golub GH, van Loan CF (1996) Matrix computations. The Johns Hopkins University Press, Baltimore
65. Gurtin ME (1981) An introduction to continuum mechanics. Academic, New York
66. Lebedev LP, Cloud MJ (2003) Tensor analysis. World Scientific, Singapore
67. Lütkepohl H (1996) Handbook of matrices. Wiley, Chichester
68. Narasimhan MNL (1993) Principles of continuum mechanics. Wiley, New York
69. Noll W (1987) Finite-dimensional spaces. Martinus Nijhoff, Dordrecht

70. Renton JD (2002) Applied elasticity: matrix and tensor analysis of elastic continua. Horwood, Chichester
71. Ruíz-Tolosa JR, Castillo (2005) From vectors to tensors. Springer, Berlin/Heidelberg/New York
72. Schade H (1997) Tensoranalysis. Walter der Gruyter, Berlin/New York
73. Schey HM (2005) Div, grad, curl and all that: an informal text on vector calculus. W.W. Norton, New York
74. Schouten JA (1990) Tensor analysis for physicists. Dover, New York
75. Šilhavý M (1997) The mechanics and thermodynamics of continuous media. Springer, Berlin/Heidelberg/New York
76. Simmonds JG (1997) A brief on tensor analysis. Springer, Berlin/Heidelberg/New York
77. Talpaert YR (2002) Tensor analysis and continuum mechanics. Kluwer, Dordrecht

Index

M. Itskov, *Tensor Algebra and Tensor Analysis for Engineers*, Mathematical Engineering,
DOI 10.1007/978-3-642-30879-6, © Springer-Verlag Berlin Heidelberg 2013